Advances in Mechanics and Mathematics

Volume 40

More information about this series at http://www.springer.com/series/5613

Aleksandra A. Bozhko • Sergey A. Suslov

Convection in Ferro-Nanofluids: Experiments and Theory

Physical Mechanisms, Flow Patterns, and Heat Transfer

Aleksandra A. Bozhko
Faculty of Physics
Perm State University
Perm, Russia

Sergey A. Suslov
Department of Mathematics
Swinburne University of Technology
Hawthorn, Victoria, Australia

ISSN 1571-8689 ISSN 1876-9896 (electronic)
Advances in Mechanics and Mathematics
ISBN 978-3-030-06849-3 ISBN 978-3-319-94427-2 (eBook)
https://doi.org/10.1007/978-3-319-94427-2

Mathematics Subject Classification: 76E06, 76E25, 76E30, 76R10, 76B70, 80A20, 82D40

This Springer imprint is published by the registered company Springer Nature Switzerland AG
The registered company address is: Gewerbestrasse 11, 6330 Cham, Switzerland

To the memory of Professor G. F. Putin, outstanding experimentalist, colleague and teacher

Preface

Scope

This book is based on the results of experimental and theoretical studies of hydrodynamic stability and heat and mass transfer processes in ferrofluids that we have been involved with over the past several decades. The main motivation for such studies has been the growing interest in the use of magnetically controllable media as heat carriers in various thermal management systems. Along with other non-gravitational mechanisms capable of inducing the motion of initially quiescent fluid such as vibrational and electroconvection thermomagnetic convection can be used to enhance heat transfer in conditions where natural convection is impossible, for example, in microgravitation conditions. The experimental investigation that is reported in this book was conducted at Perm State University, Russia, under the license from the Russian State Corporation for Space Research (RosCosmos) and formed the ground-based component of a larger programme involving experiments on board autonomous and piloted spacecrafts, the orbital station "Mir" and the International Space Station. From a fundamental point of view the book considers an intricate interaction of non-isothermal and electrically non-conducting magnetopolarisable fluid with gravitational and magnetic fields. In the absence of a magnetic field ferrofluids behave similarly to other non-magnetic nanofluids, studies of which have been growing exponentially over the past two decades due to their ever expanding applications in modern technology.

Audience

We hope that this book will be of interest to researchers and practitioners working in the areas of fluid mechanics, hydrodynamic stability and heat and mass transfer with the view of perspective applications of ferrofluids in heat management systems, in particular, in microelectronics and space technologies. The main emphasis of the book is on the influence of a uniform magnetic field on flows of non-isothermal ferrofluids and the associated heat transfer. However, we also discuss peculiar features of ferrofluid flows occurring in the absence of a magnetic field, which are shown to be drastically different from those of ordinary fluids and need to be taken into account by practitioners working with magnetic and non-magnetic nanofluids.

Content

Invention of ferrofluids, their industrial synthesis and numerous studies at micro and macro levels have been primarily motivated by their magnetic properties that are many orders of magnitude stronger than those of natural paramagnetic and diamagnetic fluids and gases. The composition of ferrofluids that defines their magnetic properties and the related mechanisms of heat and mass transfer in them are briefly reviewed in Chapter 1.

The main equations describing motion of non-isothermal ferrofluids by treating them as magnetopolarisable continuous media are summarised in Chapter 2. While such a description has its limitations that become evident when the theoretically obtained results are compared with those of experimental observations, currently, such an approximation offers the most robust way of modelling ferrofluid flows. The reasons for this are outlined in the subsequent chapters of the book. The major governing non-dimensional parameters are also defined and their physical meaning is discussed in Chapter 2.

Results of a theoretical analysis of thermomagnetic convection in geometrically simple yet practically relevant domains are presented in Chapter 3. Such an analysis sheds light on physical processes taking place in the bulk of ferrofluid offering the insight that is successfully used to guide experimental observations and measurements. In particular, the existence of thermomagnetic waves associated with the thermally induced non-uniformity of fluid magnetisation and of oscillatory regimes of convections caused by the nonlinear variation of magnetisation across a ferrofluid layer was discovered theoretically first and then was confirmed in specialised experiments. A comprehensive analysis of magnetoconvection arising in the arbitrarily oriented magnetic field in gravity-free conditions is another example of a practically important situation considered in Chapter 3 that is out of reach for ground-based laboratory experiments.

Chapter 4 contains a detailed description of experimental setups specifically designed for a comprehensive study of buoyancy and thermomagnetically driven ferrofluid flows. The details distinguishing experimental chambers and flow visualisation and heat flux measurement techniques used for working with magnetically active media from those used in experiments with non-magnetic fluids are emphasised. In particular, it is shown that the shape and size of the working chamber have a defining influence on the type of convection patterns arising in a magnetic field.

The features of thermogravitational and thermomagnetic convection arising in finite flat layers and spherical cavities filled with ferrofluids and placed in uniform gravitational and magnetic fields are detailed in Chapters 5 and 6, respectively. Notably, a strong influence of gravitational sedimentation of solid particles and their aggregates contained in ferrofluids is demonstrated experimentally. It changes qualitatively the character of convection compared to that observed in ordinary single-phase fluids. Specifically, it is shown in Chapter 5 that in the vicinity of convection threshold in ferrofluids flows become oscillatory and chaotic both in space and time. A hysteresis is observed when the onset of convection in the initially density-stratified ferrocolloid is delayed compared to that recorded for the same but pre-mixed fluid. Convection is found to arise and decay spontaneously and irregularly and this found to be related to the concentration of solid phase in experimental fluids.

The influence of magnetic fields of various orientations on ferrofluid convection and heat transfer is discussed next in Chapter 6. It is shown that such an influence is not monotonic. Depending on the values of the governing gravitational and magnetic parameters, the application of magnetic field can either enhance or suppress convection drastically changing the observed flows and offering a not-intrusive means of controlling them. The experimental evidence of the fact that conditions of a particular laboratory run, storage and past usage of a ferrofluid strongly affect its flows and performance as a heat carrier. These factors should be taken into account when interpreting physical observations of a non-isothermal ferrofluid behaviour and when using it in practical applications. Overall, the book is intended to provide a guidance to a very rich and frequently ambiguous behaviour of non-uniformly heated ferrocolloids caused by their complex composition and influenced by an external magnetic field.

Acknowledgements

This book would not be possible without the hard work of our colleagues and technical staff who were invaluable in building experimental equipment and maintaining it in working order over many months during which individual experimental runs were performed and over decades during which this research was conducted. We are especially grateful to our students T. Pilugina, D. Shupeinik, P. Bulychev, A. Sidorov, M. Krauzina, P. Krauzin, H. Rahman, P. Dey and K. Pham for their effort and time preparing and running experiments and performing computations reported in this book. We extend our gratitude to Professor A. F. Pshenichnikov and Dr. A. S. Ivanov of the Institute of Continuous Media Mechanics of the Ural Branch of the Russian Academy of Sciences for illuminating discussions of microstructure of ferrocolloids, Mr. A. N. Poludnitsyn for help with experimental photography and Dr. T. Tynjälä of Lappeenranta University of Technology, Finland, for fruitful collaboration on numerical modelling of ferrofluid flows. AAB is also grateful to the late Professors I. M. Kirko, Yu. K. Bratukhin and G. Z. Gershuni for their mentoring and help during the early years of this research.

Perm, Russia
Melbourne, Australia
April 2018

Aleksandra A. Bozhko
Sergey A. Suslov

Contents

Acronyms

CTO	Centrifugally purified transformer oil
FF	Ferrofluid
FF-PES	Ferrofluid based on polyethylsiloxane
FF-TO	Ferrofluid based on transformer oil
GCC	Gravi-concentrational convection
GS	Gravitational sedimentation
MCC	Magneto-concentrational convection
MPH	Magnetophoresis
TD	Thermodiffusion
TGC	Thermogravitational convection
TMC	Thermomagnetic convection
TO	Transformer oil

Chapter 1
Ferrofluids: Composition and Physical Processes

Abstract A brief history and an overview of the current state of knowledge of ferrofluids (also known as ferrocolloids or ferro-nanofluids) are given. Applications of ferrofluids as advanced heat carrier media in heat management systems are emphasised. It is discussed that in the absence of a magnetic field, ferrofluids can be considered as a type of synthesised nanofluids or ordinary colloids. However, when they are placed in an external magnetic field, they behave as magneto-polarisable media, the magnetic susceptibility of which is several orders of magnitude larger than that of natural fluids and gases. Various physical mechanisms of heat and mass transfer in ferrofluids are identified. It is shown that the macroscopic behaviour of ferrofluids is strongly affected by their microstructure that depends on the way they are synthesised, stored and used.

1.1 Brief History and Composition of Ferrofluids

First colloids (termed later as ferrofluids[1]) containing single-domain ferromagnetic particles with a characteristic size of 10 nm suspended in a carrier liquid were synthesised in the 1930s [85]. The interest to their technological applications was significantly boosted in the 1960s when their industrial production became possible [22, 178, 184, 244]. The presence of ferromagnetic nanoparticles with magnetic moments that are 10^3–10^4 times larger than those of ions of paramagnetic materials enables achieving ferrofluid magnetisation of up to $\sim 100\,\mathrm{kA/m}$ using external magnetic fields created by ordi-

[1] Not to be confused with magnetorheological fluids containing much larger, of the order of a micron, particles.

A. A. Bozhko, S. A. Suslov, *Convection in Ferro-Nanofluids: Experiments and Theory*, Advances in Mechanics and Mathematics 40,
https://doi.org/10.1007/978-3-319-94427-2_1

nary permanent and electromagnets. When taken out of an external magnetic field, ferrofluids lose their magnetisation due to the disorientation of magnetic moments of individual particles by Brownian motion. Because of this ferrofluids are classified as superparamagnetics [120] and are referred to as magneto-polarisable media. Fluid behaviour of electrically non-conducting ferrocolloids is similar to that of natural dia- and paramagnetic fluids but is characterised by much larger (10^4–10^6 times) magnetic susceptibility and thus magnetic forces. It is studied in a special division of hydrodynamics that is known as ferrohydrodynamics [209]. The main motivating factor for a rapid development of this area of research is a wide range of ferrofluid applications that include vibration damping, magnetic sealing, species separation as well as their use in various sensors, actuators modulators of laser radiation, MEMS and NEMS, and cancer treatment to name a few. Of the main interest in this book is the application of ferrofluids as heat carrier media in thermal management systems, for example, in power transformers and converters and solar collectors. Comprehensive reviews of heat and mass transfer processes taking place in ferrofluids can be found, for example, in [13, 26, 88, 173, 209, 215, 222].

While the main property of ferrofluids that defines their numerous applications is their ability to respond to external magnetic field, the specifics of their composition put them in a larger class of synthetic fluids known as nanofluids—fluids that contain solid particles with sizes ranging from 1 to 100 nm. The term "nanofluid" was coined relatively recently (in 1995) to denote artificial fluids created to drastically improve the performance of traditional heat carrier liquids [64] by adding solid particles. However, the field of colloidal chemistry dealing with fluids containing such small particles existed for a much longer time [268] being motivated by the existence of natural nanomaterials [107] and their use in arts, trades and industry [181, 226]. Therefore ferrofluid research benefits greatly from the knowledge and experience accumulated over decades of studies of other similar media.

Ferrofluids typically contain nanoparticles of cobalt, magnetite, hematite and various ferrites. Nanoparticles are usually obtained by ball-milling macroscopic materials [184] or by chemical precipitation [22, 139]. The latter method is frequently preferred as it results in particles with more uniform sizes. In this method magnetite particles are obtained via a chemical reaction between iron salt solutions and concentrated alcali. The magnetite sediment then is mixed with a surfactant such as oleic acid that prevents the particles from forming aggregates in a carrier fluid. Magnetic properties of such nanoparticles are defined by their size and shape, the type of a crystal lattice and its defects and the interactions of particles with molecules of a carrier liquid [49, 107].

The stability of a magnetic colloid depends on the balance between attraction and repulsion forces acting between the particles. Closely located particles can coagulate under the action of van der Waals forces. The strength of these forces reduces in inverse proportion to the sixth power of the distance between particles. Single-domain magnetic particles also experience the attraction due to magnetic dipole-dipole interaction [47, 222] that decreases with the distance slower than van der Waals forces. However, at room temperature, the dipole-dipole interaction between particles of the size of the order 10 nm is negligible [192, 222]. Electrostatic or steric repulsion is used to prevent particles from forming aggregates. Depending on the stabilisation method, ferrofluids are categorised into two main groups: ionic, where particles are surrounded by an electrically charged shell, or surfacted, where particles are coated with amphiphilic molecules. Magnetite particles are typically surfacted with oleic acid. It has long (1.8 nm) molecules that are bent in the middle due to a double bond and that attach to the particle surface with one end due to adsorption. Long molecular tails extending away from particles create steric repulsion between them.

If the balance between particle repulsion and attraction is broken, they can start forming aggregates. Particle aggregation can also be triggered by the variation of the shape of nanoparticles from spherical [49, 155], by the presence of large particles and of surface coating defects [53, 224, 266] as well as by the increase of the particle concentration [8, 118]. The appearance of aggregates strongly influences magnetic and transport properties of ferrofluids. Depending on the number of participating particles, nano- and microscopic aggregates are distinguished with characteristic sizes of tens of nanometres and microns, respectively [53, 63, 118]. Nano-sized aggregates can be quasi-spherical [53, 118, 142, 228] or chain-like [8, 192, 193]. Microscopic aggregates also known as droplet aggregates are more likely to form in a magnetic field [53, 119, 187], when the fluid is cooled or if the concentration of nanoparticles is increased [61].

Properties of ferrofluids also depend on the choice of a carrier liquid. Most common ferrofluids are based on kerosene, silicon or transformer oil and water. Organic carrier fluids themselves have a complex composition and contain molecules of different weights and sizes as well as contaminants that could lead to the formation of insoluble sediment. Given that ferrofluids also can contain unbound surfactant (i.e. free oleic acid), ferrofluids are essentially multiphase systems. Their composition is schematically shown in Figure 1.1.

Single-domain magnetic particles suspended in ferrofluids can be of spherical (1, 2) or nonspherical (3) shapes with complete (1, 2) or deficient (3) coating. The size of magnetic particles determines the prevailing mechanism of magnetic relaxation (the alignment of magnetic moments of individual particles with the applied magnetic field). For large particles (1 and 3), Brownian mechanism [222] dominates when a particle rotates in the magnetic field as a whole. In small particles (2) Neel's mechanism [166] is preferred when the magnetic moment reorients within a particle. Quasi-spherical aggregates (4) with the size of 40–90 nm are typical for ferrofluids used in the majority of experiments described in this book. They can have both Brownian and Neel's mechanisms of relaxation [53, 194].

At present even more advanced two-phase (particle-fluid) models of ferrofluids [31, 122] are incapable of fully representing their microscopic structure. Therefore, simpler models treating ferrocolloids as monofluids with continuously varying properties are still used for a theoretical description of their macroscopic flows [13], and this approach will be taken in Chapter 3 here.

1.2 Physical Processes Taking Place in Ferrofluids

Despite containing ferrous particles, ferrofluids with organic bases have a low electric conductivity. When they are nonuniformly magnetised, bulk forces can arise that are capable of exciting convective motion of a nongravitational nature as well as forces that prevent such a motion [13]. The former are due to

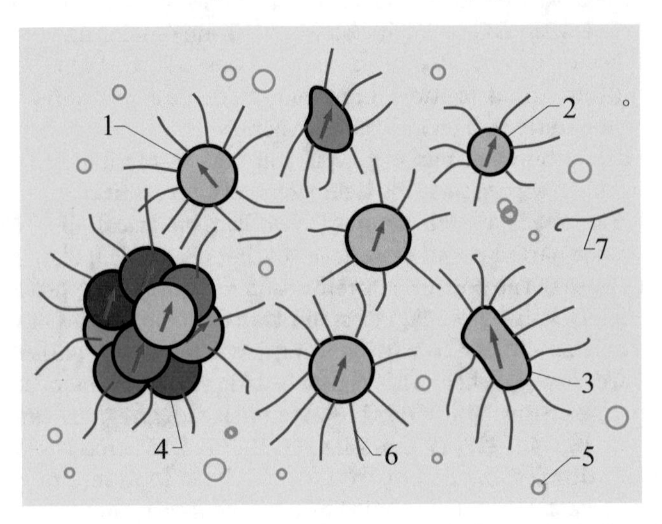

Fig. 1.1 Schematic composition diagram of a magnetic fluid: 1–3, single-domain magnetic particles of different sizes and shapes; 4, quasi-spherical aggregates; 5, molecules of a carrier liquid; 6 and 7, adsorbed and free surfactant, respectively. Arrows show magnetic moments of individual particles.

the so-called magnetic buoyancy, while the latter appear as a consequence of magnetic field distortion by the moving magnetised medium. In this context ferrohydrodynamics can be considered as the limit opposite to the inductionless limit in classical magnetohydrodynamics of electrically conducting fluids. The other distinction between magneto- and ferrohydrodynamics is that the main driving mechanism in the former is Lorentz force acting on a conducting fluid moving in a magnetic field, while in the latter it is ponderomotive Kelvin force driving stronger magnetised non-conducting fluid to regions with a stronger magnetic field. The magnetisation of ferrofluids decreases with temperature because of three reasons: Curie effect leads to demagnetisation of individual particles, more intense Brownian motion disorients their magnetic moments and thermal expansion reduces the effective concentration of magnetic phase in the fluid. Because of that ponderomotive force drives stronger magnetised cooler fluid to regions with stronger magnetic field displacing warmer fluid. Such a motion is called thermomagnetic convection. Its mechanism is shown schematically in Figure 1.2 for the cases of nonuniform and uniform external magnetic field.

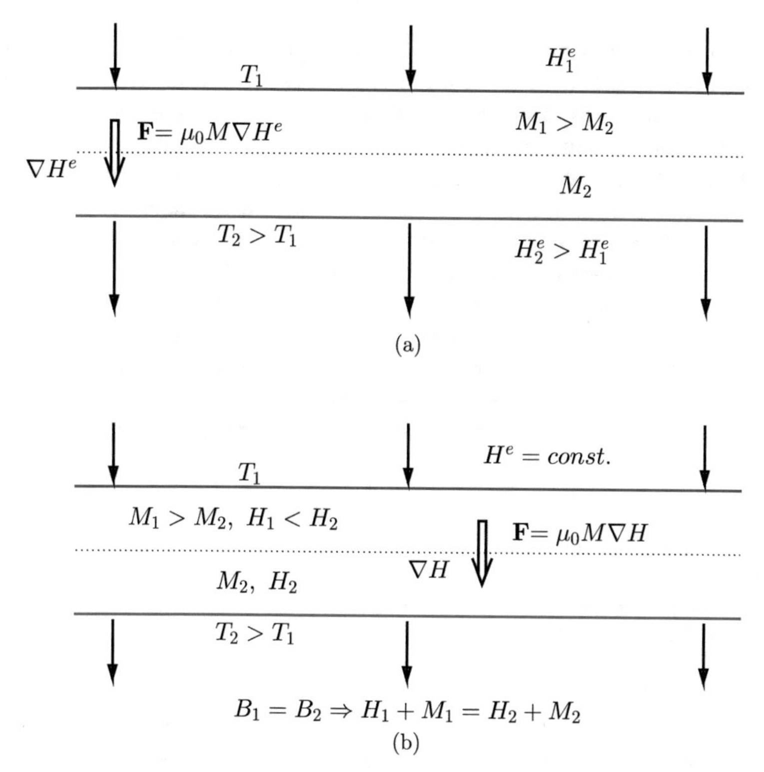

Fig. 1.2 Schematic diagram of forces driving thermomagnetic convection in (a) nonuniform and (b) uniform external magnetic field.

In the case of a nonuniform external magnetic field (such as the one created by a standard bar magnet), the motion of fluid is induced primarily by the gradient of this field (see Figure 1.2(a)). It plays the role of effective gravity, while fluid magnetisation is analogous to fluid density. Cooler fluid that has larger magnetisation (M_1) is drawn to the regions with a larger magnetic field that is in the direction of the external magnetic field gradient (∇H^e). When the nonuniformity of external magnetic field is sufficiently strong, perturbations of a magnetic field caused by the variation of fluid temperature are assumed to be negligible. Therefore, thermomagnetic convection in nonuniform external magnetic field is qualitatively similar to gravitational convection.

When the external magnetic field is uniform (e.g. magnetic field inside a solenoid), the mechanism of thermomagnetic convection is more subtle. The rigorous mathematical formulations describing forces acting in this case will be given in Chapter 2, but qualitatively their origin can be seen from Figure 1.2(b). According to Maxwell's boundary conditions at the surface separating differently magnetised media, the normal component of the magnetic induction vector **B**, which is proportional to the sum of the normal components of the magnetisation vector **M** and the magnetic field vector **H** inside the fluid, remains constant. Therefore, the magnitude of internal magnetic field (H_1) must be smaller wherever fluid is stronger magnetised. Thus, given the dependence of fluid magnetisation on the temperature discussed above, we conclude that even if the externally applied magnetic field is uniform, a cooler and stronger magnetised fluid will have a stronger "field-blocking" effect. Therefore, the internal magnetic field H there will be weaker. Hence placing a non-isothermal ferrofluid in a magnetic field parallel to the applied temperature gradient creates an inherently unstable situation, which is a necessary condition for the occurrence of thermomagnetic convection. The two approaches used to model a motion of non-isothermal magnetic fluid in an external magnetic field in the limiting cases of $\nabla H^e \gg \nabla H$ and $\nabla H^e \ll \nabla H$ are termed as inductionless (zero order) and induction (first order) approximations, respectively [12, 13].

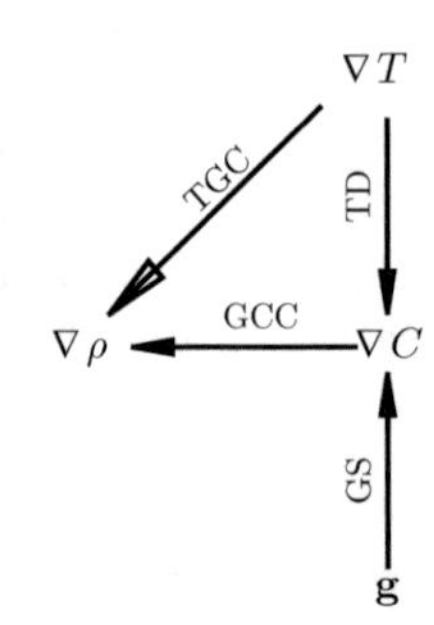

Fig. 1.3 Mechanisms of heat and mass transfer in ferrofluids in the absence of magnetic field. The acronyms denote TGC, thermogravitational convection driven by gravity **g**; GCC, gravi-concentrational convection; TD, thermodiffusion; and GS, gravitational sedimentation caused by gravity **g**. Various gradients existing in the fluid are of ∇T, temperature; $\nabla \rho$, density; and ∇C, concentration of solid phase.

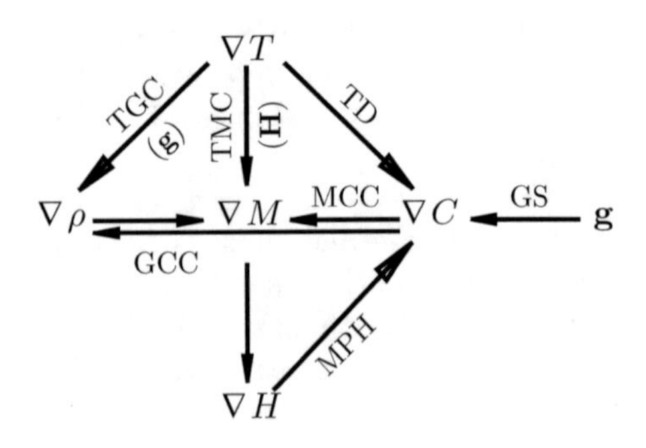

Fig. 1.4 Mechanisms of heat and mass transfer in ferrofluids in magnetic field. See caption of Figure 1.3 for the meaning of various acronyms. The additional acronyms appearing here denote TMC, thermomagnetic convection driven by the application of magnetic field **H**; MCC, magneto-concentrational convection; and MPH, magnetophoresis; ∇M and ∇H denote gradients of magnetisation and magnetic field, respectively.

It will be shown in Chapters 5 and 6 that thermoconvective instability arising in non-isothermal ferrofluids and caused by the combined action of gravitational buoyancy and ponderomotive forces is strongly influenced by the nonuniformity of solid-phase distribution in the bulk of fluid [23, 37, 38, 200]. In the absence of magnetic field, the gradient of the solid-phase concentration can be established as a result of thermodiffusion (TD) (or Soret effect) [23, 24, 75, 162, 230, 231] and gravitational sedimentation (GS) of particles and their aggregates [41, 84, 100, 141] as well as of insoluble residues [112, 161, 213] present in organic carrier fluids (see Figure 1.3). The size and weight variation of organic molecules contained in carrier fluids as well as the presence of unbound surfactant molecules can also lead to thermodiffusion [159, 190] resulting in the appearance of the concentration and thus fluid density gradient. Such a gradient can lead to gravi-concentrational convection (GCC) that may either enhance or suppress thermogravitational convection (TGC) arising due to nonuniform thermal expansion of a non-isothermal fluid.

In magnetic field, in addition to the above mechanisms, the concentration of solid phase can be affected by magnetophoresis (MPH) of magnetic particles and their aggregates [23, 197]. This may influence the local magnetisation of the fluid and lead to magneto-concentrational convection (MCC) (see Figure 1.4). These mechanisms act simultaneously with those causing thermomagnetic convection (TMC) schematically illustrated in Figure 1.2. As will be discussed in this book, the interaction of these processes leads to very complex macroscopic spatio-temporal dynamics including oscillatory convection, intermittency of convection patterns and suppression or promotion of convection onset and hysteresis.

A complex composition of ferrofluids and a wide range of microscopic processes taking place in them result in a large and frequently unmeasurable variation of fluid transport coefficients. Their values depend strongly on the conditions and history of storage and use of ferrofluids. This may lead to the unquantifiable uncertainty of the values of nondimensional control parameters that are used for identifying various flow regimes and thus to ambiguity of interpretation of experimental results. Several main reasons for that are discussed in Section 1.3.

Despite these difficulties the use of ferrofluids as magnetically driven heat carriers in conditions where gravitational convection is impossible due to the extreme congestion (microelectronics) or in the absence of gravity (spacecrafts) offers an efficient alternative to conventional heat management strategies. Laboratory experiments with ferrofluids placed in an external magnetic field are also used for physical modelling of processes taking place in the Earth's mantle, of oceanic currents and of flows arising in crystal growth applications.

1.3 Physical Properties of Ferrofluids

As noted in Section 1.1, in the absence of magnetic field, ferrofluids behave similarly to other non-magnetic nanofluids or colloids. Their effective thermal conductivity increases with concentration of nanoparticles because the nanoparticle material is a much better heat conductor than the base fluid. The thermal conductivity of a nanofluid is defined by the material and morphology of particles, the degree and type of their aggregation as well the type of the carrier liquid and the temperature. Its numerical value is well predicted by formulae first suggested by Maxwell [157] and Tareev [245] and later refined by various researchers by accounting for the shape of the particles and their aggregates, Brownian motion and thermodiffusion [103, 139, 246, 257]. The thermal conductivity of ferrofluids containing around to 10% by volume of nanoparticles increases up to 50% compared to that of a carrier liquid [26, 88, 134, 149, 188, 209]. However, in magnetic field, the value of the thermal conductivity of ferrofluid becomes a function of the field orientation relative to the direction of the temperature gradient. Depending on it the increase could vary between a few percent and several hundred percent [5, 94, 137, 149, 167, 186, 188].

Since the base liquids used for manufacturing ferrofluids are Newtonian, ferrofluids with concentration of nanoparticles up to 10% also remain Newtonian. However, as was shown by Einstein [83], the viscosity of fluids seeded with solid particles increases. Finite-size particles experience a rotating moment in a shear flow, which leads to the appearance of the so-called rotational viscosity. At present there exist around a dozen of rheological models of nanofluids that account for their viscosity dependence on nanoparticle concentration [11, 77, 122, 139, 204]. Experiments also show that the rotational viscosity depends on the size and shape of nanoparticles and their aggregates, fluid temperature and surfactants [11, 64, 77, 204, 257]. Moreover, the viscosity of ferrofluids depends on the magnitude and direction of the applied magnetic field [23, 158, 170]. The reason for that is that magnetic moments of particles with Brownian mechanism of relaxation [48] tend to align with the applied magnetic field by turning as a whole[2]. If the field is perpendicular to the local vorticity, it will hinder particle rotation, which is perceived as the increase of the fluid viscosity at a macroscopic level [108]—the so-called magnetoviscous effect. Similar to the rotational viscosity, magnetoviscosity is influenced by the local flow shear and the concentration and size of nanoparticles and their aggregates. In addition, it depends on the magnitude and direction of a magnetic field. Therefore, the value of magnetoviscosity measured in a given fluid using a viscosimeter can only give an approximation of the actual fluid viscosity in convection experiments because of the variation of flow velocity profiles from one experiment to another.

[2] Particles with Neel [166] relaxation where magnetic moments align with the field within a particle not causing its overall rotation do not lead to magnetoviscous effect.

A species separation known as Soret effect or thermodiffusion can occur in non-isothermal mixtures. In colloids this effect is two orders of magnitude stronger than in binary liquid mixtures [23, 24, 75, 232, 252]. It is characterised by the Soret coefficient, which is the ratio of the thermal and molecular diffusion coefficients. The sign of the Soret coefficient in ferrofluids depends on the concentration of nanoparticles, properties of a carrier liquid, the magnitude of the applied magnetic field and its orientation with respect to the temperature gradient [25, 231, 232, 253], and its magnitude is hard to determine due to the difficulties with measuring the coefficient of molecular diffusion [232].

One of the most important characteristics of ferrofluids placed in a magnetic field is the initial magnetic susceptibility determining the magnetic response of the ferrofluid to an applied weak magnetic field. Its magnitude also depends strongly on the microstructure of the colloid: concentration, material, morphology and size distribution of nanoparticles, types and sizes of particle aggregates, composition of the carrier fluid and the temperature of ferrofluid [104, 142, 154, 195].

To conclude, ferrofluid studies have to consider a wide range of microscopic physical processes that can take place in the bulk of a fluid and lead to its de-homogenisation. Given that convection is one of the most sensitive natural phenomena, it can easily be affected by the so-created nonuniformities so that experimental results have to be interpreted very carefully. The presence of multiple interacting microscopic transport mechanisms also makes theoretical analysis of magnetoconvection a very challenging problem. At present it can only be solved approximately by employing a number of simplifying assumptions as will be detailed in Chapter 2. The obtained theoretical and computational results thus have to be viewed in the context of the validity of the adopted simplifications, which puts an even stronger emphasis on experimental observations and quantitative measurements.

Chapter 2
Governing Equations

Abstract In this chapter the equations describing flows of non-isothermal ferrofluids and the corresponding boundary conditions are summarised. The main physical assumptions under which these equations are valid are discussed, and references to further reading are given. The constitutive equations for ferrofluid magnetisation are also reviewed. It is emphasised that commonly used Langevin's magnetisation law may be inaccurate in the case of non-isothermal ferrofluids, and thus the second-order modified mean-field model is preferred. Subsequently, the nondimensional form of equations is introduced, and the major governing nondimensional parameters in terms of which the results are presented throughout the manuscript are defined, and their physical meaning is identified.

2.1 Simplifying Physical Assumptions and Basic Equations

Equations describing flows of a ferrofluid placed in an external magnetic field were first given in [89]. If the temperature variation in the flow domain is sufficiently small, the Boussinesq approximation of the continuity, Navier-Stokes and thermal energy equations that are complemented with Maxwell's equations for the magnetic field written in the magnetostatic form due to the negligible electrical conductivity of ferrofluids [13, 209] read:

See Appendix B for the list of previously published materials reused in this chapter with permission.

A. A. Bozhko, S. A. Suslov, *Convection in Ferro-Nanofluids: Experiments and Theory*, Advances in Mechanics and Mathematics 40,
https://doi.org/10.1007/978-3-319-94427-2_2

$$\nabla \cdot \mathbf{v} = 0\,, \tag{2.1}$$

$$\rho_* \frac{\partial \mathbf{v}}{\partial t} + \rho_* \mathbf{v} \cdot \nabla \mathbf{v} = -\nabla p + \eta_* \nabla^2 \mathbf{v} + \rho \mathbf{g} + \mu_0 M \nabla H\,, \tag{2.2}$$

$$\frac{\partial T}{\partial t} + \mathbf{v} \cdot \nabla T = \kappa_* \nabla^2 T\,, \tag{2.3}$$

$$\nabla \times \mathbf{H} = 0\,,\ \nabla \cdot \mathbf{B} = 0\,, \tag{2.4}$$

where

$$\mathbf{B} = \mu_0(\mathbf{M} + \mathbf{H})\,, \quad \mathbf{M} = \frac{M(H,T)}{H}\mathbf{H}\,. \tag{2.5}$$

In the above equations, $\mathbf{v}$ is the velocity vector with the respective components (u, v, w) in the x, y and z directions, t is time, T is the temperature, p is the pressure, $\mathbf{g}$ is the gravity acceleration, $\mathbf{B}$ is the magnetic flux density, ρ_* is the fluid density, η_* is the dynamic viscosity, κ_* is the thermal diffusivity of the fluid and $\mu_0 = 4\pi \times 10^{-7}$ H/m is the magnetic constant. The internal magnetic field inside the fluid domain is $\mathbf{H}$ such that $|\mathbf{H}| = H$. It induces fluid magnetisation $\mathbf{M}$ such that $|\mathbf{M}| = M$, which is assumed to be codirected with the magnetic field: $\mathbf{M} = \chi_* \mathbf{H}$, where χ_* is the integral magnetic susceptibility of the fluid. As discussed, for example, in [172] and references therein, this is true if the magnetic particle size does not exceed $d_p \sim 13$ nm. In this case the ratio of the Brownian particle magnetisation relaxation time $\tau_B = (4\pi d_p^3 \eta_*)/(k_B T)$, where $k_B = 1.38 \times 10^{-23}$ J/K is Boltzmann constant, to the viscous time $\tau_v = \rho_* d^2/\eta_*$ characterising the macroflow development is $\tau_B/\tau_v \sim 10^{-5}$. Thus it is safe to assume that the orientation of the magnetic moments of individual particles and thus of the fluid magnetisation follows the direction of a local magnetic field. However, the orientation of magnetic particle aggregates can in principle be affected by the mechanical torque due to the local shear of the flow so that they can misalign with the local magnetic field. Yet the experiments reported in [174] show that such a misalignment only becomes noticeable for shear rates exceeding 15 s^{-1}, while the shear rate for typical convection flows that are of interest here is of the order of 0.1 s^{-1} or smaller. Therefore the misalignment of the magnetisation and magnetic field vectors can be safely neglected. It is also common to assume that the fluid magnetisation depends only on field and temperature, which is the case when the concentration of magnetic particles remains uniform in experiments (this assumption is not always valid in reality due gravitational sedimentation, thermo- or magnetophoresis of solid particles, which we discuss in detail in Section 6.3.4).

The subscript $*$ in the governing equations denotes the values of the fluid properties evaluated at the reference temperature T_* and reference internal magnetic field H_*. In writing Equation (2.2), we assume that the fluid remains Newtonian. It has been found in experiments of [29] that this is a reasonable approximation for fluids with the concentration of solid phase not exceeding

$f = 0.1$. The more recent measurements reviewed in [171, Ch. 4] have indicated that ferrofluids placed in the magnetic field can also behave as Bingham fluids with a non-zero yield-stress that increases approximately quadratically with the applied magnetic field. However, the yield-stress magnitude remains very small for the field strength range relevant to thermomagnetic flows of interest here so that the Newtonian fluid approximation is well justified.

As evidenced by numerous studies [13, 86, 170–172, 174, 175, 193, e.g.], the viscosity of concentrated ferrofluids depends on the applied magnetic field, and the local flow shear that influences the concentration of aggregates formed as a result of a dipole interaction between magnetised particles. In general, both the average and local values of viscosity can vary. Even though multiple experiments aiming at quantifying such a dependence have been reported in literature [171, 174, 193, e.g.], the data collected in these measurements cannot be used directly to model flows in geometries and conditions that are significantly different from those of rheological experiments. However, in theoretical studies the reference average fluid viscosity only enters the nondimensional governing equations in combination with other fluid properties forming magnetic Rayleigh or Grashof numbers (e.g. see Section 2.2). In parametric studies their values are typically allowed to vary over a wide range, which effectively includes all experimental conditions even though the exact value of magnetoviscosity remains unknown. The unknown variation of the local viscosity and other fluid properties subject to the action of the locally varying magnetic field and shear presents a more daunting problem. It is well known [239, 240, e.g.] that if sufficiently large such a variation can strongly influence the structure of the flow and its stability. Yet to make analytical progress in absence of a quantitative rheological model, one is forced to neglect these spatial variations of fluid properties in Equation (2.2). This is consistent with a widely used Boussinesq thermal approximation adapted for magnetic fluids [13] and is reasonable if the temperature and magnetic field variation across the domain occupied by fluid remain small. The qualitative agreement between the computational results and the experimental observations reported in [238, e.g.] indicates that indeed such a simplification preserves sufficient accuracy of the model and makes it tractable. Further discussion and a quantitative justification of this simplification will be provided for the specific case of a vertical fluid layer in Sections 2.2 and 3.2.

The last term in Equation (2.2) represents a ponderomotive (Kelvin) force that acts on a magnetised fluid in a nonuniform magnetic field and drives it towards regions with a stronger magnetic field as discussed in [13, 144]. To close the problem, thermal and magnetic equations of state are required, which are assumed to be in the simplest linear form that is valid for small temperature and field variations within the layer,

$$\rho = \rho_* - \beta_*(T - T_*)\,, \tag{2.6}$$

$$M = M_* + \chi\Delta H - K\Delta T\,, \quad \Delta H \equiv H - H_*\,, \quad \Delta T \equiv T - T_*\,. \tag{2.7}$$

Here $\beta_* = -\left.\frac{\partial\rho}{\partial T}\right|_P$ is the coefficient of thermal expansion of the fluid at $T = T_*$, H_* and $M_* = \chi_* H_*$ are the magnitudes of the magnetic field, and the magnetisation at the location with temperature T_*, $\chi = \partial M/\partial H|_{(H_*,T_*)}$ is the differential magnetic susceptibility and $K = -\partial M/\partial T|_{(H_*,T_*)}$ is the pyromagnetic coefficient. Using Equation (2.7) it is possible to rewrite (2.5) as

$$\mathbf{M} = \frac{\chi H + (\chi_* - \chi)H_* - K\Delta T}{H}\mathbf{H}\,. \tag{2.8}$$

Subsequently, eliminating the magnetisation in favour of the magnetic field, one obtains from the second of Equation (2.4)

$$(1+\chi)\nabla\cdot\mathbf{H} + \frac{(\chi_* - \chi)H_* - K\Delta T}{H}(\nabla\cdot\mathbf{H} - \nabla H\cdot\mathbf{e}) - K\nabla T\cdot\mathbf{e} = 0\,, \tag{2.9}$$

where $\mathbf{e} = (e_1, e_2, e_3) \equiv \mathbf{H}/H$ is the unit vector in the direction of the magnetic field. This equation shows that thermomagnetic coupling occurs mostly when the magnetic field and the temperature gradient have components in the same direction.

It is convenient to redefine pressure p entering the momentum equation (2.2) so that it includes both a hydrostatic component and a Kelvin force potential (see also [170, pp. 86, 87]). Upon using Equation (2.7), one writes

$$\begin{aligned}\mu_0 M\nabla H &= \mu_0[M_* + \chi\Delta H - K\Delta T]\nabla H \\ &= \mu_0\nabla[M_* H + \frac{1}{2}\chi\Delta H^2] - \mu_0 K\Delta T\nabla H\,.\end{aligned}$$

Upon introducing the modified pressure

$$P = p - \rho_*(\mathbf{r}\cdot\mathbf{g}) - \mu_0\left[M_* H + \frac{1}{2}\chi\Delta H^2\right]\,, \tag{2.10}$$

where $\mathbf{r} = (x, y, z)$ is the position vector equation (2.2) is written as

$$\rho_*\frac{\partial\mathbf{v}}{\partial t} + \rho_*\mathbf{v}\cdot\nabla\mathbf{v} = -\nabla P + \eta_*\nabla^2\mathbf{v} - \rho_*\beta_*(T - T_*)\mathbf{g} - \mu_0 K\Delta T\nabla H\,. \tag{2.11}$$

The governing Equations (2.1)–(2.4) require a set of appropriate boundary conditions. At the solid boundaries limiting the domain occupied by fluid, its velocities vanish and the temperature is assumed to be known:

$$\mathbf{v} = \mathbf{0}\,, \quad T = T_b\,. \tag{2.12}$$

The applied magnetic field must satisfy magnetic boundary conditions:

$$(\mathbf{H}^e - \mathbf{H})\times\mathbf{n} = \mathbf{0}\,,\ (\mathbf{B}^e - \mathbf{B})\cdot\mathbf{n} = 0\,, \tag{2.13}$$

where superscript e denotes external fields and $\mathbf{n}$ is the normal vector to the boundaries. Using Equation (2.9), the second of the conditions in Equation (2.13) can be rewritten as

$$[((1+\chi)H + (\chi_* - \chi)H_* \pm K\Theta)\mathbf{e} - \mathbf{H}^e] \cdot \mathbf{n} = 0 \tag{2.14}$$

where it is assumed for definiteness that the temperatures at the opposite surfaces bounding a ferrofluid layer deviate from the average value T_{av} and are $T_{av} \pm \Theta$. This completes the formulation of a main set of governing equations.

2.2 Nondimensionalisation and Governing Parameters

The governing equations and boundary conditions are nondimensionalised using

$$(x, y, z) = d(x', y', z')\,,\ \mathbf{v} = \frac{\kappa_*}{d}\mathbf{v}'\,,\ t = \frac{d^2}{\kappa_*}t'\,,\ P = \frac{\rho_* \kappa_*^2}{d^2}P'\,, \tag{2.15}$$

$$T - T_* = \Theta\theta'\,,\ \mathbf{H} = \frac{K\Theta}{1+\chi}\mathbf{H}'\,,\ H = \frac{K\Theta}{1+\chi}H'\,, \tag{2.16}$$

$$\mathbf{M} = \frac{K\Theta}{1+\chi}\mathbf{M}'\,,\ M = \frac{K\Theta}{1+\chi}M'\,, \tag{2.17}$$

where d is the characteristic size of the domain occupied by fluid and Θ is the characteristic temperature difference across the domain. Then omitting primes for simplicity of notation, one obtains nondimensional governing equations

$$\nabla \cdot \mathbf{v} = 0\,, \tag{2.18}$$

$$\frac{\partial \mathbf{v}}{\partial t} + \mathbf{v} \cdot \nabla \mathbf{v} = -\nabla P + \mathrm{Pr}\,\nabla^2 \mathbf{v} - \mathrm{Ra}\,\mathrm{Pr}\,\theta \mathbf{e}_g - \mathrm{Ra}_m\,\mathrm{Pr}\,\theta \nabla H\,, \tag{2.19}$$

$$\frac{\partial \theta}{\partial t} + \mathbf{v} \cdot \nabla \theta = \nabla^2 \theta\,, \tag{2.20}$$

$$\nabla \times \mathbf{H} = \mathbf{0}\,, \tag{2.21}$$

$$(1+\chi)(\nabla \cdot \mathbf{H} - \nabla\theta \cdot \mathbf{e}) + \frac{(\chi_* - \chi)\mathrm{N} - (1+\chi)\theta}{H}(\nabla \cdot \mathbf{H} - \nabla H \cdot \mathbf{e}) = 0\,, \tag{2.22}$$

$$\mathbf{M} = [\chi H + (\chi_* - \chi)\mathrm{N} - (1+\chi)\theta]\mathbf{e} \tag{2.23}$$

with the boundary conditions

$$[((1+\chi)(H \pm 1) + (\chi_* - \chi)\mathrm{N})\mathbf{e} - \mathbf{H}^e] \cdot \mathbf{n} = 0\,, \tag{2.24}$$

$$\mathbf{v} = \mathbf{0}\,,\ \theta = \theta_b \tag{2.25}$$

along the solid boundary. The dimensionless parameters appearing in equations are

$$\mathrm{Ra} = \frac{g\beta\Theta d^3}{\eta_* \kappa_*}\,,\ \mathrm{Ra}_m = \frac{\mu_0 K^2 \Theta^2 d^2}{\eta_* \kappa_* (1+\chi)}\,,\ \mathrm{Pr} = \frac{\eta_*}{\rho_* \kappa_*}\,,\ \mathrm{N} = \frac{H_*(1+\chi)}{K\Theta}\,. \tag{2.26}$$

The gravitational and magnetic Rayleigh numbers Ra and Ra_m characterise the importance of buoyancy and magnetic forces, respectively, Prandtl number Pr characterises the ratio of viscous and thermal diffusion transport and parameter N represents the nondimensional magnetic field at the reference location. Note that while magnetic Rayleigh number is the main nondimensional parameter characterising pure magnetoconvection, in laboratory experiments, the influence of gravitational convection usually cannot be neglected, and in studies of mixed gravitational and magnetic convection flows, it is traditional to use a nondimensionalisation based on "viscous speed" $\frac{\eta_*}{\rho_* d}$ rather than on "thermal speed" $\frac{\kappa_*}{d}$ used in (2.15). In this case the nondimensional momentum and thermal energy equations (2.19) and (2.20) take a slightly different form and read

$$\frac{\partial \mathbf{v}}{\partial t} + \mathbf{v} \cdot \nabla \mathbf{v} = -\nabla P + \nabla^2 \mathbf{v} - \mathrm{Gr}\theta \mathbf{e}_g - \mathrm{Gr}_m \theta \nabla H\,, \tag{2.27}$$

$$\frac{\partial \theta}{\partial t} + \mathbf{v} \cdot \nabla \theta = \frac{1}{\mathrm{Pr}} \nabla^2 \theta\,, \tag{2.28}$$

where gravitational and magnetic Grashof numbers are related to gravitational and magnetic Rayleigh numbers as

$$\mathrm{Gr} = \frac{\mathrm{Ra}}{\mathrm{Pr}} \text{ and } \mathrm{Gr}_m = \frac{\mathrm{Ra}_m}{\mathrm{Pr}}\,, \tag{2.29}$$

respectively.

Among other important physical quantities characterising the magnetic properties of the fluid are differential and integral magnetic susceptibilities χ and χ_* and pyromagnetic coefficient K that depend on the applied magnetic field and the temperature. The pyromagnetic coefficient K only enters the governing equations as an element of the nondimensional groups (2.26) so that its exact value is not required for the analysis. However, the magnitude of K (and thus of parameter N) can be conveniently used to distinguish between paramagnetic and ferromagnetic fluids. It is small in the former case and typically is of the order of 10^2 in the latter. At the same time, the values of magnetic susceptibilities χ and χ_* are important parameters entering the governing equations directly. It is a common practice to estimate them from Langevin magnetisation law that reads [144, 209]

$$M_L(H) = M_\infty L(\xi)\,, \quad L(\xi) = \coth \xi - \frac{1}{\xi}\,, \quad \xi = \mu_0 \pi \frac{M_s d_p^3 H}{6 k_B T}\,, \tag{2.30}$$

where M_∞ is the experimentally measured saturation magnetisation of the fluid, $L(\xi)$ is Langevin's function and ξ is Langevin's parameter. Here M_s is the saturation magnetisation of a magnetic phase at a given temperature, and d_p^3 is the average cube of the diameter of a magnetised core of solid particles. Due to Curie effect [144] (demagnetisation of a ferromagnetic with the increasing temperature) and thermal expansion of a carrier fluid , the saturation magnetisation of magnetic material and ferrofluid vary as

$$M_s = M_{s*} \frac{1 - \beta_2 T^2}{1 - \beta_2 T_*^2}\,, \tag{2.31}$$

$$M_\infty = M_{\infty *} \frac{1 - \beta_2 T^2}{1 - \beta_2 T_*^2} (1 - \beta_* (1 - f)(T - T_*))\,, \tag{2.32}$$

where β_* is the coefficient of thermal expansion of a carrier fluid, β_2 is Curie coefficient and f is the volume fraction of the magnetic phase. For example, magnetite frequently used to manufacture nanoparticles for ferrofluids has saturation magnetisation $M_{s*} = 480$ kA/m and Curie coefficient $\beta_2 = 8 \times 10^{-7}$ K^{-2} at the reference temperature $T_* = 293\,\text{K}$.

However, both experimental measurements and molecular dynamics simulations show that the magnetisation law of a realistic ferrofluid deviates significantly from the Langevin dependence. The main reason for this is that Langevin's law assumes no interparticle interactions, which is not the case for experimental fluids with magnetic phase concentration as high as $f = 0.1$. A comprehensive review of this issue is given in [117]. There the authors showed that the significant improvement of the accuracy of the magnetisation law for a ferrofluid is obtained via the use of the so-called second-order modified mean-field (MMF2) model that is essentially a two-term expansion of Weiss mean-field model [247, 258]. It is obtained by replacing the Langevin parameter ξ with

$$\bar{\xi} = \mu_0 \pi \frac{M_s d_p^3 \bar{H}}{6 k_B T}\,, \quad \bar{H} = H + \frac{1}{3} M_L(H) \left(1 + \frac{1}{48} \frac{dM_L(H)}{dH}\right)\,. \tag{2.33}$$

Here $\bar{H}$ is the effective magnetic field that takes into account mean magnetic interactions between particles in a concentrated magnetic fluid.

The only physical quantity that remains unknown in the above formulae is the average cube of the diameter of magnetised particle cores. This quantity depends strongly on the (unknown) size dispersion of nanoparticles and their aggregates present in the fluid. Thus in practice d_p^3 is determined by matching the predictions of the initial differential magnetic susceptibility $\lim_{H \to 0} \chi(H)$ of the fluid with its experimentally determined value [145, 196].

Even though the MMF2 model is shown to produce accurate values for magnetic properties of a ferrofluid, it requires the value of a local magnetic field H as an input. However, this quantity depends on the geometry of the considered problem. For example, it will be shown in Chapter 3 that when an external magnetic field H^e is applied perpendicularly to an infinite differentially heated layer of a ferrofluid bounded by parallel non-magnetic plates, the internal magnetic field in the midplane of the layer is $H_* = H^e/(1+\chi_*)$. The values of H_* and the corresponding fluid magnetisation M_* are shown in Figure 2.1(c) for the range of external magnetic field $0 \leq H^e \leq 35$ kA/m that corresponds to experiments discussed in Chapter 6. The comparison of

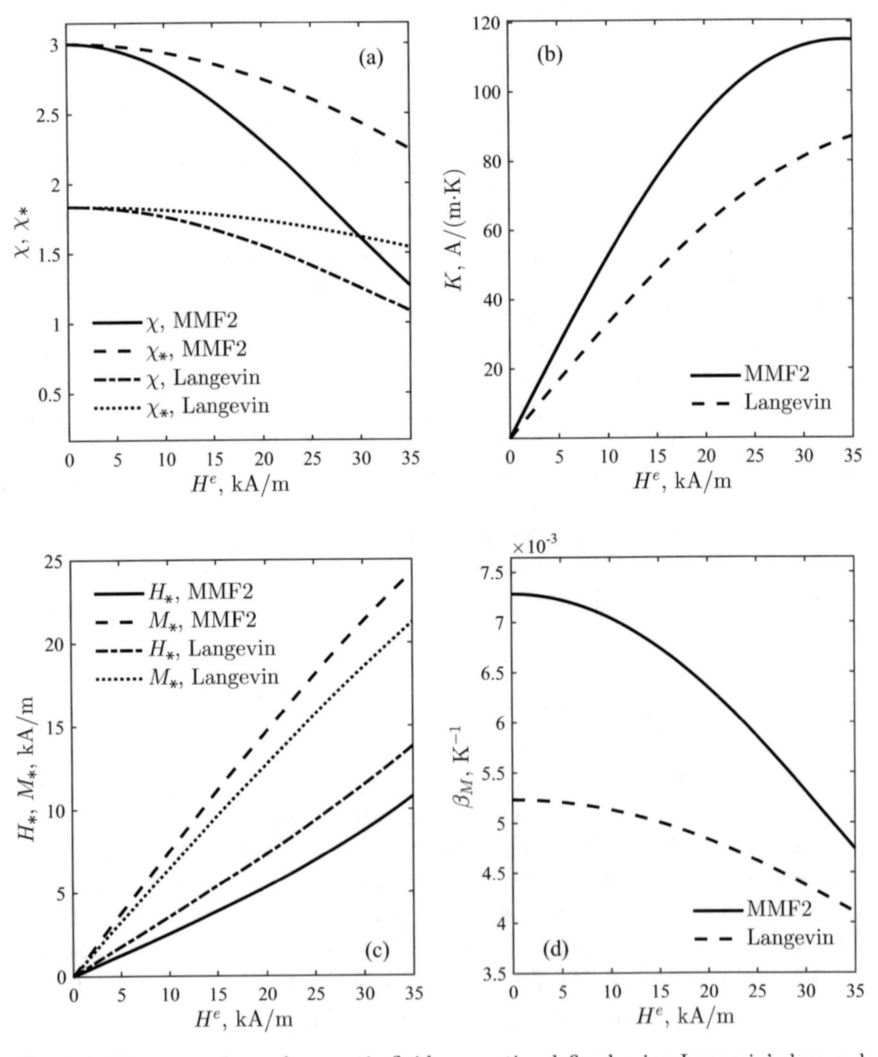

Fig. 2.1 The comparison of magnetic fluid properties defined using Langevin's law and MMF2 model [117] with $T_* = 293$ K.

relevant quantities obtained using Langevin and MMF2 models given in Figure 2.1 demonstrates that Langevin's law systematically underestimates the values of the magnetic susceptibilities and the pyromagnetic coefficient.

Note also that as discussed in [89] and [234], $\chi = \chi_*$ along the linear segment of the magnetisation curve. However, $\chi < \chi_*$ when the fluid's magnetisation approaches saturation. Therefore, by choosing different values of the differential and integral magnetic susceptibilities, one can investigate the effect of nonlinearity of the magnetisation law while still using the linearised magnetic equation of state (2.7).

Finally, note from Figure 2.1(b) that the value of the pyromagnetic coefficient K varies strongly with magnetic field. Therefore, when the comparison of theoretical and experimental results is desired, it may be more convenient to define the magnetic Rayleigh (or Grashof) number in terms of the relative pyromagnetic coefficient $\beta_M = K/M = K/(\chi_* H)$:

$$\mathrm{Ra}_m = \frac{\mu_0 \beta_M^2 \chi_*^2 H_*^2 \Theta^2 d^2}{\eta_* \kappa_* (1+\chi)} . \tag{2.34}$$

As seen from Figure 2.1(d) the variation of β_M with the applied magnetic field is not as pronounced, and thus it is seen that the value of Ra_m is approximately proportional to the square of a magnitude of the applied magnetic field.

Chapter 3
Theory of Thermogravitational and Thermomagnetic Convection in an Infinite Vertical Layer of Homogeneous Ferrofluid

Abstract The chapter discusses results of hydrodynamic stability analysis of non-isothermal ferrofluid flow arising between two parallel infinite plates maintained at different temperatures and placed in a uniform external magnetic field of various orientations. To distinguish between gravitational and magnetic buoyancy effects in the most straightforward way, the vertical layer configuration with the downward gravity is chosen while magnetic field is applied in the direction perpendicular to the layer or under a small angle with respect to the normal to the layer. In the absence of the gravity, such a configuration reduces to magnetic Rayleigh-Bénard problem. Comprehensive linear and weakly nonlinear stability results are presented. The existence of multiple three-dimensional convection patterns is demonstrated, and symmetry breaking effects of nonlinear fluid magnetisations are emphasised.

3.1 Introduction

Differentially heated flat vertical fluid layer is one of standard geometric configurations that has been used for convection studies for many decades, see [14, 18, 45, 97, 109, 126, 131, 211, 240, 251] and references therein for a comprehensive introduction to the field and historical overview. There are two main reasons for this. Firstly, such geometry is a practically relevant prototype for various realistic heat exchange systems. Secondly, this configuration is relatively simple to recreate experimentally. In this chapter we will also use a simplifying assumption that the lateral dimensions of a fluid layer are much larger than its thickness so that the influence of the layer edges can be ignored

See Appendix B for the list of previously published materials re-used in this chapter with permission.

A. A. Bozhko, S. A. Suslov, *Convection in Ferro-Nanofluids: Experiments and Theory*, Advances in Mechanics and Mathematics 40,
https://doi.org/10.1007/978-3-319-94427-2_3

in the middle part of the fluid domain. This assumption enables one to derive analytical expressions for the distributions of physical quantities of interest such as the temperature, velocity, magnetic field and fluid magnetisation across the layer and determine unambiguously the physical factors influencing the appearance and properties of thermomagnetic convection. However, prior to investigating the influences of a magnetic field on magnetically active medium contained in a layer, we first summarise the basic features of convection that occurs in such a geometry in ordinary non-magnetic fluids.

When the vertical walls of a layer are heated to different temperatures, Archimedean force occurs caused by a nonuniform thermal expansion that drives fluid up the warm wall and down the cool one. In an ideal fluid with constant transport properties, the vertical velocity profile arising in an infinitely tall and wide layer has a centrosymmetric cubic shape [14, 211, 251, e.g.]. If the fluid properties vary with temperature, then the symmetry is broken [240, e.g.], yet the flow remains parallel. The speed of such a flow is proportional to the applied cross-layer temperature difference or, nondimensionally, to the Grashof number. When it exceeds the critical value, this parallel flow becomes unstable with respect to periodic perturbations that are responsible for the onset of a cross-layer convection. The corresponding

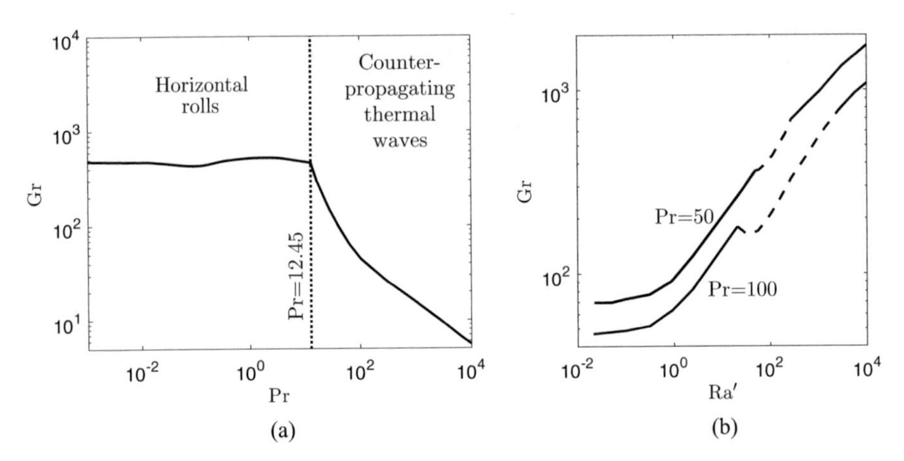

Fig. 3.1 Linear stability diagrams for a flow in a differentially heated infinite vertical fluid layer when (a) the walls are maintained at different but constant values [98] and (b) when a constant vertical temperature gradient is applied at the differentially heated walls [18]. The solid and dashed segments of stability boundaries in panel (b) correspond to thermal waves and stationary horizontal rolls, respectively.

stability diagrams are shown in Figure 3.1. The spatio-temporal patterns observed after the basic parallel flow loses its stability depend on thermoviscous properties of a fluid. In liquid metals and simple gases that have low viscosity and high thermal conductivity and thus are characterised by small Prandtl numbers, the instability is associated with the inflection point of the basic flow profile and is of hydrodynamic nature (shear instability). This instability exists above the mostly horizontal segment of the stability boundary in Figure 3.1(a). The convection patterns in this case consist of stationary horizontal rolls with axes in the midplane of the layer. If the symmetry of the problem is broken by the variation of fluid properties with temperature, the axes of such horizontal rolls shift away from the centreplane towards one of the vertical walls. As a result, the rolls start drifting in the direction of the local basic flow [239, 240]. Such instability is referred to as oscillatory. In viscous fluids with low thermal conductivity characterised by large Prandtl numbers Pr $>$ 12.45, the dominant instability mode leading to the cross-layer convection is of thermal nature. The perturbations are located near the heated and cooled boundaries and counter-propagate in the directions of the local basic flow: up near the warm wall and down near the cool one. Such patterns are termed as thermal waves. Unlike for the shear instability, the critical value of Grashof number for thermal waves decreases rapidly with Prandtl number as seen from Figure 3.1(a). Therefore, thermogravitational cross-layer convection in ferrofluids based on large-Prandtl-number organic carrier fluids is expected to take the form of vertically propagating waves.

While the infinite layer assumption enables one to make a significant analytical progress in convection studies, in practice containers of finite height are always used. Therefore fluid flowing vertically in the middle part of a layer necessarily turns and starts flowing horizontally near the top and bottom (usually adiabatic) edges of a finite layer. This leads to two major effects. The first is that since the fluid is forced to flow across the layer between the differentially heated vertical walls, the overall heat transfer across the finite layer always contains a near-edge convection component. Its intensity is approximately proportional to the speed of a vertical flow in the middle of the layer. The second effect caused by a turning flow is the formation of warm and cool regions near the top and bottom edges, respectively, and the establishment of the associated vertical average temperature stratification in a vertical layer of a finite size. They are caused by the rising warm jet turning from the warm to cool wall and then a cool jet impinging the bottom boundary. The investigation of such turning flows cannot be given fully analytically and typically requires a numerical consideration, see, for example, [131, 147]. However, a reasonable compromise can be achieved for finite layers of large aspect ratio by replacing the presence of solid top and bottom boundaries with specifying a vertical temperature gradient within a framework of an infinitely tall layer as was suggested in [18]. As seen from Figure 3.1(b), the vertical temperature stratification expressed in terms of the vertical Rayleigh number

$$\mathrm{Ra}' = 16\frac{g\beta d^4}{\eta_* \kappa_*}\frac{dT}{dy}$$

(refer to Figure 3.2 for the meaning of geometric parameters) stabilises a conduction state in a layer. Moreover, there exists a finite range of stratification values for which the thermal waves give way to stationary roll convection even in large-Prandtl-number fluids. Such conclusions based on a linear stability analysis of the basic conduction state have been also confirmed experimentally [126].

Keeping this in mind, in the subsequent sections, we will focus on the analysis of thin ferrofluid layers where the stratification effects are less pronounced. This is required to make sure that magnetic rather than gravitational buoyancy effects play a dominant role in determining the overall flow structure as this is the main focus of the current study.

3.2 Problem Definition and Basic Flow Solutions

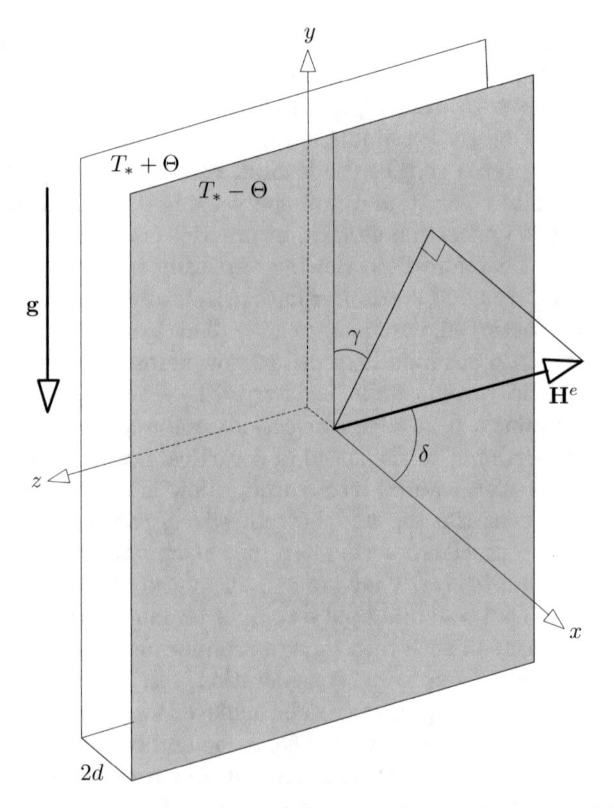

Fig. 3.2 Sketch of the problem geometry. The vector of external magnetic field, $\mathbf{H}^e$, forms angles δ and γ with the coordinate axes.

Consider a layer of a ferromagnetic fluid that fills a gap between two infinitely long and wide parallel non-magnetic plates as shown in Figure 3.2. The plates are separated by the distance $2d$ and are maintained at constant different temperatures $T_* \pm \Theta$. An external uniform magnetic field, $\mathbf{H}^e = (H^e_x, H^e_y, H^e_z)$ such that $|\mathbf{H}^e| = H^e$, where $H^e_x = H^e \cos\delta$, $H^e_y = H^e \sin\delta\cos\gamma$ and $H^e_z = H^e \sin\delta\sin\gamma$, is applied at an arbitrary inclination to the layer. We choose the right-hand system of coordinates (x, y, z) with the origin in the midplane of the layer in such a way that the plates are located at $x = \pm d$ and the y and z axes are parallel to the plates.

The governing Equations (2.18), (2.21)–(2.25), (2.27) and (2.28) admit steady solution of the form

$$\mathbf{v}_0 = \mathbf{0}\,,\ \theta_0 = \theta_0(x)\,,\ P_0 = P_0(x)\,,\ \mathbf{H}_0 = (H_{x0}(x), H_{y0}(x), H_{z0}(x))\,.$$

They satisfy the reduced equations

$$DP_0 = -\mathrm{Gr}_m \theta_0 e_{10} DH_{x0}\,,\ D^2 v_0 = -\mathrm{Gr}\theta_0\,,\ D^2\theta_0 = 0\,, \tag{3.1}$$

$$D\left(\left(1 + \frac{M_0}{H_0}\right) H_{x0}\right) = 0\,, \tag{3.2}$$

$$DH_{y0} = 0\,,\ DH_{z0} = 0\,, \tag{3.3}$$

and the boundary conditions representing no-slip condition for the fluid velocity and Maxwell's conditions for magnetic field (see Equations (2.13) and (2.14))

$$\mathbf{v}_0 = \mathbf{0}\,,\ \theta_0 = \pm 1\,,\ H_{y0} = H_y^e\,,\ H_{z0} = H_z^e \tag{3.4}$$

$$\left(1 + \frac{M_0}{H_0}\right) H_{x0} = H_x^e \text{ at } x = \mp 1\,, \tag{3.5}$$

where $H_0 \equiv \sqrt{H_{x0}^2 + H_{y0}^2 + H_{z0}^2}$, $M_0 \equiv \sqrt{M_{x0}^2 + M_{y0}^2 + M_{z0}^2}$ and $D \equiv \mathrm{d}/\mathrm{d}x$. Upon introducing the unit vector

$$\mathbf{e}_0(x) \equiv (e_{10}(x), e_{20}(x), e_{30}(x)) = \left(\frac{H_{x0}}{H_0}, \frac{H_{y0}}{H_0}, \frac{H_{z0}}{H_0}\right)$$

in the direction of the magnetic field, the basic flow solutions of Equation (3.1) are written as

$$\theta_0 = -x\,,\ v_0 = \frac{\mathrm{Gr}}{6}(x^3 - x)\,,\ \text{and } P_0 = \mathrm{Gr}_m \int_0^x x e_{10} DH_{x0}\,\mathrm{d}x + C\,, \tag{3.6}$$

where C is an arbitrary constant.

Equation (3.3) along with boundary conditions (3.4) results in the expressions for tangential components of the magnetic field that are constant inside the fluid layer $H_{y0}(x) = H_y^e$ and $H_{z0}(x) = H_z^e$. In view of (2.23) and (3.5), Equation (3.2) is integrated to obtain a nonlinear algebraic equation for the x-component of the unperturbed magnetic field

$$((1+\chi)(H_0 - \theta_0) + (\chi_* - \chi)\mathrm{N})H_{x0} = H_x^e H_0\,, \tag{3.7}$$

In particular, for a perpendicular field when $\mathbf{e}_0 = (1, 0, 0)$, this equation is integrated to produce an exact solution

$$H_0 = H_{x0} = \mathrm{N}_0 - x \tag{3.8}$$

for the basic flow component of magnetic field in the x direction across the layer, where N_0 is defined by (3.9) [234]

$$H^e = (1+\chi_*)\mathrm{N}_0 \tag{3.9}$$

However, when the external field is applied obliquely, a nonlinear Equation (3.7) does not have a closed-form solution and has to be solved numerically. Yet by evaluating this equation at the reference position $x = 0$, we obtain exact expressions for the magnetic field and its x-component there:

$$H_{x0}(0) = \mathrm{N}_0 \cos\delta\,, \quad \mathrm{N} \equiv H_0(0) = \mathrm{N}_0 \cos\delta \sqrt{1+(1+\chi_*)^2 \tan^2\delta}\,. \tag{3.10}$$

It follows from Equations (3.10) and (3.9) that the magnetic field in the mid-plane of the layer monotonically increases from N_0 for a normal field to H^e when the applied field is tangential to the layer.

For practical ferrofluids $\mathrm{N}_0 \sim 10^2 \gg x \sim 1$ and the solution of (3.7) can also be written asymptotically as

$$\begin{aligned}\frac{H^a_{x0}}{\mathrm{N}_0\cos\delta} &= 1 - \frac{1+\chi}{(1+\chi_*)^3\sin^2\delta + (1+\chi)\cos^2\delta}\frac{\mathrm{N}}{\mathrm{N}_0}\frac{x}{\mathrm{N}_0} \\ &+ \sin^2\delta\frac{(1+\chi)^2(1+\chi_*)^3\left(\frac{\mathrm{N}^2}{\mathrm{N}_0^2}+\frac{1}{2}\frac{\chi_*-\chi}{1+\chi_*}\cos^2\delta\right)}{((1+\chi_*)^3\sin^2\delta+(1+\chi)\cos^2\delta)^3}\frac{x^2}{\mathrm{N}_0^2} + o\left(\frac{x^2}{\mathrm{N}_0^2}\right)\,.\end{aligned} \tag{3.11}$$

A similar asymptotic approach was used in [111] to determine the approximate expression for the magnetic field in a layer of magnetic fluid subject to longitudinal temperature gradient (in contrast to the transverse gradient considered here).

If the magnetisation law is linear, that is if $\chi = \chi_*$, the above expression simplifies to

$$\frac{H^a_{x0}}{\mathrm{N}_0\cos\delta} = \left(1-\frac{x}{\mathrm{N}}\right) + (1+\chi)^2\sin^2\delta\frac{\mathrm{N}_0^2}{\mathrm{N}^2}\frac{x^2}{\mathrm{N}^2} + o\left(\frac{x^2}{\mathrm{N}^2}\right)\,. \tag{3.12}$$

The first two terms in the asymptotic solution (3.12) are equivalent to expression (9) given in [113] for a magnetic field in a layer of paramagnetic fluid with $\chi = \chi_* \ll 1$. However, the nonlinearity of the magnetic field in the layer was fully neglected in [113]. As seen from expressions (3.11) and (3.12), this is only true when $\mathrm{N}_0 \to \infty$ that is when pyromagnetic coefficient $K \to 0$; see Definition (2.26). This is shown to be a good approximation in the case of paramagnetic fluids [113]. However, K is large for ferrofluids; see Figure 2.1(b). Therefore N_0 is finite, and the nonlinearity of the magnetic field inside the layer of ferromagnetic fluid cannot be ignored. This is confirmed by fully nonlinear numerical solutions for the magnetic field shown for the finite values of $\chi = \chi_* = 3$ in Figure 3.3. The degree of nonlinearity is stronger for thermomagnetically more sensitive fluids. The thermomagnetic sensitivity of ferrofluids is characterised by the pyromagnetic coefficient K; see Figure 2.1(b). Nondimensionally, this is accounted for by the values of parameter N defined in (2.26) or, equivalently, by the magnitude of the nondimensional applied magnetic field H^e; see Figure 2.1(d). Since both N and H^e are inversely proportional to K, more thermomagnetically sensitive fluids are characterised by the smaller values of these parameters. To compare the behaviour of fluids with different thermomagnetic sensitivities below, the results are reported for the representative magnitudes of the nondimensional external magnetic fields $H^e = 100$ and $H^e = 10$. The first value corresponds to experiments discussed in Chapter 6, while the second is chosen consistently with [202, 203] to highlight the effects caused by the nonlinearity of magnetic field within the fluid layer. The three-term asymptotic solution (3.11) remains robust providing the accuracy within 1–2% of the numerically computed values even in the latter strongly nonlinear case.

Importantly, Figure 3.3 demonstrates that the relative deviation of the magnetic field within the layer from its average value cannot exceed $1/N_0$. Using the data presented in Figure 2.1(d), one then concludes that the field varies within the layer by less than 4%. This is the natural measure of the error that is introduced in the considered model by assuming that the field-dependent fluid properties remain constant in Equation (2.2).

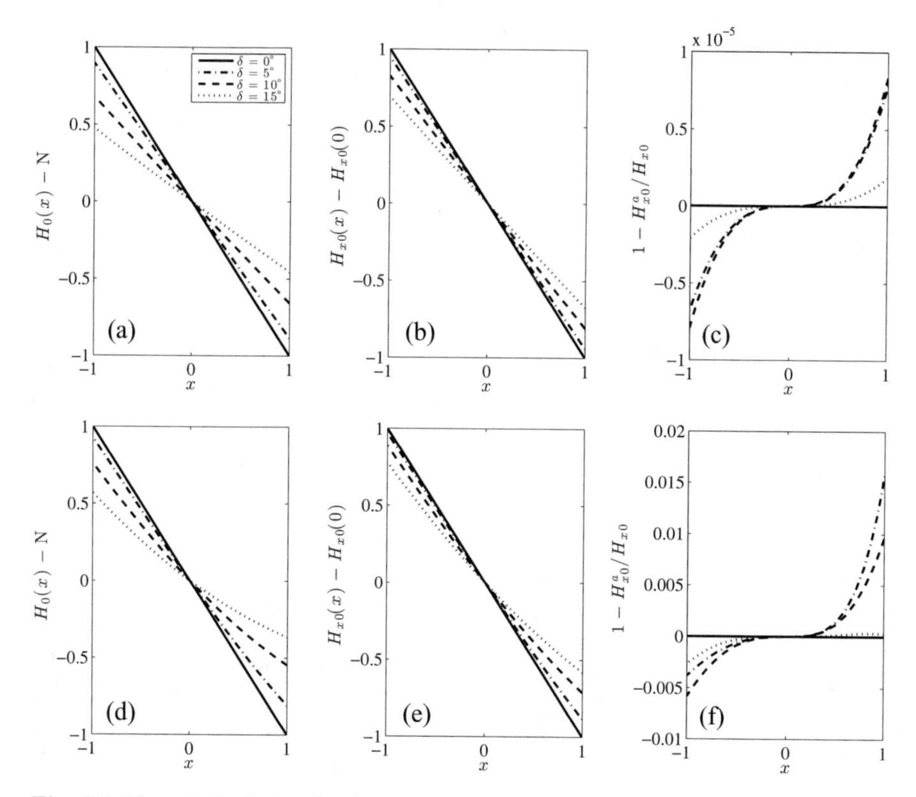

Fig. 3.3 Numerical solution for the magnitude H_0 of the undisturbed magnetic field ((a) and (d)) and its cross-layer component H_{x0} ((b) and (e)) for $H^e = 100$ (top row), $H^e = 10$ (bottom row), $\chi = \chi_* = 3$ and various field inclination angles δ. Plots (c) and (f) show the corresponding relative error of the asymptotic solution (3.11) [202].

Once the magnetic field within the layer is determined, the unperturbed fluid magnetisation is computed using

$$M_0 \equiv \sqrt{M_{x0}^2 + M_{y0}^2 + M_{z0}^2} = \chi H_0 + (\chi_* - \chi)\mathrm{N} - (1+\chi)\theta_0\,. \qquad (3.13)$$

The typical distributions of the magnetisation and magnetic pressure across the layer are shown in Figure 3.4. Both Figures 3.3 and 3.4 demonstrate that the field inclination leads to a noticeably asymmetric difference of the magnitudes of the cross-layer components of magnetic and magnetisation fields and their full magnitudes from their respective values at the midplane of the layer. These differences are more pronounced near the hot wall. The resulting asymmetry in the basic flow fields will be shown to influence the stability results qualitatively. The other observation is that the magnitude of the external magnetic field influences the fields inside the layer: weaker oblique external fields lead to a stronger nonlinearity of internal fields (compare the

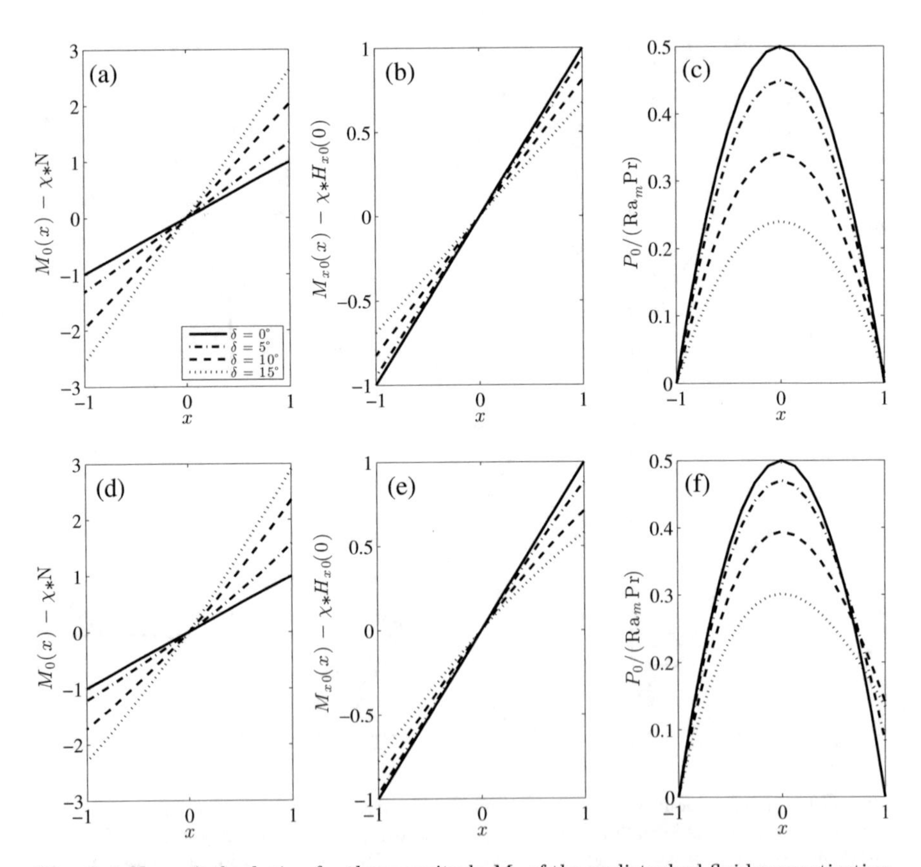

Fig. 3.4 Numerical solution for the magnitude M_0 of the undisturbed fluid magnetisation ((a) and (d)), its cross-layer component M_{x0} ((b) and (e)) and magnetic pressure P_0 ((c) and (f)) for $H^e = 100$ (top row), $H^e = 10$ (bottom row), $\chi = \chi_* = 3$ and various field inclination angles δ [202].

top and bottom rows in Figures 3.3 and 3.4). The symmetry-breaking effect of the field inclination is also evident in the behaviour of the magnetic pressure P_0 shown in Figure 3.4(c) and (f). As the field inclination angle increases, the pressure near the cold wall grows with respect to that near the hot wall. As will be discussed later, this leads to the preferential shift of instability structures towards the hot wall, which introduces a further asymmetry and qualitative change in stability characteristics compared to the normal field case considered in [89] and [234].

The behaviour of magnetic field lines inside the layer of a ferrofluid is shown in Figure 3.5(a) and (c). In contrast to the case of a normal field considered in [89] and [234], the nondimensional magnitude (relative to the fluid magnetisation) of the obliquely applied field strongly affects the geometry of magnetic lines. The curvature of magnetic lines is especially pronounced in stronger magnetisable fluids (plot (c) as contrasted to plot (a)). This has a profound effect on the distribution of the normal non-potential component $F_{K0} = -\theta_0 \frac{\mathrm{d}H_0}{\mathrm{d}x}$ of the nondimensional Kelvin force that can be viewed as a

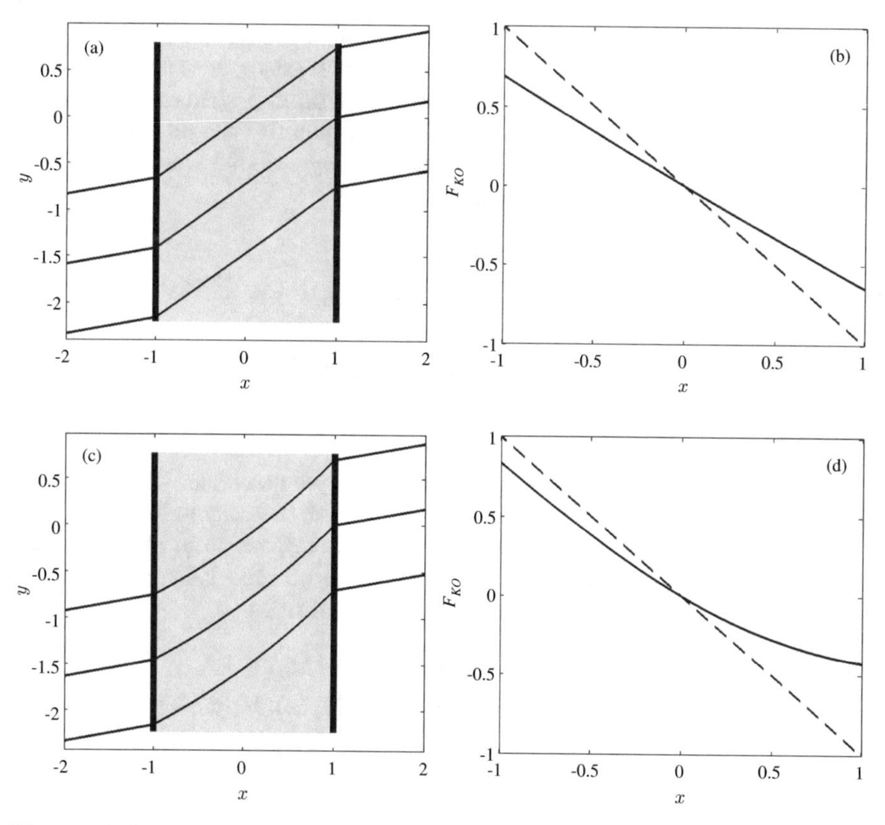

Fig. 3.5 Refraction of magnetic lines ((a) and (c)) and the distribution of Kelvin force ((b) and (d)) in a layer of magnetic fluid heated from the left for the field inclination angles $\delta = 10°$ (solid line), $\delta = 0°$ (dashed line), $\gamma = 0°$, $\chi = \chi_* = 3$, $H^e = 100$ ((a) and (b)) and $H^e = 10$ ((c) and (d)) [202].

magnetic buoyancy force. It is shown in Figure 3.5(b) and (d). Such a force is positive near the left wall and negative near the right wall, which corresponds to an inherently unstable situation when hot fluid near $x = -1$ is forced to flow towards the cold wall at $x = 1$ and vice versa. This situation is similar to an unstably stratified layer of a regular fluid heated from below in the downward gravitational field. Yet such similarity is complete only if the external magnetic field is normal to the layer. In this case, similar to its gravitational counterpart, the magnetic buoyancy is a linear function of the cross-layer coordinate x, and its nondimensional value is independent of the strength of the applied field; see dashed lines in Figure 3.5(b) and (d). However, when the oblique field of the same magnitude is applied to the layer, at least three qualitative differences arise due to the nonlinearity of the induced internal field. Firstly, the magnetic buoyancy force becomes more uniform across the layer so that the onset of thermomagnetic instability is expected to be delayed compared to the normal field situation. Secondly, the magnetic buoyancy force becomes a function of the magnitude of the applied magnetic field. Thirdly, and most importantly, the nonlinearity of the internal magnetic field leads to the situation when the unstably stratified with respect to magnetic buoyancy layer is effectively reduced to a sublayer in the vicinity of a hot wall; see Figure 3.5(d). Therefore in contrast to the case of a normal field, the cross-layer symmetry of the arising instability structures is broken. In the following sections, we will establish and quantify the physical features of instability patterns that arise in a normal magnetic field and are brought about by the inclination of an external field.

3.3 Flow Patterns in a Normal Magnetic Field

3.3.1 Linearised Equations for Infinitesimal Perturbations

We start with considering a vertical ferrofluid layer placed in a uniform magnetic field normal to it ($\delta = 0°$, see Figure 3.2). We investigate linear stability of the basic state discussed in Section 3.2 with respect to infinitesimal disturbances, which are periodic in the y and z directions. We use a standard normal mode approach and write perturbed quantities as

$$\begin{aligned}(\mathbf{v}, P, \theta, \mathbf{H}, H, \mathbf{M}, M) &= (\mathbf{v}_0, P_0, \theta_0, \mathbf{H}_0, H_0, \mathbf{M}_0, M_0) \\ &+ [(\mathbf{v}_1(x), P_1(x), \theta_1(x), \mathbf{H}_1(x), H_1(x), \mathbf{M}_1(x), M_1(x)) \\ &\times \exp(\sigma t + i\alpha y + \mathrm{i}\beta z) + c.c.] \,,\end{aligned}$$

where $\sigma = \sigma^R + \mathrm{i}\sigma^I$ is the complex amplification rate, α and β are real wavenumbers in the y and z directions, respectively, and $c.c.$ denotes the complex conjugate of the expression in brackets. To satisfy Equation (2.21) identically it is convenient to introduce perturbation $\phi_1(x)\exp(\sigma t + i\alpha y + \mathrm{i}\beta z)$ of a magnetic potential so that

$$\begin{aligned}
\mathbf{H}_1 &= (D\phi_1, \mathrm{i}\alpha\phi_1, \mathrm{i}\beta\phi_1)^T\,, \quad H_1 = D\phi_1\,,\\
\mathbf{M}_1 &= [\chi_* D\phi_1 + (\chi - \chi_*)H_1 - (1+\chi)\theta_1, \mathrm{i}\alpha\chi_*\phi_1, \mathrm{i}\beta\chi_*\phi_1]^T\,, \qquad (3.14)\\
M_1 &= \chi H_1 - (1+\chi)\theta_1\,.
\end{aligned}$$

The linearisation of Equations (2.18)–(2.23) about the basic state leads to

$$Du_1 + \mathrm{i}(\alpha v_1 + \beta w_1) = 0\,, \qquad (3.15)$$

$$\sigma u_1 + \left(\alpha^2 + \beta^2 + i\alpha v_0 - D^2\right) u_1 + DP_1 + \mathrm{Gr}_m DH_0\,\theta_1 + \mathrm{Gr}_m\theta_0 D^2\phi_1 = 0\,, \qquad (3.16)$$

$$\sigma v_1 + Dv_0\,u_1 + \left(\alpha^2 + \beta^2 + \mathrm{i}\alpha v_0 - D^2\right) v_1 + \mathrm{i}\alpha P_1 - \mathrm{Gr}\theta_1 + \mathrm{i}\alpha\mathrm{Gr}_m\theta_0 D\phi_1 = 0\,, \qquad (3.17)$$

$$\sigma w_1 + \left(\alpha^2 + \beta^2 + \mathrm{i}\alpha v_0 - D^2\right) w_1 + \mathrm{i}\beta P_1 + \mathrm{i}\beta\mathrm{Gr}_m\theta_0 D\phi_1 = 0\,, \qquad (3.18)$$

$$\sigma\theta_1 + D\theta_0 u_1 + \left(\frac{\alpha^2 + \beta^2 - D^2}{\mathrm{Pr}} + \mathrm{i}\alpha v_0\right)\theta_1 = 0\,, \qquad (3.19)$$

$$(D^2 - \alpha^2 - \beta^2)\phi_1 - \left(\frac{\chi_* - \chi}{1+\chi}\mathrm{N}_0 - \theta_0\right)\left(\alpha^2 + \beta^2\right)\frac{\phi_1}{H_0} - D\theta_1 = 0\,. \qquad (3.20)$$

The disturbance velocity and temperature fields are subject to standard homogeneous boundary conditions

$$u_1 = v_1 = w_1 = \theta_1 = 0 \quad \text{at } x = \pm 1\,. \qquad (3.21)$$

As was discussed in [89] in the case of non-magnetic boundaries, a perturbation of a magnetic field within a fluid layer causes perturbation of the external field. If there are no induced currents outside the layer and the ambient space is filled with non-magnetic medium (air), then the external magnetic field has a potential $\phi_1^e(x)\exp(\sigma t + \mathrm{i}\alpha y + \mathrm{i}\beta z)$. As follows from Equations (2.4) and (2.5), it must satisfy Laplace's equation

$$(D^2 - \alpha^2 - \beta^2)\phi_1^e = 0$$

in regions $x < -1$ and $x > 1$. A physically relevant bounded solution then is given by

$$\phi_1^e(x) = \begin{cases} C_1 e^{\sqrt{\alpha^2+\beta^2}x}\,, & x < -1 \\ C_2 e^{-\sqrt{\alpha^2+\beta^2}x}\,, & x > 1 \end{cases}\,. \qquad (3.22)$$

Linearisation of the first of conditions (3.4) and boundary condition (2.24) then leads to

$$\phi_1^e = \phi_1\,, \quad D\phi_1^e = (1+\chi)D\phi_1 \quad \text{at } x = \pm 1\,. \qquad (3.23)$$

Eliminating $C_{1,2}$ from (3.22) and (3.23), the mixed boundary conditions for ϕ_1 are finally obtained

$$(1+\chi)D\phi_1 \pm \sqrt{\alpha^2+\beta^2}\phi_1 = 0 \quad \text{at } x = \pm 1\,. \qquad (3.24)$$

Further, the generalised Squire's transformation

$$
\begin{aligned}
&x = \widetilde{x}\,,\ \alpha y + \beta z = \widetilde{\alpha}\widetilde{y}\,,\ w = \widetilde{z}\,,\ \sigma = \widetilde{\sigma}\,,\ \alpha^2 + \beta^2 = \widetilde{\alpha}^2\,,\ \beta = \widetilde{\beta}\,,\\
&u_1 = \widetilde{u}\,,\ \alpha v_1 + \beta w_1 = \widetilde{\alpha}\widetilde{v}\,,\ w_1 = \widetilde{w}\,,\ \theta_1 = \widetilde{\theta}\,,\ P_1 = \widetilde{P}\,,\ \phi_1 = \widetilde{\phi}\,,\\
&\alpha \mathrm{Gr} = \widetilde{\alpha}\widetilde{\mathrm{Gr}}\,,\ \mathrm{Gr}_m = \widetilde{\mathrm{Gr}_m}\,,\ \mathrm{Pr} = \widetilde{\mathrm{Pr}}\,,\ \chi = \widetilde{\chi}\,,\ \chi_* = \widetilde{\chi}_*
\end{aligned}
\tag{3.25}
$$

can be introduced. Upon noting that $\alpha v_0 = \widetilde{\alpha}\widetilde{v}_0$, where $\widetilde{v}_0 = \widetilde{\mathrm{Gr}}(\widetilde{x}^3 - \widetilde{x})/6$, $\theta_0 = \widetilde{\theta}_0$, $H_0 = \widetilde{H}_0$ and keeping tildes to denote only nontrivially transformed quantities, Equations (3.15)–(3.20) are rewritten as

$$Du + \mathrm{i}\widetilde{\alpha}\widetilde{v} = 0\,, \tag{3.26}$$

$$\sigma u + \left(\widetilde{\alpha}^2 + \mathrm{i}\widetilde{\alpha}\widetilde{v}_0 - D^2\right) u + DP + \mathrm{Gr}_m DH_0\,\theta + \mathrm{Gr}_m\theta_0 D^2\phi = 0\,, \tag{3.27}$$

$$\sigma\widetilde{v} + D\widetilde{v}_0\, u + \left(\widetilde{\alpha}^2 + \mathrm{i}\widetilde{\alpha}\widetilde{v}_0 - D^2\right)\widetilde{v} + \mathrm{i}\widetilde{\alpha}P - \widetilde{\mathrm{Gr}}\theta + \mathrm{i}\widetilde{\alpha}\mathrm{Gr}_m\theta_0 D\phi = 0\,, \tag{3.28}$$

$$\sigma w + \left(\widetilde{\alpha}^2 + \mathrm{i}\widetilde{\alpha}\widetilde{v}_0 - D^2\right) w + \mathrm{i}\beta P + \mathrm{i}\beta\mathrm{Gr}_m\theta_0 D\phi = 0\,, \tag{3.29}$$

$$\sigma\theta + D\theta_0 u + \left(\frac{\widetilde{\alpha}^2 - D^2}{\mathrm{Pr}} + \mathrm{i}\widetilde{\alpha}\widetilde{v}_0\right)\theta = 0\,, \tag{3.30}$$

$$(D^2 - \widetilde{\alpha}^2)\phi - \left(\frac{\chi_* - \chi}{1+\chi}\mathrm{N}_0 - \theta_0\right)\widetilde{\alpha}^2\frac{\phi}{H_0} - D\theta = 0\,, \tag{3.31}$$

where we also suppress subscript 1 to simplify the notation. Equation (3.28) is obtained in a standard way by multiplying Equation (3.17) by α, Equation (3.18) by β, adding them together and dividing the result by $\widetilde{\alpha}$. Note that only Equation (3.29) contains w, and thus it can be solved for w for any specified β after σ, P and ϕ are found from Equations (3.26)–(3.28), (3.30) and (3.31) which comprise of an equivalent two-dimensional problem.

Consider two limiting cases. If $\beta = 0$, i.e. if the solution is periodic only in the vertical direction, Equation (3.29) is fully decoupled from the rest of the system and becomes

$$\sigma w = \left(D^2 - \widetilde{\alpha}^2 - \mathrm{i}\widetilde{\alpha}\widetilde{v}_0\right) w\,. \tag{3.32}$$

It has a trivial solution $w = 0$ unless σ is an eigenvalue of (3.32). Upon multiplying Equation (3.32) by w^*, the complex conjugate of w, integrating it across the fluid layer by parts and taking the real part of the resulting expression, we obtain

$$\sigma^R = -\widetilde{\alpha}^2 - \frac{\int_{-1}^{1}|Dw|^2\mathrm{d}x}{\int_{-1}^{1}|w|^2\mathrm{d}x}$$

which is always negative. Therefore any aperiodic motion in the horizontal z direction must decay exponentially quickly due to viscous dissipation so that for $\beta = 0$ the asymptotic solution $w = 0$ holds.

Note that in the other limit of $\alpha = 0$ when the solution is periodic in the horizontal z, but not vertical y direction, the Squire transformation requires that $\widetilde{\mathrm{Gr}} = 0$ for any finite Gr. This means that thermal convection characterised by Grashof number plays no role at all in establishing a periodic flow pattern in the horizontal direction. It is fully defined by the thermomagnetic effects and by the value of Gr_m.

Upon discretisation and exclusion of Equation (3.29), system (3.26)–(3.31) results in a generalised algebraic eigenvalue problem for the complex amplification rate σ

$$(\mathsf{A} + \sigma \mathsf{B})\mathbf{q} = \mathbf{0}\,, \tag{3.33}$$

where $\mathsf{A}=\mathsf{A}(\widetilde{\alpha}, \widetilde{\mathrm{Gr}}, \mathrm{Gr}_m, \mathrm{Pr}, \chi, \chi_*)$ and B are matrices obtained after discretisation of the perturbation equations and eigenvector $\mathbf{q}$ contains discretised components of $(u, \widetilde{v}, P, \theta, \phi)^T$. Once both σ and $\mathbf{q}$ are found, equation

$$\left(D^2 - \sigma - \widetilde{\alpha}^2 - \mathrm{i}\widetilde{\alpha}\widetilde{v}_0\right) w = \mathrm{i}\beta(P + \mathrm{Gr}_m \theta_0 D\phi) \tag{3.34}$$

is solved for w. The inverse Squire transformation (3.25) then recovers full three-dimensional solution for perturbations.

3.3.2 Stability Results for an Equivalent Two-Dimensional Problem

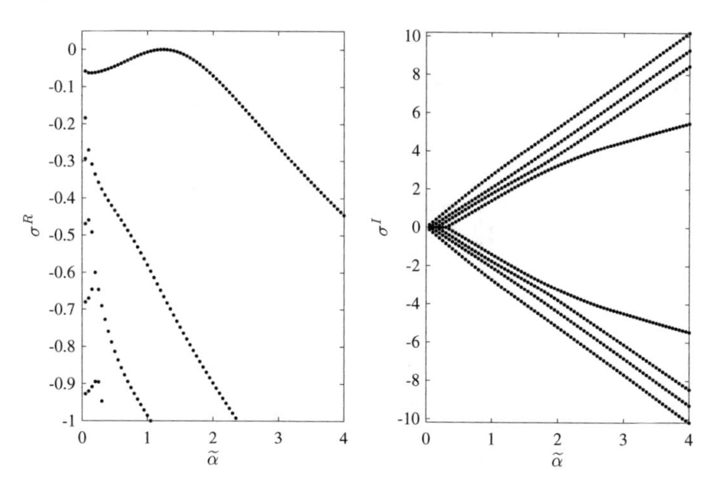

Fig. 3.6 Leading disturbance temporal amplification rates σ^R (left) and frequencies σ^I (right) as functions of the combined wavenumber $\widetilde{\alpha}$ for $(\mathrm{Gr}_m, \widetilde{\mathrm{Gr}}) = (0, 40.974)$ (the onset of thermogravitational convection).

To identify all possible physical mechanisms leading to the onset of convection in an infinite differentially heated flat layer of ferrofluid placed in a uniform external magnetic field normal to the layer, we first discuss the stability results for an equivalent two-dimensional problem given by Equations (3.26)–(3.28), (3.30) and (3.31). To obtain such results for each set of physical governing parameters, the problem is discretised using the Chebyshev pseudo-spectral collocation method as outlined in Appendix A and solved for a range of wavenumbers $\widetilde{\alpha}$ to locate the maximum of the disturbance amplification rate σ^R; see, for example, the left plot in Figure 3.6. Then the values of Gr_m or $\widetilde{\mathrm{Gr}}$ are iteratively changed until the set of values is found such that the maximum of $|\sigma^R|$ becomes smaller than the given threshold. This set of parameters then gives a point on the marginal stability boundary. A full stability diagram as presented in Figure 3.7(a) is obtained by repeating this computational process for a prescribed range of governing parameters. The stability region for an equivalent two-dimensional problem is bounded by three lines each representing different type of instability char-

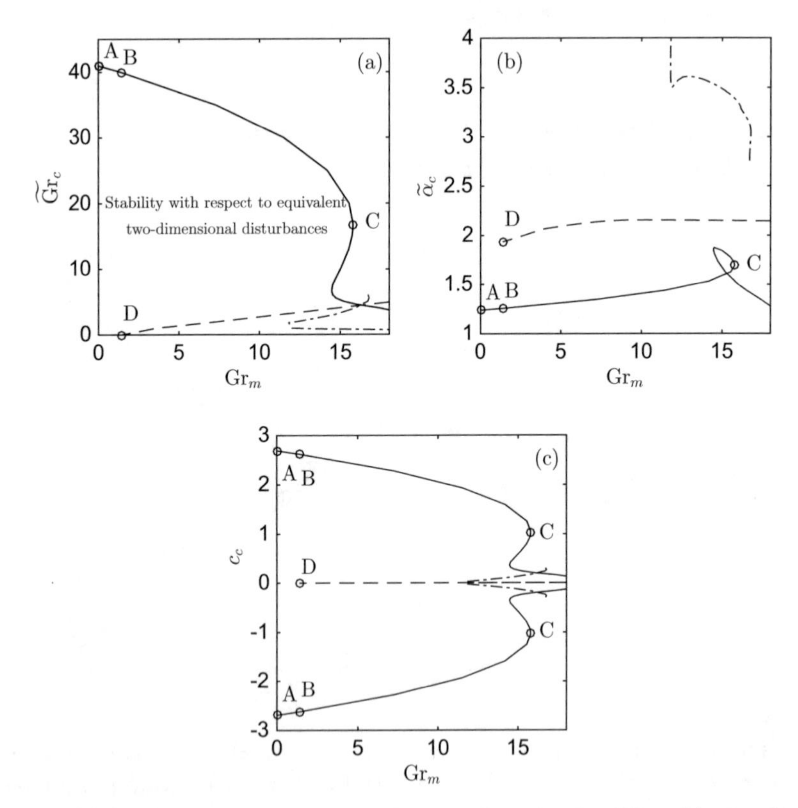

Fig. 3.7 (a) Stability diagram for an equivalent two-dimensional problem, (b) critical combined wavenumber $\widetilde{\alpha}$ and (c) the corresponding wave speeds along the stability boundaries shown in plot (a). Parameter values for points A–D are given in Table 3.1.

Table 3.1 Selected critical values for Pr $= 130$ and $\chi = \chi_* = 5$. Here $\widetilde{v}_{0max}$ is the maximum speed of the basic flow, and $\widetilde{c} \equiv -\frac{\sigma^I}{\widetilde{\alpha}_c}$ is the disturbance wave speed. The corresponding points A–D are shown in Figure 3.7.

	$\widetilde{\mathrm{Gr}}_c$	Gr_{mc}	$\widetilde{\alpha}_c$	$\widetilde{c}$	$\widetilde{v}_{0max}$
A	40.974	0	1.2384	±2.692	2.628
B	39.976	1.398	1.2563	±2.622	2.564
C	16.69	15.775	1.6964	±1.033	1.071
D	0	1.398	1.9365	0	0

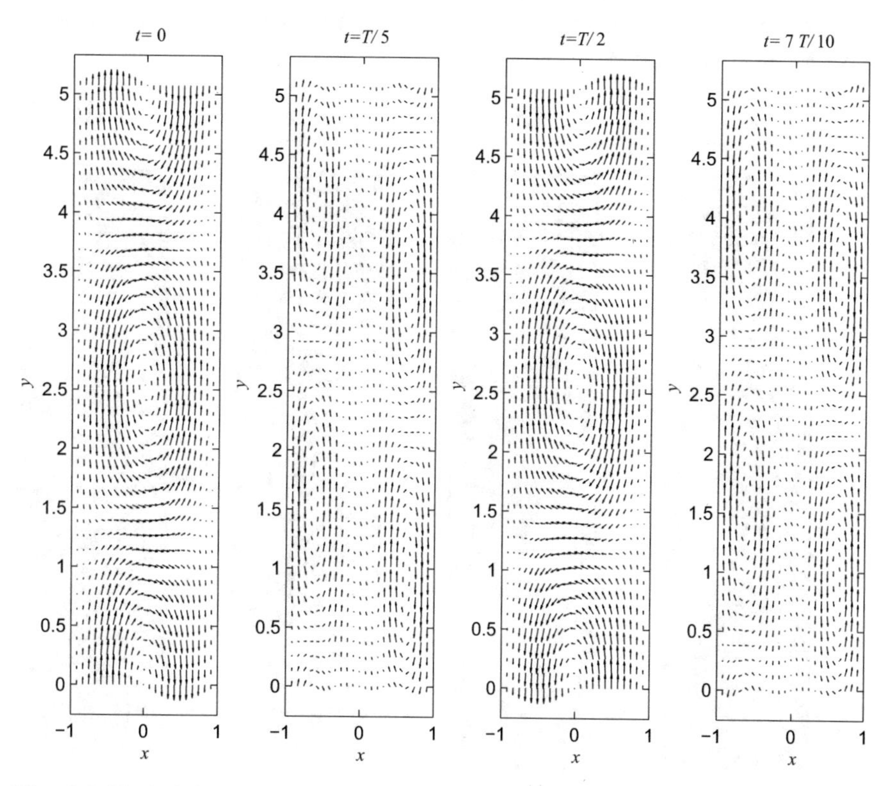

Fig. 3.8 Typical disturbance velocity fields for thermogravitational instability (point A in Figure 3.7) for $(\mathrm{Gr}_m, \widetilde{\mathrm{Gr}}) = (0, 40.974)$. Snapshots are for the indicated times where $T \approx 1.88$ is the period of oscillations.

acterised by its own wavenumber; see Figure 3.7(b). The solid line originates at $\mathrm{Gr}_m = 0$ and thus corresponds to the onset of thermogravitational convection. In subsequent sections, we will also refer to this type of convection as Type I instability. For large-Prandtl-number fluid, this type of convection is represented by two counter-propagating thermal waves [55, 98, 256]. Numerically, this is seen from the right plot in Figure 3.6: the eigenvalues of the linearised problems appear as the complex conjugate pairs (the σ^I

values have equal magnitudes but opposite signs). In turn the corresponding wave speeds $\widetilde{c}$ shown by solid lines in Figure 3.7(c) have the opposite signs as well. It is noteworthy that the wave speeds of disturbance thermal waves leading to thermogravitational convection are larger than the maximum speed of the basic flow; see points A and B in Table 3.1. On the other hand, when the role of magnetic effects increases, the disturbance wave speed decreases; see point C in Table 3.1. Such a wave speed behaviour is indicative of the gradual change of the instability mechanism from thermogravitational to thermomagnetic as the ratio $\widetilde{\mathrm{Gr}_m/\mathrm{Gr}}$ increases. The perturbation energy balance discussed in Section 3.3.3 will be used to determine where exactly such a transition occurs.

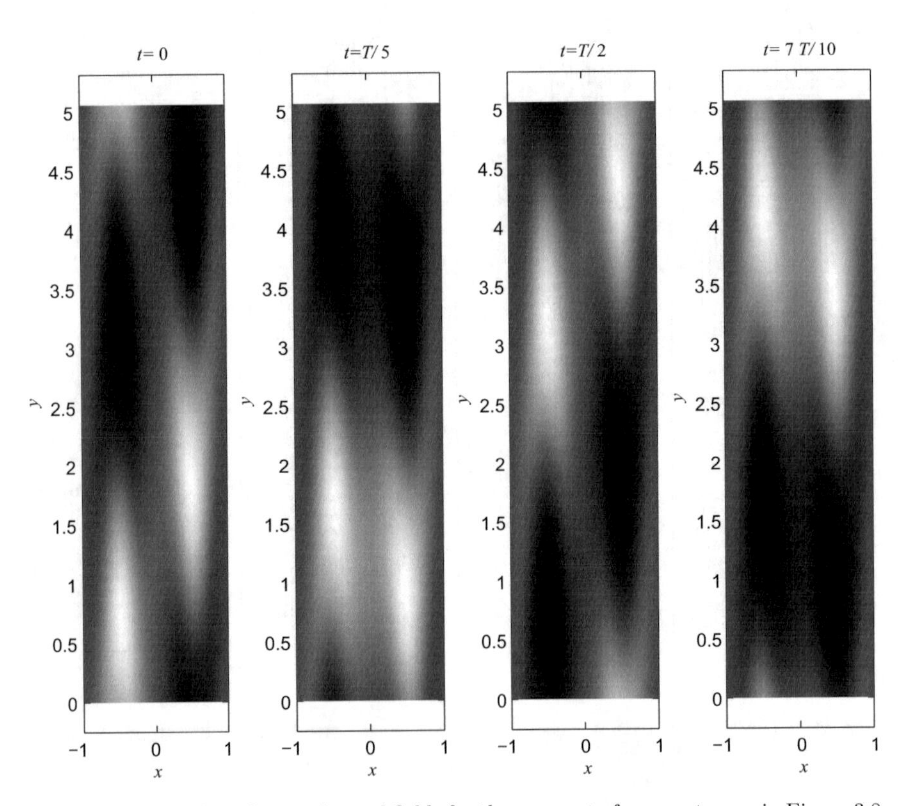

Fig. 3.9 Typical disturbance thermal fields for the same set of parameters as in Figure 3.8 (point A in Figure 3.7). Lighter areas correspond to higher temperature.

Typical disturbance velocity and thermal fields for thermogravitational convection (Type I instability) are shown as a series of snapshots in Figures 3.8 and 3.9. The fields are composed of two counter-propagating waves whose amplitudes (which are undetermined in the linear stability framework) are assumed to be equal so that the resulting pattern remains symmetric. As noted in [98], such a symmetry is indeed observed in fully nonlinear simulations provided that the boundary conditions are also symmetric. The instability velocity pattern consists of stationary counter-rotating vortices centered along the midplane of the fluid layer. However, they regularly change the direction of rotation as seen in a series of snapshots in Figure 3.8. During the transition, central vortices are destroyed and replaced by pairs of short-lived vortices, which appear near the walls and propagate in the direction of the local basic flow before they are replaced by central vortices rotating in the opposite direction. As expected, the direction of the vortex rotation is correlated with the location of warmer fluid regions: less dense warmer fluid (see light areas in Figure 3.9) tends to rise confirming the thermogravitational buoyancy-driven nature of this instability. In contrast to the stationary central vortex system seen in Figure 3.8, the thermal field snapshots clearly demonstrate the presence of two counter-propagating thermogravitational waves: the alternating warm and cold fluid regions shift upwards along the hot left wall and downwards along the cold right wall.

It will be shown in Section 3.3.3 that when the value of $\widetilde{\mathrm{Gr}}$ decreases along the solid instability boundary approximately below point C in Figure 3.7(a), the thermogravitational instability waves (Type I instability) are replaced with thermomagnetic waves. While the transition between the two types of waves is continuous along the solid line in Figure 3.7(a) and their appearances are similar, the physical mechanisms causing them are quite different. The thermogravitational waves are due to a buoyancy force acting in the vertical direction, while thermomagnetic waves are caused by a magnetic force acting across the gap. To avoid any ambiguity, the nature of this force is discussed in detail in the following paragraph in the context of stationary pure thermomagnetic convection. However, it is important to keep in mind that the physical mechanism discussed next is dominant for both stationary and wave-like instability regimes occurring at the relatively small values of $\widetilde{\mathrm{Gr}}$; see [236].

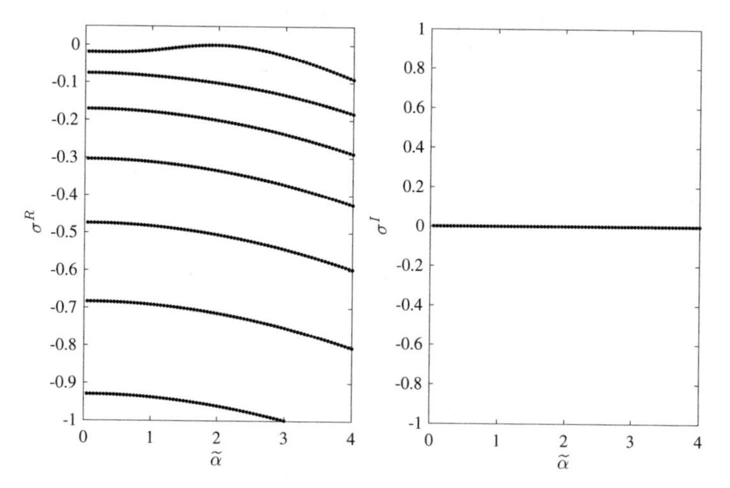

Fig. 3.10 Same as Figure 3.6 but for $(\mathrm{Gr}_m, \widetilde{\mathrm{Gr}}) = (1.398, 0)$ (the onset of stationary magnetoconvection, point D in Figure 3.7).

The stability region in Figure 3.7(a) is bounded from below by the dashed line which originates at $\widetilde{\mathrm{Gr}} = 0$ and therefore corresponds to the onset of magnetoconvection (also referred to as Type II instability in the subsequent sections). As seen from Figures 3.10 and 3.7(b), magnetoconvection is characterised by stationary patterns since the corresponding $\sigma^I = \widetilde{c} = 0$. This observation is consistent with findings of [89] and [229]. As seen from Figure 3.7(b), stationary magnetoconvection patterns have larger wavenumbers than those of thermogravitational convection. This distinction is most evident for smaller values of the magnetic Grashof number when the thermomagnetic convection rolls have a characteristic size about 1.5 times smaller than their thermogravitational counterparts. Typical disturbance fields for magnetoconvective instability are shown in Figure 3.11. Similar to thermogravitational instability, this instability leads to the appearance of central vortices seen in the leftmost plot in Figure 3.11. While these vortices are also stationary, they do not change the direction of their rotation. Since the regions of warm and cool fluid are centrally located above each other, they cannot cause buoyancy-driven motion of fluid and thus cannot be the reason for the vortex's appearance. Instead the pressure gradient directed from dark to light regions in the third plot is responsible for the vortical motion. For example,

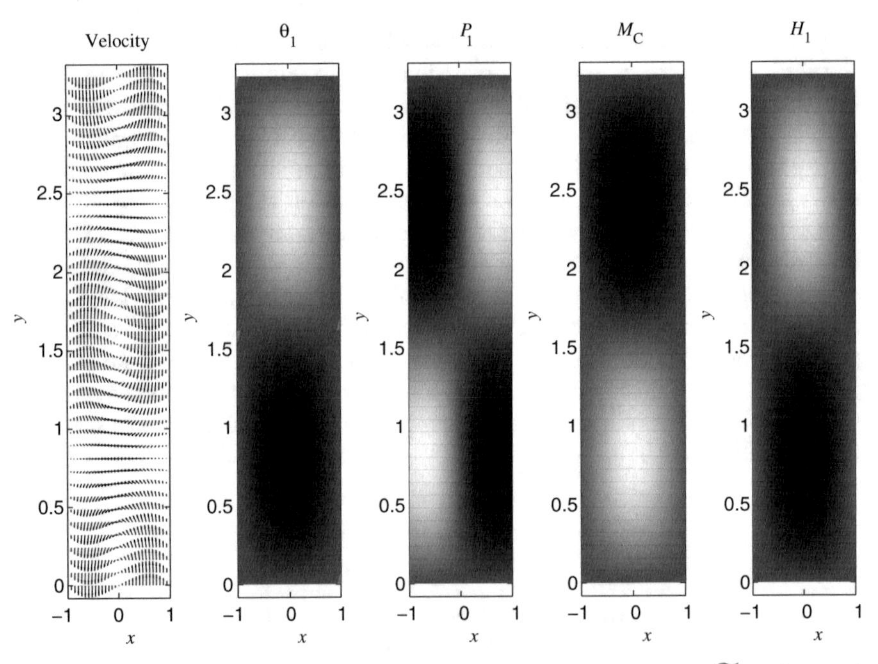

Fig. 3.11 Disturbance fields for magnetoconvective instability for $(\mathrm{Gr}_m, \widetilde{\mathrm{Gr}}) = (1.398, 0)$ (point D in Figure 3.7).

consider the region near the hot left wall near $\widetilde{y} = 1.5$[1]. As seen from the third plot in Figure 3.11, the value of the $\widetilde{y}$ component of the pressure gradient $-\frac{\partial P}{\partial \widetilde{y}}$ attains its positive maximum at this location. Responding to such a driving force, the fluid moves upwards here as illustrated in the leftmost plot. At the same time, the value of $-\frac{\partial P}{\partial \widetilde{y}}$ reaches its negative minimum near the cold wall around $\widetilde{y} = 1.5$ driving the fluid downwards. As a result, a vortex is formed which is seen near $\widetilde{y} = 1.5$ in the leftmost plot.

While providing the most obvious explanation for the disturbance flow pattern, the above discussion of the driving pressure gradient gives only a secondary reason for the existence of magnetoconvection. Indeed the definition of pressure P given by (2.10) is introduced for mathematical convenience and contains a combination of thermodynamic, hydrostatic and magnetic components. It is not directly seen which of them dominates in any particular regime. To disclose the true nature of the primary physical mechanism driving convection, we refer to the last term in Equation (2.2). It represents the force which acts on a magnetised fluid driving it to the regions with a stronger magnetic field. The primary magnetic field (3.8) weakens with the distance away from the hot wall. Therefore stronger magnetised fluid particles tend to move towards the hot wall. According to (3.14), the disturbance magnetisation field is affected by two factors: the induced disturbance magnetic field $\mathbf{H}$ and the disturbance thermal field (more precisely, the negative of it). A comparison of the second and fourth plots in Figure 3.11 shows that in the thermomagnetic convection regime fluid magnetisation is strongly correlated with the disturbance thermal field: highly magnetised regions correspond to cooler locations. Therefore, a larger thermal gradient leads to more pronounced variations in the local magnetisation. On the other hand, the induced magnetic field (the rightmost plot) tends to weaken the local magnetisation gradient and thus thermomagnetic effects. The disturbance energy analysis presented in [236] confirms that the induction of the magnetic field by the displacement of magnetised particles always plays a stabilising role. Yet the magnitude of this effect is always smaller than the destabilisation due to the thermomagnetic effects. Therefore the details of thermomagnetic convection mechanism are as follows. Thermal perturbations lead to the formation of cooler stronger magnetised regions in the flow domain (light areas in the fourth plot in Figure 3.11). The fluid in these regions is then driven towards the hot wall, where the basic magnetic field is stronger (see the major flow direction between $\widetilde{y} = 0.5$ and $\widetilde{y} = 1$ in the leftmost plot). This impinging jet of magnetised fluid hits the wall and creates a high-pressure region there; see the third plot. The so-created vertical pressure gradient is, subsequently, responsible for driving the fluid away from the impingement point. The fluid conservation then requires that warmer less magnetised fluid is displaced towards the cold wall thus creating the vortices seen in the leftmost plot in Figure 3.11.

[1] Note that in the context of this section dealing with the equivalent two-dimensional problem $\widetilde{y}$ denotes any direction in the plane of the layer. The three-dimensional unfolding of results presented here will be discussed in Section 3.3.4.

There also exists a small parametric region in the lower right corner of Figure 3.7(a) (see also the close-up in Figure 3.21(b)) where the stability region is bounded from below by the dash-dotted line. Unlike the solid and dashed lines corresponding to the thermogravitational/thermomagnetic waves (Type I instability) and stationary magnetoconvection (Type II instability), respectively, the nature of this dash-dotted boundary is not apparent from Figure 3.7(a) alone. Thus we refer to the growth rate spectrum plots in Figure 3.12 to identify it. It follows from Figure 3.7(b) that the dash-dotted line corresponds to the rightmost top maximum in the left plot in Figure 3.12. It is also seen from Figure 3.12 that this maximum is continuously connected to the leftmost maximum which, according to the wavenumber and wave speed data in Figure 3.7(b) and (c), corresponds to the onset of thermomagnetic wave instability (see the discussion of solid lines in Figure 3.7 above). Therefore we conclude by continuity that the dash-dotted segment of the stability boundary in Figures 3.7(a) and 3.21(b) still corresponds to thermomagnetic waves. However, they are characterised by a larger wavenumber, and we refer to them as Type III instability below. Thus it is expected that at larger values of the magnetic Grashof number, the unsteady thermomagnetic convection may reveal itself as a combination of two pairs of counter-propagating

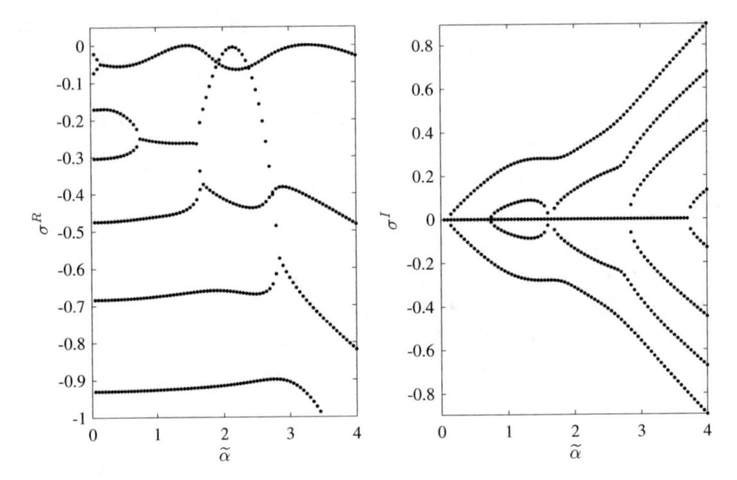

Fig. 3.12 Same as Figure 3.6 but for $(\mathrm{Gr}_m, \widetilde{\mathrm{Gr}}) = (16.2, 4.4)$. In the left plot, the left and right maxima correspond to small- and large-wavenumber thermomagnetic waves, respectively, and the middle maximum corresponds to stationary thermomagnetic rolls.

thermomagnetic waves (see solid and dash-dotted lines in Figure 3.7(c)) with different wavenumbers and wave speeds. The disturbance fields corresponding to the left and right maxima in Figure 3.12 are qualitatively similar to those shown in Figures 3.8 and 3.9 and are not presented here.

Similarly, we deduce that the middle maximum in the left plot in Figure 3.12 corresponds to a stationary magnetoconvection mode (the corresponding values of $\sigma^I = \widetilde{c} = 0$, Type II instability) with disturbance fields similar to those shown in Figure 3.11.

It is worthwhile noting that unlike the dashed and solid marginal stability lines in Figure 3.7(a), the dash-dotted segment of the stability boundary has a finite extent. The reasons for this become clear from Figures 3.13 and 3.14. At the right end of the dash-dotted segment, the σ^R maximum seen in Figure 3.12 for large wavenumbers shifts to the left until it disappears blending with the right "slope" of the maximum existing for smaller wavenumbers. As a result, the range of unstable disturbance wavenumbers widens forming a plateau seen in the left plot in Figure 3.13, and the distinction between short- and long-wavelength modes becomes obscured. All disturbances with wavenumbers $1.5 \lesssim \widetilde{\alpha} \lesssim 3$ have approximately the same growth rate and cannot be distinguished on this ground alone. This is in contrast to the regime depicted in Figure 3.12 where disturbances corresponding to the three discrete wavenumbers (three maxima) are expected to dominate the instability pattern.

At the left end of the dash-dotted stability boundary segment, the middle maximum seen in the left plot of Figure 3.12 dominates. This maximum corresponds to a stationary magnetoconvection instability mode. It leaves little room for disturbance waves with large wavenumbers and suppresses

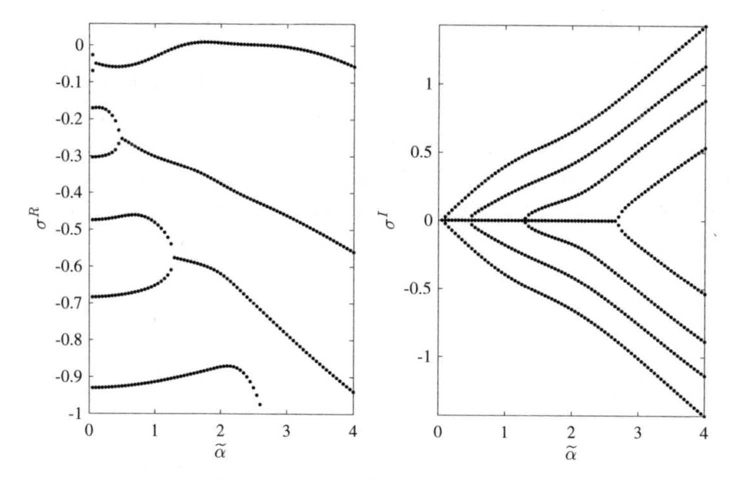

Fig. 3.13 Same as Figure 3.6 but for $(\mathrm{Gr}_m, \widetilde{\mathrm{Gr}}) = (16.7, 6.5)$ (thermomagnetic waves with a wide wavenumber range).

waves with small wavenumbers completely; see Figure 3.14. Two real eigenvalues σ (one of which corresponds to a stationary thermomagnetic instability mode) collide at $\widetilde{\alpha} \approx 3.6$ (see the left plot in Figure 3.14) and form a pair of complex conjugate eigenvalues which correspond to two counter-propagating thermomagnetic waves. The real part σ^R of these complex conjugate eigenvalues remains negative for $\mathrm{Gr}_m \lesssim 11.8$ which corresponds to the left end of

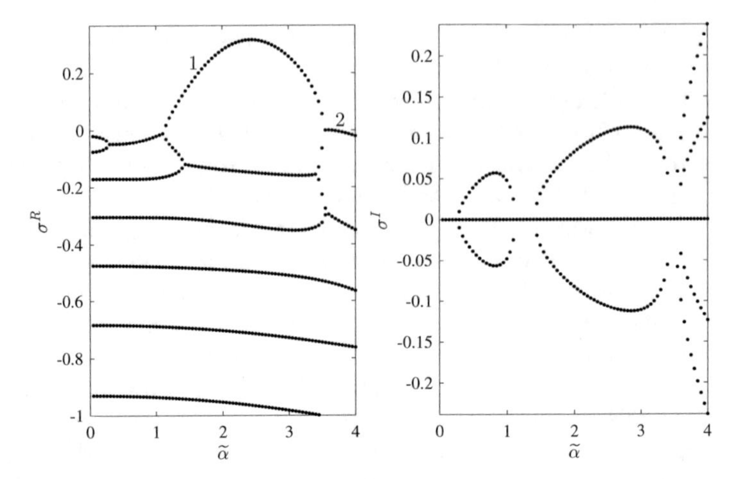

Fig. 3.14 Same as Figure 3.6 but for $(\mathrm{Gr}_m, \widetilde{\mathrm{Gr}}) = (11.8, 1.8)$. In the left plot: modes 1 and 2 correspond to stationary magnetoconvection rolls and large-wavenumber thermomagnetic waves, respectively.

the dash-dotted line in Figure 3.7(a). This is the smallest value of Gr_m for $Pr = 130$ for which thermomagnetic waves can exist.

Overall, the computational stability results show that magnetic effects play a strongly destabilising role in thermogravitational convection. Critical Grashof number decreases rapidly with the increasing magnetic Grashof number; see the solid line in Figure 3.7(a). A stronger magnetic field also slows down thermal waves as seen from Figure 3.7(c) and Table 3.1. On the other hand, thermogravitational effects tend to inhibit magnetoconvection: a stronger magnetic field is required for the onset of magnetoconvection when $\widetilde{\mathrm{Gr}}$ is increased. This effect was also confirmed by the disturbance energy analysis in [236].

3.3.3 Perturbation Energy Balance

Linearised perturbation energy balance analysis offers a computationally straightforward way of providing a further quantitative insight into the roles various instability mechanisms play in defining the observed flow patterns [238]. To derive the disturbance kinetic energy equation, we multiply the momentum Equations (3.27) and (3.28) by the complex conjugate velocity components $\bar{u}$ and $\bar{v}$, respectively, add them together, integrate by parts across the layer using boundary conditions (3.21) and the continuity equation (3.26) and take the real part of the result to obtain

$$\sigma^R \Sigma_k = \Sigma_{uv} + \Sigma_{\mathrm{Gr}} + \Sigma_{vis} + \Sigma_{m1} + \Sigma_{m2} \,, \tag{3.35}$$

where

$$\Sigma_k = \int_{-1}^{1} \left(|u|^2 + |\widetilde{v}|^2\right) \, dx > 0 \,, \tag{3.36}$$

$$\Sigma_{uv} = -\int_{-1}^{1} D\widetilde{v}_0 \Re(u\bar{\widetilde{v}}) \, dx \,, \tag{3.37}$$

$$\Sigma_{\text{Gr}} = \int_{-1}^{1} \underbrace{\widetilde{\text{Gr}}\Re(\theta\bar{\widetilde{v}})}_{E_{\text{Gr}}} \, dx \,, \tag{3.38}$$

$$\Sigma_{vis} = -\widetilde{\alpha}^2 \Sigma_k - \int_{-1}^{1} \left(|Du|^2 + |D\widetilde{v}|^2\right) \, dx = -1 \,, \tag{3.39}$$

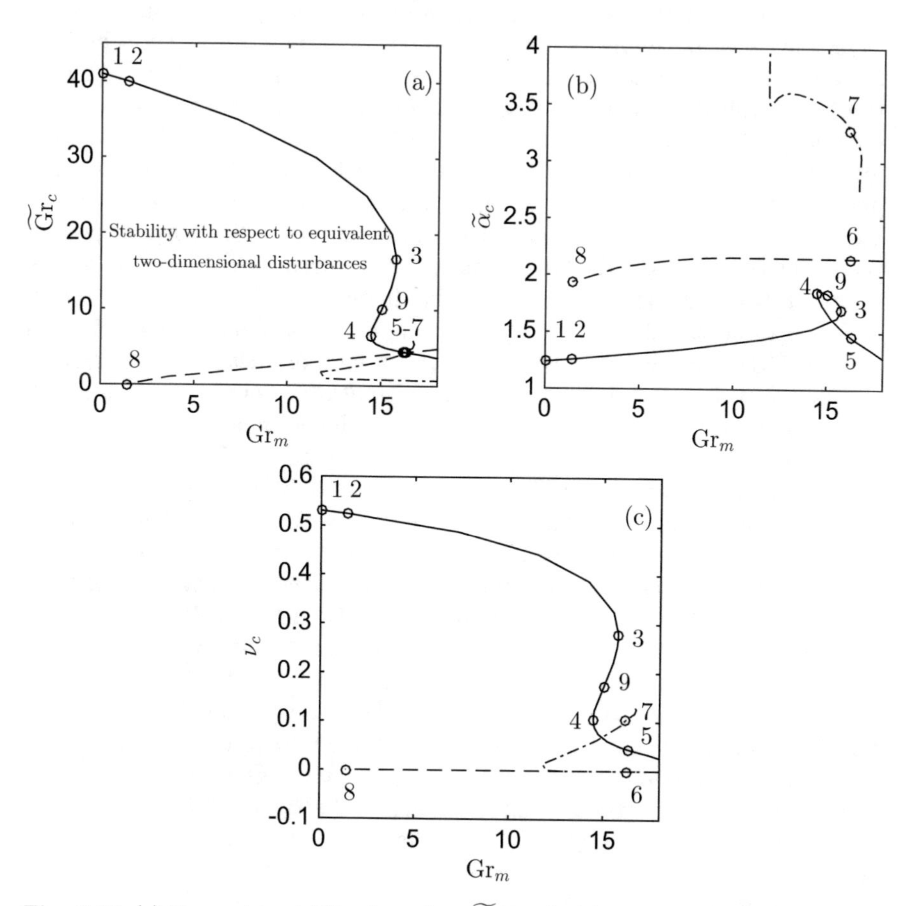

Fig. 3.15 (a) Parametric stability boundary $\widetilde{\text{Gr}}(\text{Gr}_m)$, (b) nondimensional critical disturbance wavenumber $\widetilde{\alpha}_c(\text{Gr}_m)$ and (c) nondimensional disturbance frequency $\nu_c(\text{Gr}_m) \equiv \sigma_c^I/(2\pi)$ (c) for equivalent two-dimensional problem at $\text{Pr} = 130$ [234, 238].

$$\Sigma_{m1} = \int_{-1}^{1} \underbrace{-\mathrm{Gr}_m DH_0 \,\Re(\theta \bar{u})}_{E_{m1}} \,\mathrm{d}x \,, \tag{3.40}$$

$$\Sigma_{m2} = \int_{-1}^{1} \underbrace{\mathrm{Gr}_m D\theta_0 \,\Re(D\phi \,\bar{u})}_{E_{m2}} \,\mathrm{d}x \,. \tag{3.41}$$

Equation (3.35) does not involve any approximations and is exact along the marginal stability boundary $\sigma^R = 0$. It should not be confused with fully nonlinear yet approximate energy stability estimates discussed, for example, in [227]. Equation (3.35) is simply an integrated form of the linearised Equations (3.26)–(3.28), but it enables one to do what cannot be achieved using the original equations, namely, to directly and unambiguously determine the main instability mechanism. Here this will be done simply by inspecting the signs and relative magnitudes of terms entering Equation (3.35) and judging their direct contribution to the disturbance amplification rate σ^R. Importantly, all information required to make physical conclusions based on this equation is local, i.e. the physical flow instability is classified directly at a given parametric point without any need to refer to a global stability diagram such as the one shown in Figure 3.15 (which is similar to Figure 3.7 but contains more details). Each of the integral terms (3.39) has a distinct physical meaning: Σ_k is the kinetic energy of perturbations; Σ_{uv} is the energy exchange between the basic flow and the disturbance velocity fields; Σ_{Gr} is the energy contribution due to buoyancy; and Σ_{vis} is the negatively defined viscous dissipation. Because within the framework of linearised disturbances the eigenfunctions are defined up to a multiplicative constant, without loss of generality the perturbation kinetic energy balance Equation (3.35) can be normalised so that $\Sigma_{vis} = -1$. The remaining two terms, Σ_{m1} and Σ_{m2}, are both due to magnetic effects. However, they have a different physical nature. Σ_{m1} accounts for the disturbance kinetic energy variation due to the thermal disturbances in the field. They influence the degree of local fluid magnetisation and the appearance of a magnetic force driving cool strongly magnetised fluid particles to the regions of stronger magnetic field. On the other hand, Σ_{m2} contains perturbations of a magnetic field itself that are caused by the cross-layer motion of a nonuniformly heated fluid. This is essentially an energy perturbation due to the magnetic induction. Given that the disturbance kinetic energy Σ_k is positively defined, the basic flow can only be unstable, i.e. $\sigma^R > 0$, if the sum of terms in the right-hand side of Equation (3.35) is positive. In other words, positive terms in the right-hand side of (3.35) unambiguously identify physical effects that lead to flow destabilisation, and the comparison of magnitudes of these terms determines the relative strength of various physical influences.

While results presented in Figure 3.15 provide a comprehensive stability map of the flow, there is still a question awaiting clarification is the distinction between two magnetically driven modes shown by the dashed and dash-dotted lines in Figure 3.15. The physical behaviour of a flow along the segment of the instability boundary shown by the solid line in Figure 3.15 also requires a clarification. It connects, in non-monotonous but continuous way, the limiting regimes of pure gravitational convection ($\mathrm{Gr}_m = 0$, $\widetilde{\mathrm{Gr}} \neq 0$, point 1) and magnetically driven flow ($\mathrm{Gr}_m \neq 0$, $\widetilde{\mathrm{Gr}} \to 0$, the solid line segment beyond point 5). Clearly, the physical nature of instability has to change along this curve, but where exactly this happens cannot be determined based on modal analysis of infinitesimal perturbations alone. Thus we consider a number of representative points 1–8 on the marginal stability boundary and investigate perturbation energy balance at these points to answer aforementioned questions.

The values of various energy balance terms computed at selected marginal stability ($\sigma^R = 0$) points marked by circles in Figure 3.15 are presented in Table 3.2. From the analysis of the obtained computational data, the following conclusions are made:

- The contribution of the basic flow velocity into the disturbance energy balance, Σ_{uv}, can be either slightly positive or negative, but it remains close to zero in all regimes; see the dashed line in Figure 3.16. Therefore the interaction of the disturbance velocity field with the basic flow is weak. This confirms that instability associated with the presence of an inflection point in the basic flow velocity profile, which is the dominant instability in similar flows of low-Prandtl-number fluids, see, for example, [239, 240], does not occur in typical ferrofluids characterised by large values of Pr.
- The thermogravitational contribution, Σ_{Gr}, depends strongly on the values of both $\widetilde{\mathrm{Gr}}$ and Gr_m and can be either positive or negative. It is strongly

Table 3.2 Disturbance energy integrals for selected marginal stability points shown by circles in Figure 3.15.

	Gr_m	$\widetilde{\mathrm{Gr}}$	$\widetilde{\alpha}_c$	Σ_{uv}	Σ_{Gr}	Σ_{m1}	Σ_{m2}
1	0	40.974	1.238	−0.006	1.006	0	0
2	1.398	39.976	1.256	−0.007	1.005	0.004	−0.002
3	15.775	16.690	1.696	−0.011	0.795	0.302	−0.086
4	14.468	6.6	1.853	−0.003	0.180	1.141	−0.318
5	16.353	4.4	1.463	0.001	−0.036	1.795	−0.759
6	16.239	4.4	2.147	0.002	−0.219	1.812	−0.595
7	16.189	4.4	3.278	−0.003	−0.163	1.516	−0.350
8	1.398	0	1.936	0	0	1.584	−0.584
9	15.020	10.052	1.843	−0.008	0.504	0.677	−0.173

destabilising in the absence of a magnetic field, i.e. for $\mathrm{Gr}_m \sim 0$, but becomes stabilising for the larger values of Gr_m when the motion caused by the actions of the vertical buoyancy force and the horizontal magnetic ponderomotive force start competing with each other. The peculiar Z-shape of the stability boundary shown by the solid line in Figure 3.15(a) is the consequence of this competition: for larger values of $\widetilde{\mathrm{Gr}}$, both magnetic and thermogravitational mechanisms play a destabilising role, and their combination leads to a reduction in the parametric area of the stability region (the solid line bends to the left). However, for small values of $\widetilde{\mathrm{Gr}}$, the buoyancy starts playing a stabilising role; see the dash-dotted line in Figure 3.16. Figure 3.17 offers a possible explanation of this fact. Note that the horizontal ponderomotive magnetic force effectively drives the convection motion only during the first and third quarters of a wave period when it is aligned with the perturbation velocity; see the snapshots for $t = 0$ and $t = T/2$. At these instances, cool (warm) and stronger (weaker) magnetised fluid is pushed towards the hot (cool) wall where the basic magnetic field given by Equation (3.8) is stronger (weaker). It is mostly the buoyancy force that drives convection over the second and fourth quarters of the period; see the snapshots for $t = T/4$ and $t = 3T/4$. At these moments, the warm (cool) fluid disturbance velocity has components up (down) and towards the warm (cool) wall. Thus the work is done by the gravity force to enhance convective motion. However, the work is also done by the fluid against the horizontal ponderomotive magnetic force, and this reduces the fluid's kinetic energy. At relatively large values of $\widetilde{\mathrm{Gr}}$, the gravity work exceeds that done by fluid against the ponderomotive force so that the overall buoyancy effect is destabilising. For small values of $\widetilde{\mathrm{Gr}}$, the situation is reversed, and the buoyancy contribution to disturbance energy balance becomes negative.

- The first of the two magnetic contributions to the energy balance, Σ_{m1}, is always non-negative. This term represents a ponderomotive force that drives stronger magnetised cooler fluid particles into the regions of a stronger basic magnetic field (i.e. from the cool wall towards the warm one, see expression (3.8)). Therefore in the considered configuration, the dependence of fluid magnetisation on the temperature always plays a destabilising role ultimately leading to the onset of a thermomagnetic convection.
- In contrast, the second magnetic term, Σ_{m2}, remains negative. It represents the induction of a disturbance magnetic field by the displaced ferromagnetic fluid particles. The corresponding modification of the basic magnetic field (an analogy with an electromagnetic transmitter that requires energy supply for its operation might be helpful here) always absorbs energy and thus plays a stabilising role hindering the change in the primary magnetisation field. However this magnetic stabilisation effect is always weaker than the thermomagnetic destabilisation characterised by Σ_{m1}. Therefore the overall magnetic influence in the considered geometry is always destabilising.

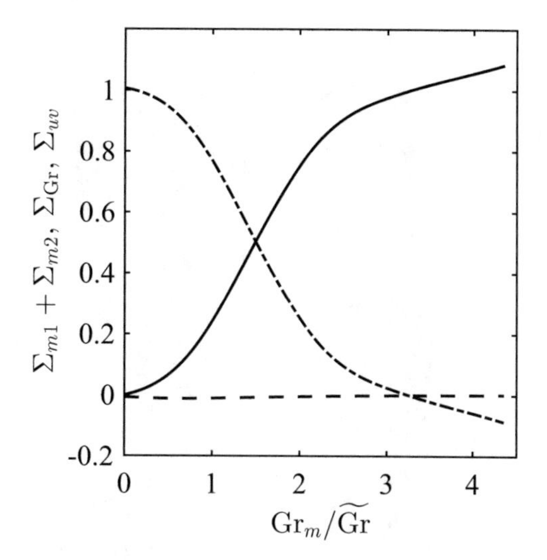

Fig. 3.16 The perturbation energy integrals characterising thermomagnetic ($\Sigma_{m1} + \Sigma_{m2}$, solid line) and thermogravitational (Σ_{Gr}, dash-dotted line) mechanisms of convection and exchange with the basic flow (Σ_{uv}, dashed line) as functions of the ratio $\mathrm{Gr}_m/\widetilde{\mathrm{Gr}}$ along the stability boundary shown by the solid line in Figure 3.15(a).

The above observations enable us to draw the general conclusion that the destabilisation of the primary parallel flow along the solid line in Figure 3.15(a) is due to two competing physical mechanisms: the action of ponderomotive magnetic and buoyancy forces. They also define the nature of the instability whose boundaries are shown by the dashed (points 6 and 8) and dash-dotted (point 7) lines in Figure 3.15(a): purely magnetic ponderomotive force. Yet the details of these instability modes are not made clear so far. Therefore next we consider the spatial distribution of the three destabilising integrands, E_{m1}, E_{m2} and E_{Gr}, defined in (3.39) and plotted in Figure 3.18 for points 1–8 marked by circles in Figure 3.15.

For small values of Gr_m and large values of $\widetilde{\mathrm{Gr}}$ (points 1 and 2), the thermogravitational instability mechanism dominates; see the dash-dotted line in plots 1 and 2 in Figure 3.18. The energy integrand E_{Gr} has two well-defined symmetric maxima near the walls. This is a reflection of the well-known fact that in large-Prandtl-number fluids such as a typical kerosene-based ferrocolloid, the thermogravitational instability takes the form of two waves counter-propagating in the wall regions [55, 98, 256]. Computations show that they are almost insensitive to a magnetic field and exist even when the magnetic Grashof number is significantly increased; see plots for points 3 and 4 in Figure 3.18. However as the ratio $\mathrm{Gr}_m/\widetilde{\mathrm{Gr}}$ increases, the thermomagnetic effects quantitatively characterised by $E_{m1} + E_{m2}$ intensify significantly, while the role of gravitational buoyancy, quantified by Σ_{Gr}, weakens. This is demonstrated in Figure 3.16. The dominating role of gravitational instability mechanism is transferred to thermomagnetic mechanism at point 9 in Figure 3.15 (which also corresponds to the intersection of the solid and dash-dotted lines in Figure 3.16), where the condition

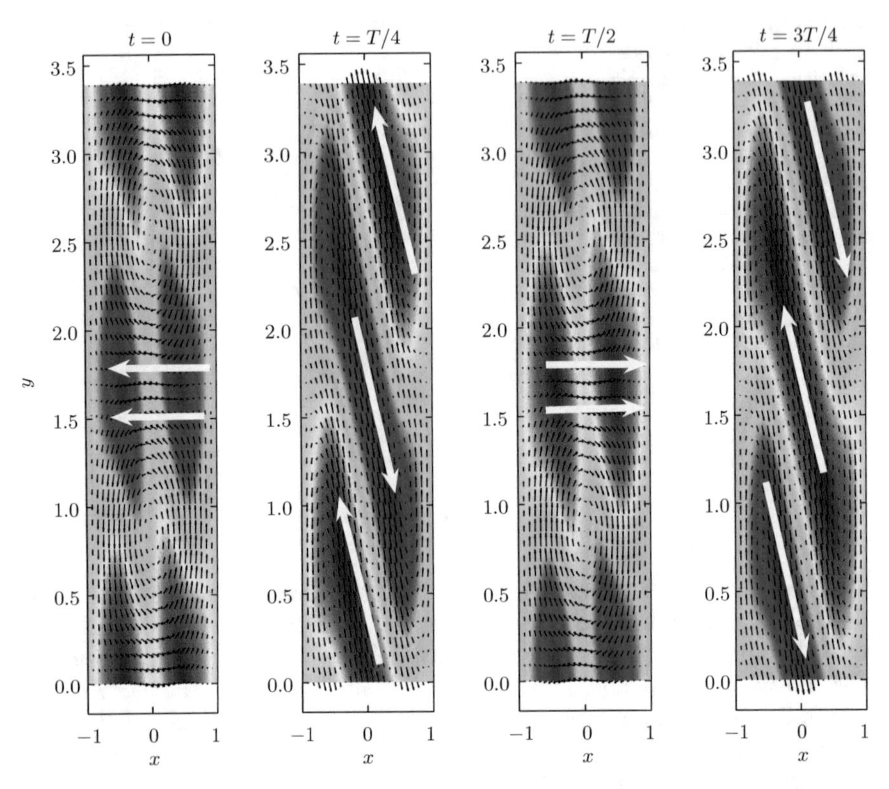

Fig. 3.17 Snapshots of thermomagnetic waves (point 4 in Figure 3.15) over the wave period T. Colour represents a thermal field (red, warm, less dense and weaker magnetised fluid; blue, cool, denser and stronger magnetised fluid). White arrows show the dominant direction of disturbance flow driven by the ponderomotive magnetic force.

$$\Sigma_{\mathrm{Gr}} = \Sigma_{m1} + \Sigma_{m2} \tag{3.42}$$

is satisfied. This transition in dominance from one physical mechanism of convection to another still results in a pair of waves propagating up/down along the hot/cold wall. Despite a clear difference in driving physical mechanisms, the transition between the two types of waves occurs in a continuous way; see the solid line in Figure 3.15. The comparison of Figures 3.19 and 3.17 shows that the dominant component of the thermogravitational perturbation velocity field is vertical (along the layer, parallel to the gravity vector) and of thermomagnetic waves is horizontal (across the layer, parallel to the applied magnetic field). Another distinguishing feature is the behaviour of the disturbance wavenumbers: when $\widetilde{\mathrm{Gr}}_m$ increases, so does the wavenumbers of thermogravitational waves. However, this trend is reversed once they are replaced with thermomagnetic waves; see the solid line in Figure 3.15(b).

When $\widetilde{\mathrm{Gr}}$ decreases, the wave speeds (see Figure 3.7(c)) also decrease, the instability pattern becomes nearly stationary and its maximum shifts

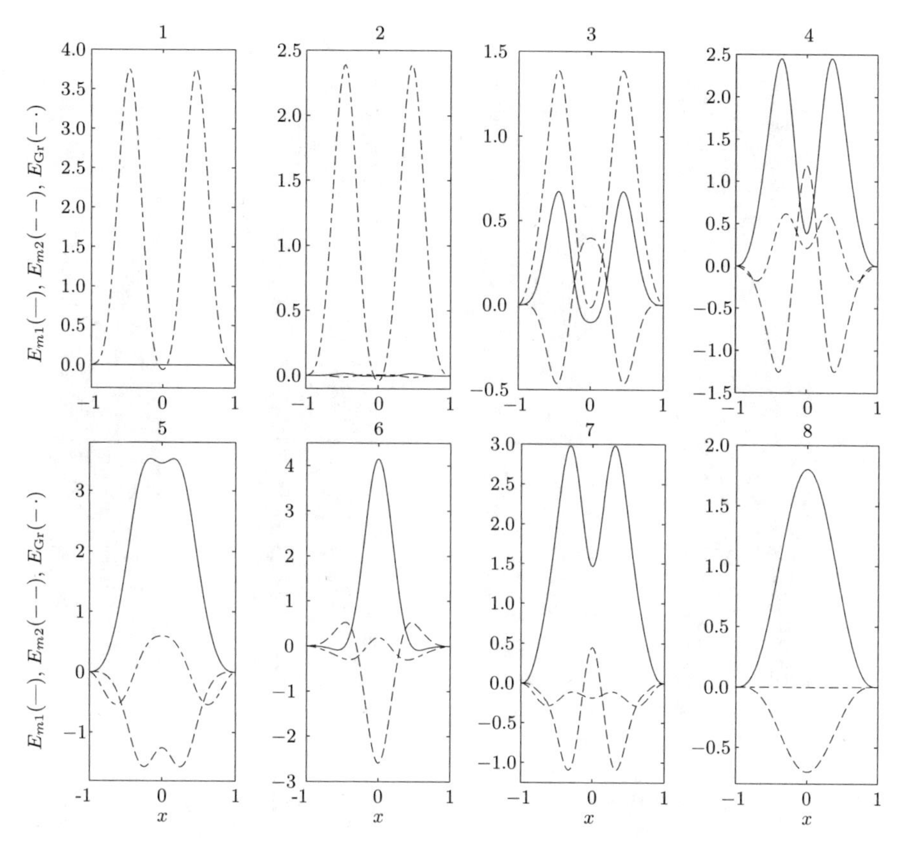

Fig. 3.18 Selected disturbance energy integrands for points denoted by circles in Figure 3.15.

from wall regions towards the centre of the layer; see the plot for point 5 in Figure 3.18. The thermogravitational convection mechanism continues to play a destabilising role in the centre of the layer, but its influence in the wall regions becomes stabilising. A shift of the instability production region to the centre of the layer has a profound effect on the characteristic wavenumber of perturbations: it quickly decreases; see the right end of the solid line in Figure 3.15(b). This has a straightforward explanation: the disturbance structures in the centre of the layer near the inflection point of the basic flow velocity profile are subject to large shear forces. These forces "stretch" convection rolls decreasing their wavenumber. The centrally located instability structures elongated by the shear forces then become so large that they cause a strong "flow blocking" effect. Eventually they are destroyed by the basic flow giving way to much shorter structures; see the dash-dotted lines containing point 7 in Figure 3.15. Plot 7 in Figure 3.18 shows that the physical mechanism generating these flow structures is indeed the same as that for the thermomagnetic waves discussed above. Their characteristic length scale is

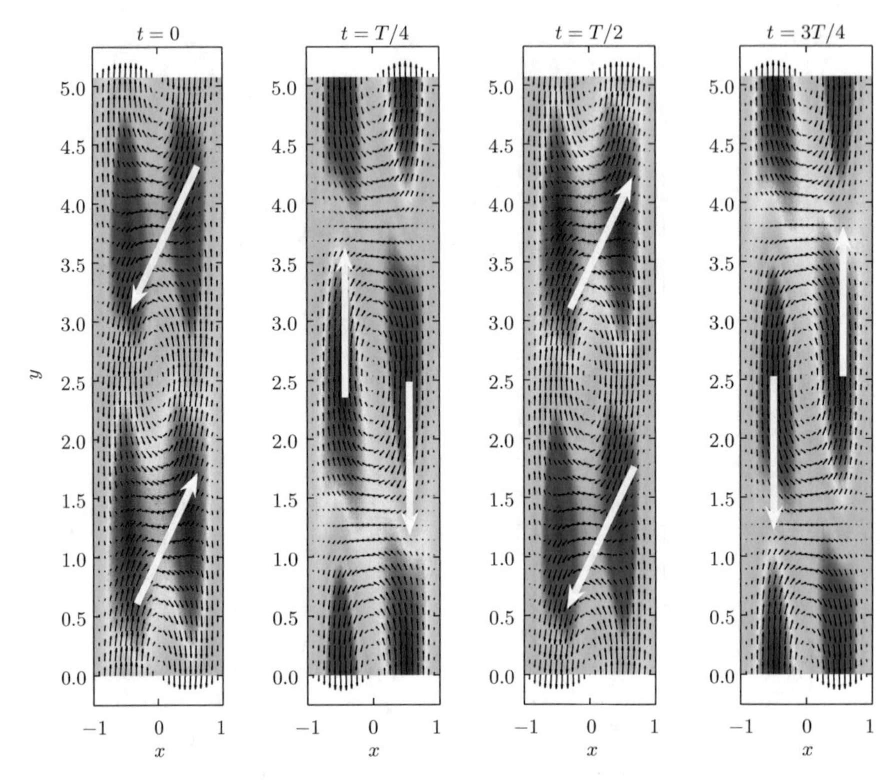

Fig. 3.19 Snapshots of thermogravitational waves (point 1 in Figure 3.15) over the wave period T. Colour represents a thermal field (red—warm and less dense fluid; blue—cool and denser fluid). White arrows show the dominant directions of disturbance flow driven by the buoyancy force.

sufficiently small (the wavenumber is large) so that the basic flow "blocking effect" is reduced and the two disturbance waves propagating along the opposite walls reappear. We also note that although overall magneto-induction effect Σ_{m2} is always stabilising, the energy integrand E_{m2} for points 3, 4 and 7 is positive in the centre of the layer (between the counter-propagating thermomagnetic waves). Thus it contributes to the local destabilisation near the midplane of the layer.

Points 6 and 8 in Figure 3.15 belong to the third type of stability boundary which is disjoint from the two segments discussed so far. The physical mechanism causing this instability is of purely thermomagnetic type: E_{m1} is strongly positive, while E_{Gr} is close to zero. Therefore the gravitational buoyancy plays no essential role in these regimes of convection. The major destabilisation occurs near the middle of the layer where basic flow velocity is zero. As a consequence, the corresponding instability patterns are stationary [234]. They take the form of vertical rolls.

3.3.4 Three-Dimensional Results

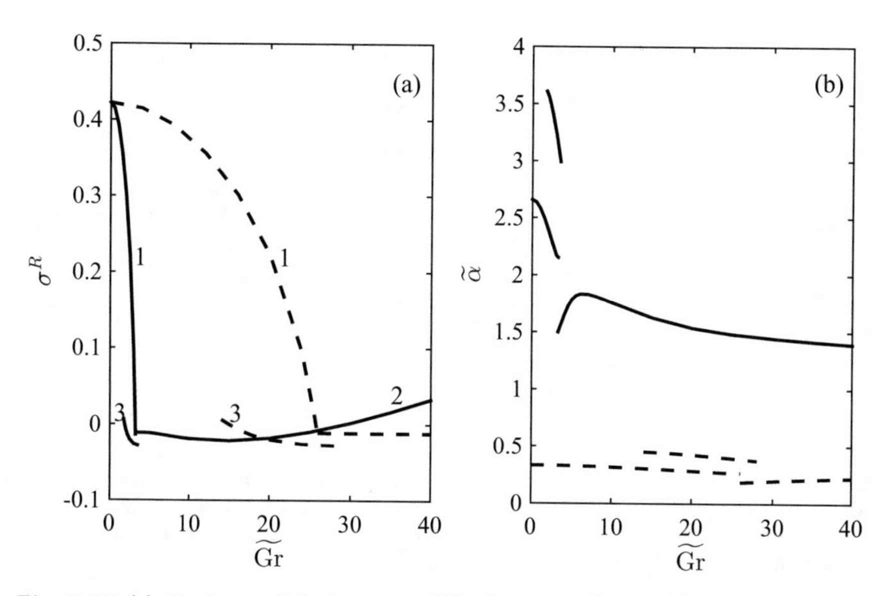

Fig. 3.20 (a) Maximum disturbance amplification rate σ^R and (b) the corresponding vertical wavenumbers for horizontal (solid lines) and oblique (dashed lines) convection patterns with the inclination angle $\psi = \arccos\frac{1}{8}$ with respect to the horizontal direction for $\mathrm{Gr}_m = 12$. Labels in plot (a) denote: 1, stationary magnetoconvection rolls; 2, thermogravitational waves; 3, thermomagnetic waves.

In Sections 3.3.2 and 3.3.3, the stability results obtained by solving an equivalent two-dimensional problem (3.26)–(3.28), (3.30) and (3.31) were discussed. While the physical reasons for the detected instabilities do not depend on whether two- or three-dimensional patterns have been considered, the parametric stability region is very sensitive to the spatial orientation of perturbation patterns. In this section we will systematically describe various regions of three-dimensional instability which unfold from the two-dimensional stability diagram of Figure 3.7(a) upon the inverse transformation (3.25). The most essential feature of this transformation is that the magnetic Grashof number remains invariant, while the thermogravitational Grashof number for three-dimensional patterns is necessarily larger than its equivalent two-dimensional counterpart. Namely,

$$\mathrm{Gr} = \frac{\widetilde{\alpha}}{\alpha}\widetilde{\mathrm{Gr}}, \quad \alpha = \sqrt{\widetilde{\alpha}^2 - \beta^2}. \tag{3.43}$$

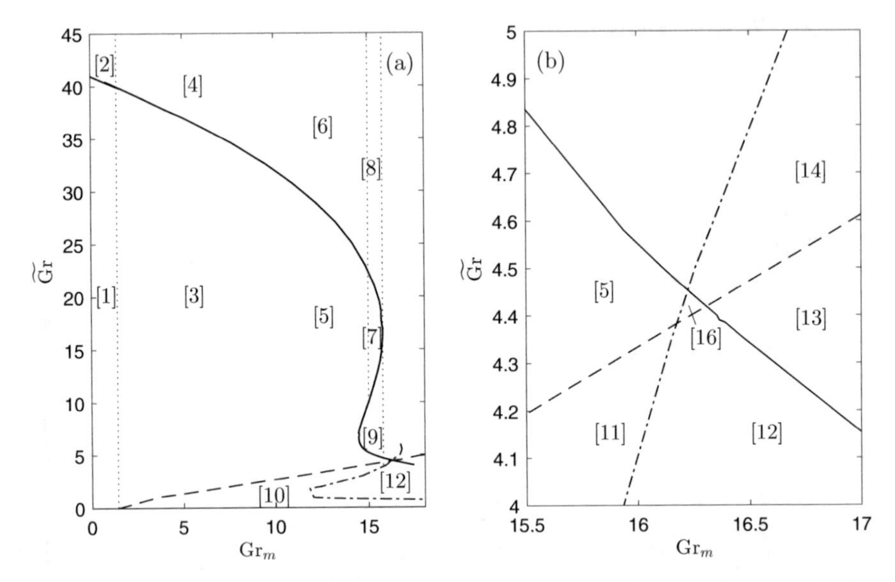

Fig. 3.21 (a) Three-dimensional unfoldings of the stability diagram for an equivalent two-dimensional problem; (b) close-up of a region with three types of instability modes. Line types are the same as in Figure 3.7 (discussed in text). To improve readability, small parametric regions [13]–[16] are only marked in the close-up.

Geometrically, this means that under the inverse transformation, all lines not crossing the horizontal axis in the equivalent two-dimensional stability diagram of Figure 3.7(a) will be shifted upwards, and only points with $\widetilde{\mathrm{Gr}} = 0$ will remain fixed. Therefore if a line separates the stability region below and the instability region above, then upon transformation the stability region will be enlarged. Equivalently this means that two-dimensional disturbance structures consisting of horizontal convection rolls are most dangerous. Conversely, if a line separates an instability region below and a stability region above, then under the inverse transformation the instability region will be enlarged meaning that oblique or vertical convection patterns are more dangerous than horizontal convection rolls. Thus the instability observed experimentally will be represented by three-dimensional patterns.

Figure 3.20 provides another illustration for the meaning of transformation (3.43). Note that $\psi = \pm \arccos \frac{\sqrt{\widetilde{\alpha}^2 - \beta^2}}{\widetilde{\alpha}} = \pm \arccos \frac{\widetilde{\mathrm{Gr}}}{\mathrm{Gr}}$ is the inclination angle of convection rolls with respect to the horizontal direction so that $\beta = 0$ (or $\alpha = \widetilde{\alpha}$) corresponds to horizontal patterns (solid lines). Within a linear analysis framework, both positive and negative inclination angles are equally possible. For $\beta = 0$, the amplification rate $\sigma^R < 0$ and the basic flow is stable with respect to vertically periodic two-dimensional patterns in the range $3.2 \lesssim \mathrm{Gr} \lesssim 30$; see solid lines crossing the $\sigma^R = 0$ level in Figure 3.20(a). However if inclined three-dimensional patterns are considered, the $\sigma^R > 0$ region bounded from the left by the $\mathrm{Gr} = 0$ axis and from the right by the solid line 1 expands to the right (see the dashed line 1). In the limit of $\alpha = 0$, $\psi = \pm 90°$ (i.e. the pattern consists of the vertical magnetoconvection rolls) and the $\sigma^R > 0$ region bounded by the dashed line 1 region will cover the stability region found for two-dimensional disturbances completely. The solid line 3 corresponding to thermomagnetic waves will also move to the right into the two-dimensional stability region. This means that three-dimensional oblique thermomagnetic waves and vertical stationary magnetoconvection rolls are more dangerous than two-dimensional vertically propagating thermomagnetic waves and horizontal magnetoconvection rolls. On the other hand, under transformation (3.43), the instability region found for vertically propagating thermogravitational waves at $\widetilde{\text{Gr}} \gtrsim 30$ (solid line 2 in Figure 3.20(a)) will move towards the larger values of Gr. This implies that the two-dimensional vertically propagating thermogravitational waves are more dangerous than similar three-dimensional oblique waves.

Such a geometrical consideration enables us to identify 16 regions shown in Figure 3.21 which are characterised by distinct instability patterns (or combinations of patterns). These parametric regions are separated by the corresponding segments of the two-dimensional marginal stability boundaries and vertical dotted lines. They are described in detail next:

[1] This is the only region of true linear stability. It corresponds to $\text{Gr}_m \lesssim 1.398$ and $\text{Gr} \lesssim 40.974$. Neither thermogravitational nor magnetic convection can develop for this range of parameters so that the basic flow remains parallel with the linear temperature profile.

[2] Here the thermogravitational convection sets. The most dangerous disturbances correspond to two counter-propagating waves similar to those found in natural convection of large-Prandtl-number fluids [55, 98, 256]. They lead to a formation of waves whose snapshot is shown as a set of horizontal convection rolls in Figure 3.22. The observation plane is chosen to be at $x = 0.95$ near the cold wall in order to simplify comparison with future experiments: in practice, thermal fields are observed through a transparent cold wall, while the hot wall is attached to a non-transparent heater. The thermal pattern seen in the right plot of Figure 3.22 propagates downwards, while its counterpart near the hot wall propagates upwards. In contrast, the velocity field shown in the left snapshot in Figure 3.22 consists of rolls which do not propagate but rather periodically change the direction of their rotation. This confirms that the combination of two counter-propagating thermogravitational waves is responsible for the formation of a standing velocity wave. The application of a magnetic field does not change the orientation of convection rolls. However it has an overall destabilising effect reducing the value of the critical Grashof number below $\text{Gr} = 40.974$.

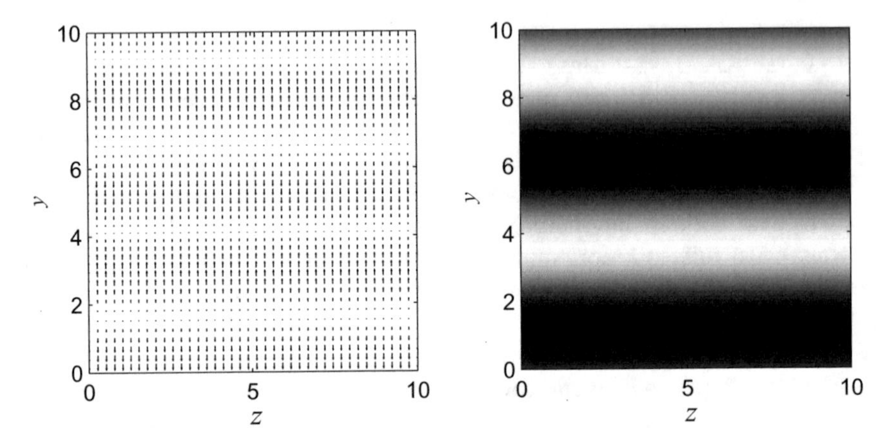

Fig. 3.22 Three-dimensional instantaneous disturbance velocity (left) and temperature (right) fields near the cold wall in the plane $x = 0.95$ for $(\mathrm{Gr}_m, \mathrm{Gr}) = (0, 41)$. Thermogravitational instability. Lighter areas correspond to warmer fluid.

[3] In this region only a magnetoconvective instability is present. As discussed in Section 3.3.2, increasing the value of $\widetilde{\mathrm{Gr}}$ tends to inhibit magnetoconvection. Therefore the fastest amplification of magnetoconvective disturbances is achieved along the horizontal axis of Figure 3.7(a). According to transformations (3.25), this amplification rate will be observed for three-dimensional disturbances at any value of Grashof number defined by (3.43). Since $\widetilde{\mathrm{Gr}} \to 0$ for the most dangerous disturbances, the finite value of Gr can only be obtained if $\alpha \to 0$ and $\widetilde{\alpha} \to \beta$. This means that in region [3], the vertical stationary magnetoconvection rolls are expected to dominate the instability pattern, yet weaker inclined rolls may also exist.

[4] The instability pattern here is a combination of horizontal rolls resulting from a pair of thermogravitational waves and stationary vertical magnetoconvection rolls.

[5] Region [5] is a stability region for horizontal convection rolls. However as follows from the discussion of Figure 3.20 above, the parallel basic flow here is unstable with respect to the stationary vertical and oblique magnetoconvection rolls. This region is similar to region [3], but oblique thermomagnetic instability waves may be found here which correspond to the large-wavenumber instability (dash-dotted line in Figure 3.21). Yet as seen from Figure 3.20(a), the amplification rate of this mode is significantly smaller than that of stationary vertical rolls. Therefore the large-wavenumber thermomagnetic pattern might not be easy to observe experimentally. Yet it should be possible to detect its unsteady signature since it corresponds to a pair of thermomagnetic waves propagating over a stationary vertical magnetoconvection background.

[6] In this region three instability patterns co-exist: stationary vertical magnetoconvection rolls, vertically propagating thermogravitational waves and larger-wavenumber oblique thermomagnetic waves. Again stationary vertical magnetoconvection rolls are expected to be the most prominent feature of the resulting mixed instability pattern at least near the lower boundary of this region since the corresponding disturbances have the largest amplification rate.

[7] This region corresponds to basic flow which is stable with respect to the horizontal convection rolls. In order to better understand its three-

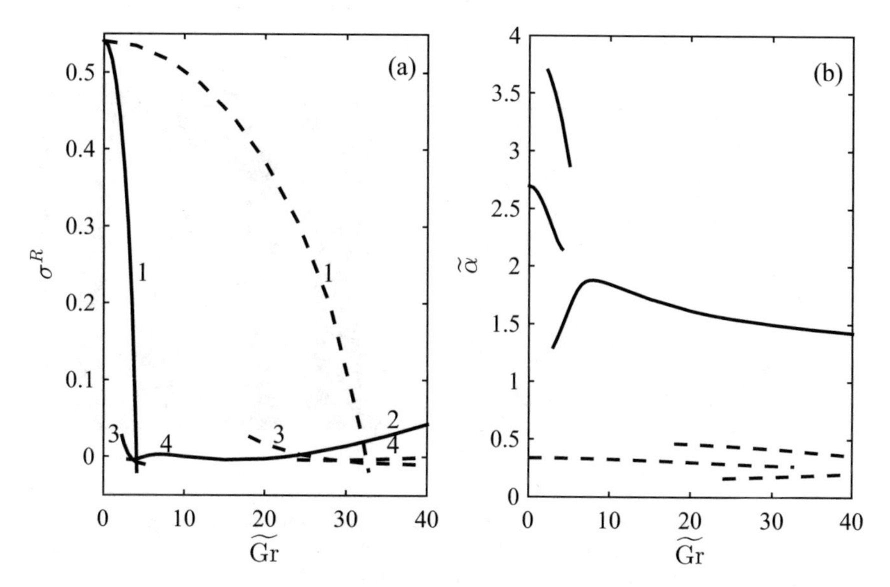

Fig. 3.23 Same as Figure 3.20 but for $\mathrm{Gr}_m = 15$. Labels in plot (a) denote: 1, stationary magnetoconvection rolls; 2, thermogravitational waves; 3, large-wavenumber thermomagnetic waves; 4, small-wavenumber thermomagnetic waves.

dimensional stability features, refer to Figure 3.23. As an example, let us consider the point $(\mathrm{Gr}_0, \mathrm{Gr}_m) = (15, 15)$ which belongs to this parametric domain. Note that the solid $\sigma^R(\mathrm{Gr})$ line in Figure 3.23(a) which shows the maximum possible linear amplification rates for various two-dimensional instability modes at $\mathrm{Gr}_m = 15$ has three maxima for $\mathrm{Gr} < 15$: at $\mathrm{Gr}_1 = 0$, $\mathrm{Gr}_2 \approx 2$ and $\mathrm{Gr}_3 \approx 7$. It is straightforward to deduce from Figures 3.7(c) and 3.23(b) that these maxima correspond to stationary magnetoconvection and large- and small-wavenumber thermomagnetic waves, respectively. Using transformation (3.43), we then deduce that despite the fact that the flow is stable with respect to any two-dimensional vertically periodic disturbance pattern, it is unstable with respect to stationary thermomagnetic rolls with the inclination angle of $\psi_1 = \pm \arccos \frac{\mathrm{Gr}_1}{\mathrm{Gr}_0} = \pm 90°$, i.e. vertical rolls (see Figure 3.24), and large- and small-wavenumber counter-propagating thermomagnetic waves forming rolls inclined at $\psi_2 = \pm \arccos \frac{\mathrm{Gr}_2}{\mathrm{Gr}_0} \approx \pm 82°$ and $\psi_3 = \pm \arccos \frac{\mathrm{Gr}_3}{\mathrm{Gr}_0} \approx \pm 62°$, respectively; see Figures 3.25 and 3.26. Note also that the larger the value of Gr in region [7] the larger the inclination angle of thermomagnetic waves. For example, the upper boundary of region [7] at $\mathrm{Gr}_m = 15$ corresponds to $\mathrm{Gr} \approx 23$ (at which vertically propagating thermogravitational waves appear, see the solid line 2 in Figure 3.23(a)). The inclination angles for this parametric regime are then expected to be $\psi_1 = \pm 90°$ (vertical stationary magnetoconvection rolls), $\psi_2 \approx \pm 85°$ (large-wavenumber thermomagnetic

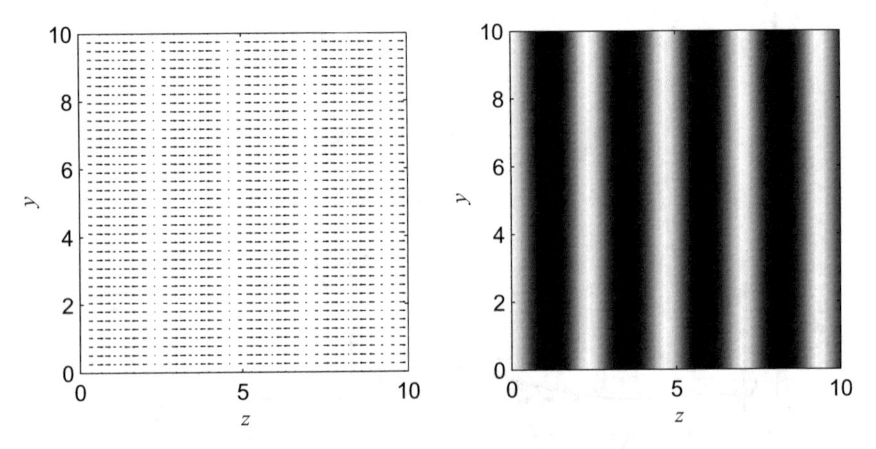

Fig. 3.24 Three-dimensional disturbance velocity (left) and temperature (right) fields near the cold wall in the plane $x = 0.95$ for $(\mathrm{Gr}_m, \mathrm{Gr}) = (15, 15)$. Stationary magnetoconvective instability. Lighter areas correspond to warmer fluid.

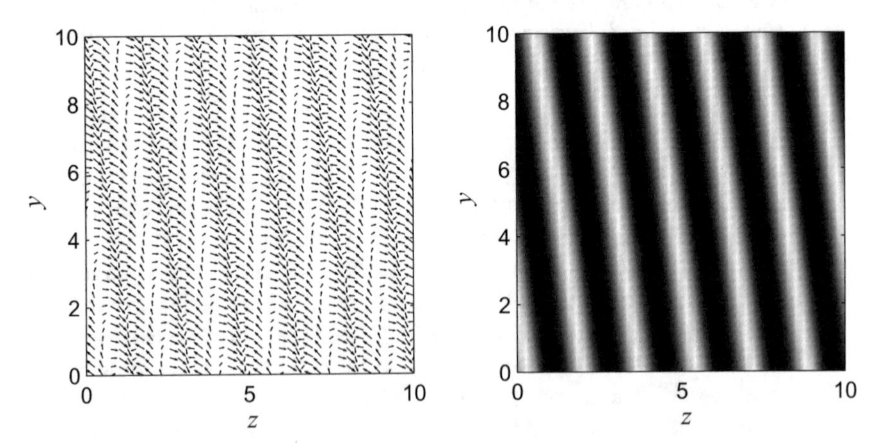

Fig. 3.25 Same as Figure 3.24 but for large-wavenumber thermomagnetic waves. The shown patterns correspond to $\psi \approx -82°$.

waves) and $\psi_3 \approx \pm 72°$ (small-wavenumber thermomagnetic waves). It is also seen from Figure 3.23(a) that the amplification rates of

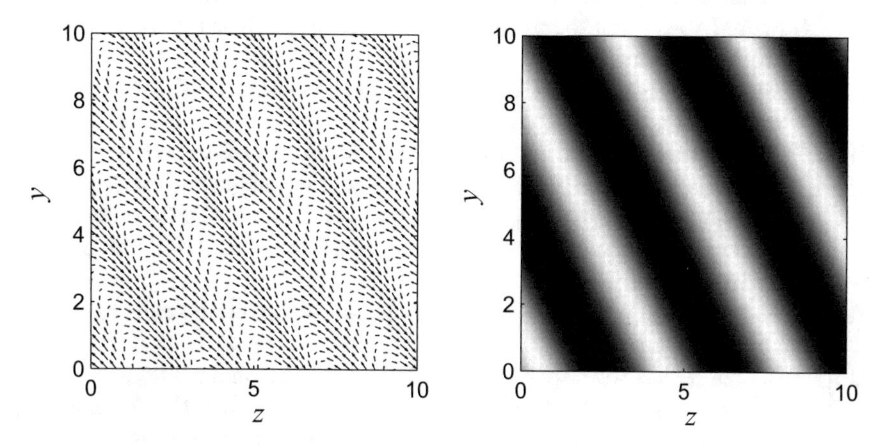

Fig. 3.26 Same as Figure 3.24 but for a small-wavenumber thermomagnetic waves. The shown patterns correspond to $\psi \approx -62°$.

stationary thermomagnetic convection rolls (lines 1) are significantly larger than that for thermomagnetic waves (lines 3 and 4). Therefore vertical rolls are expected to dominate the convection pattern. The presence of oblique thermomagnetic waves should be possible to detect by observing the relatively weak "blinking" superposed on developed vertical rolls. Such blinking was indeed seen in experiments described in Section 6.3.2.

[8] If Grashof number is increased so that one enters region, [8] then, in addition to three vertical and near vertical oblique convection patterns described above, a pair of thermogravitational vertically propagating waves also appear. As seen from Figure 3.23(a) for Gr $\gtrsim$ 30, the amplification rate of such a horizontal pattern becomes larger than that of oblique patterns. Therefore it is expected that the overall flow pattern will primarily be a combination of stationary vertical rolls and horizontal rolls formed by vertically propagating thermogravitational waves. However this pattern will be less regular than in region [6] due to the presence of weak oblique waves.

[9] In this region vertically propagating thermomagnetic waves lead to the appearance of horizontal convection rolls. However their amplification rate is significantly smaller than that of oblique waves and of vertical magnetoconvection rolls (see Figure 3.23(a)), which are expected to dominate the flow.

[10] Here the instability is caused by the magnetoconvection mechanism alone. The orientation of the resulting stationary convection rolls can be arbitrary; however for any non-zero value of Gr, the basic flow tends to cause the vertical alignment of the disturbance patterns.

[11] This region is similar to [5] with the difference that large-wavenumber thermomagnetic waves are expected to propagate at a smaller inclination angle. Vertical stationary magnetoconvection rolls are still expected to dominate, while small-wavenumber thermomagnetic waves are not present.

[12] Similar to region [11] but with large-wavenumber thermomagnetic waves propagating vertically.

[13] The instability pattern here consists of the small-wavenumber thermomagnetic waves propagating vertically, slightly oblique large-wavenumber thermomagnetic waves and predominantly vertical stationary magneto-convection rolls which can form oblique or even horizontal patches.

[14] Same as region [13] but stationary magnetoconvection rolls cannot be horizontal here.

[15] In this region the superposition of vertical stationary magnetoconvection rolls, vertically propagating small-wavenumber thermomagnetic waves and weak obliquely or vertically propagating large-wavenumber thermomagnetic waves is expected to define the flow pattern.

[16] This is the smallest parametric region in the diagram. It is characterised by the presence of the predominantly vertically propagating large-wavenumber thermomagnetic waves and mostly vertical stationary magnetoconvection rolls. Small-wavenumber thermomagnetic waves are not present.

To conclude, the above analysis demonstrates that magnetoconvection plays a major destabilising role for $\mathrm{Gr}_m > 1.4$. Stationary vertical convection rolls should be experimentally observable in parametric regions [3]–[16]. Vertically counter-propagating thermogravitational waves should be visible in regions [2], [4], [6] and [8], while oblique thermomagnetic waves should be seen in regions [5], [7], [9] and [11]–[16].

3.3.5 Symmetry-Breaking Effects of Nonuniform Fluid Magnetisation

Table 3.3 The critical values of Grashof number $\widetilde{\mathrm{Gr}}$, wavenumber $\widetilde{\alpha}$ and disturbance wave speed $\widetilde{c} = -\sigma^I/\widetilde{\alpha}$ for the two leading waves of magnetogravitational convection in a normal magnetic field ($\delta = 0°$) for $\mathrm{Gr}_m = 15$, $H^e = 100$, $\mathrm{Pr} = 55$ and various values of χ and χ_*.

		Wave propagating downward			Wave propagating upward		
χ	χ_*	$\widetilde{\alpha}_c$	$\widetilde{\mathrm{Gr}}_c$	$\widetilde{c}_c$	$\widetilde{\alpha}_c$	$\widetilde{\mathrm{Gr}}_c$	$\widetilde{c}_c$
5	5	1.219	57.39	−3.655	1.215	57.60	3.670
3	5	1.240	54.88	−3.486	1.236	55.15	3.504
3	3	1.220	56.89	−3.622	1.218	57.01	3.631
1.5	2.5	1.239	54.58	−3.467	1.237	54.70	3.476
0.5	1.5	1.250	52.78	−3.347	1.248	52.86	3.353

We have seen in the previous sections that in the absence of a magnetic field (or, equivalently, when $\mathrm{Gr}_m = 0$), the problem under consideration possesses symmetry that results in thermogravitational instability in the form of two thermal waves with equal linear growth rates propagating with equal speeds in the opposite directions. However, the computational data given in Table 3.3 demonstrate that the application of normal magnetic field is capable of breaking the symmetry between the two counter-propagating waves. This effect is traced back to the magnetic potential Equation (3.31) that can be written as

$$\left(D^2 - \frac{1+\chi_*}{1+\chi}\frac{\widetilde{\alpha}^2}{1-x/\mathrm{N}}\right)\phi - D\theta = 0 \tag{3.44}$$

taking into account expressions (3.6) and (3.8) for the basic temperature and magnetic field distributions. The asymmetry is caused by the presence of the term $-x/\mathrm{N}$ in the denominator. As has been discussed in [202], in weakly magnetisable fluids, this term is of the order of 10^{-2} and can be considered small. However, in stronger magnetisable fluids, this term is not small, and thus the symmetry-breaking effects cannot be neglected. The representative computational results for $\mathrm{Gr}_m = 15$ and $\mathrm{Pr} = 55$, corresponding to fluid used in experiments that we discuss in Chapter 6, demonstrating the symmetry-breaking effect of a nonlinear magnetisation distribution are given in Table 3.3. These data show that the basic flow becomes unstable with respect to the downward wave at smaller values of $\widetilde{\mathrm{Gr}}$. At their respective onsets, the downward waves propagate with a slightly smaller wave speed and have a slightly shorter wavelengths than their upward counterparts. This is in contrast to the completely symmetric thermal waves observed at $\mathrm{Gr}_m = 0$ and characterised by the critical values of $(\widetilde{\mathrm{Gr}}, \widetilde{\alpha}, \widetilde{c}) = (65.34, 1.127, \pm 4.202)$ at $\mathrm{Pr} = 55$. In addition to the symmetry-breaking effect of the fluid's magnetisation, the data in Table 3.3 reveal that the application of a normal magnetic field always leads to the reduction of the critical value of the Grashof number. Thus the normal magnetic field plays a destabilising role. Somewhat counter-intuitively, the data also show that the basic flow of stronger magnetisable fluids with a larger differential magnetic susceptibility χ remains more stable indicating a subtle interplay between the fluid magnetisation on one hand and its ability to "screen" the applied magnetic field on the other. The instability is also promoted as the fluid approaches magnetic saturation (when both χ and χ_* decrease and χ becomes smaller than χ_*).

3.3.6 Variation of Stability Characteristics and Summary of Results for Convection in Normal Field

The stability results discussed in Sections 3.3.2–3.3.4 have been obtained for a particular fluid with $\mathrm{Pr} = 130$ and $\chi = \chi_* = 5$. Figure 3.27(a) that is similar to Figure 3.7(a), but computed for $\mathrm{Pr} = 55$ and $\chi = \chi_* = 3$, demonstrates that the dependence of the flow stability characteristics on thermoviscous properties of the fluid placed in the normal magnetic field and on its magnetic susceptibilities is just quantitative. Thus the purpose of this section is to use results presented in Figure 3.27 to briefly summarise the major findings

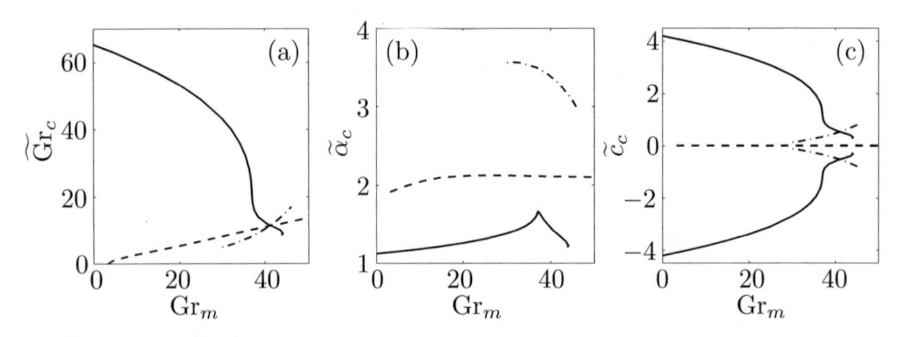

Fig. 3.27 (a) Stability diagram, (b) wavenumbers and (c) wave speeds for a normal magnetic field $H^e = 100$ and $\mathrm{Pr} = 55$ and $\chi = \chi_* = 3$.

discussed in Sections 3.3.2–3.3.4 before qualitatively new effects associated with magnetic field inclination are considered in the subsequent sections.

The stability diagram in Figure 3.27(a) still consists of three lines each representing a different type of instability characterised by its own wavenumber as follows from Figure 3.27(b). The solid line in Figure 3.27(a) starts from $\mathrm{Gr}_m = 0$, which corresponds to the threshold of a thermogravitational convection instability [55, 98, 126, 256]. As discussed earlier, this is the Type I instability characterised by two counter-propagating waves. The basic flow is subject to such an instability above the solid line in Figure 3.27(a). The Type I instability corresponds to vertically propagating patterns with $\alpha = \widetilde{\alpha}$ and $\beta = 0$ that are y-periodic and uniform in the horizontal z-direction.

The dashed line in Figure 3.27(a) starts from $\widetilde{\mathrm{Gr}} = 0$ and therefore corresponds to the threshold of magnetoconvection. In this case, the disturbance amplification rate σ^R is real. As follows from an earlier discussion, this is the Type II instability that is stationary in the normal field (see also [17, 89]). The basic flow is unstable below the dashed line in Figure 3.27(a), and therefore an additional analysis of the inverse Squire's transformation is required to determine the spatial orientation of such patterns. To perform it, refer to Figure 3.28(a), where the linear amplification rate σ^R is plotted as the function of the Squire-transformed ("two-dimensional") Grashof number $\widetilde{\mathrm{Gr}}$. The maximum amplification rate of the Type II instability (the dashed line) is detected when $\widetilde{\mathrm{Gr}} \to 0$. The inverse Squire's transformation then states that the disturbance amplification rate of three-dimensional perturbations, which is invariant under Squire's transformation, will be observed at any value of a non-transformed ("three-dimensional") Grashof number Gr related to $\widetilde{\mathrm{Gr}}$ as

$$\alpha \mathrm{Gr} = \sqrt{\alpha^2 + \beta^2}\,\widetilde{\mathrm{Gr}}\,.$$

This means that the maximum amplification rate observed for the Type II instability when $\widetilde{\mathrm{Gr}} \to 0$ will be observed at an arbitrary value of the non-transformed Grashof number Gr provided that $\alpha \to 0$. That is $\beta \to \widetilde{\alpha}$, where $\widetilde{\alpha}$ is the leftmost value along the dashed line in Figure 3.28(b). In other

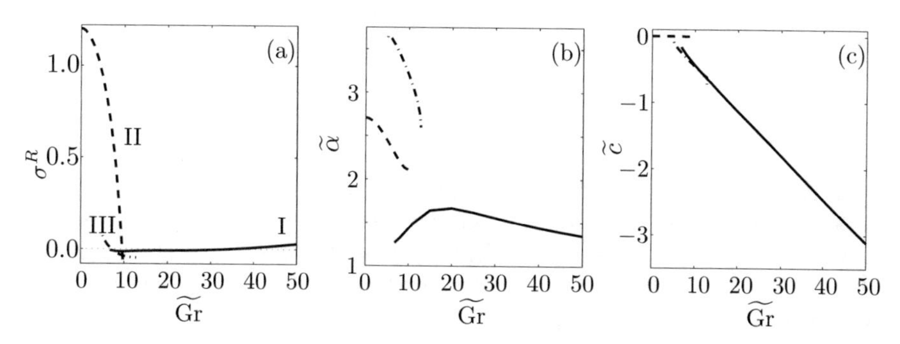

Fig. 3.28 (a) Maximum amplification rate for an equivalent two-dimensional problem, (b) the corresponding wavenumbers $\widetilde{\alpha}$ and (c) wave speeds for $\mathrm{Gr}_m = 35$, $H^e = 100$, $\mathrm{Pr} = 55$ and $\chi = \chi_* = 3$ in a normal magnetic field.

words, the Type II instability in the form of vertical magnetoconvection rolls will arise for arbitrary values of Gr once Gr_m exceeds the critical value corresponding to the leftmost point along the dashed line in Figure 3.27(a). For any values of Gr exceeding those corresponding to the solid line in Figure 3.27(a), the stationary vertical rolls of the Type II instability will overlap with the vertically propagating Type I instability waves.

Given that the inverse Squire's transformation (3.43) indicates that for sufficiently large values of Gr_m the flow is always unstable with respect to vertical thermomagnetic rolls regardless of the value of Gr, the physical meaning of the dashed line in Figure 3.27(a) needs to be clarified. When the gravity is absent and $\mathrm{Gr} = 0$, the arising magnetoconvection rolls can be arbitrarily oriented as all directions in the fluid layer plane are equivalent. When the gravity is introduced and Gr becomes non-zero, the basic gravitational convection flow arises and removes the spatial degeneracy so that the vertically oriented rolls are preferred. Yet it is clear that at small values of Gr, rolls of all other orientations still can exist even though vertical rolls now have a larger growth rate. As the value of the Grashof number increases, the growth rate of the vertical rolls remains the same at fixed Gr_m, but the growth rate of nonvertical rolls becomes smaller. Eventually, when the value corresponding to the dashed line in Figure 3.27(a) is reached, horizontal rolls cannot grow anymore and disappear. Above the dashed line, there exists a maximum roll inclination angle beyond which the Type II instability cannot be observed. To clarify this, consider the following example. The presented computational results show that in the normally applied field, the flow stabilisation occurs when the value of the Squire-transformed Grashof number

$\widetilde{\text{Gr}}$ exceeds 9.55 for $\text{Gr}_m = 35$ and $\chi = \chi_* = 3$ (this corresponds to the point in Figure 3.28(a) where the dashed line crosses zero and to the respective critical point $(\text{Gr}_{mc}, \widetilde{\text{Gr}}_c) = (35, 9.55)$ on the dashed line in Figure 3.27(a)). Say, the experimental value of interest is $\text{Gr} = 15$. Then we conclude that in such experimental conditions, it is expected that the instability patterns will be in the form of stationary vertical rolls that however could be modulated by weaker rolls with the axes forming the angle of up to

$$\sin^{-1}\frac{\alpha}{\widetilde{\alpha}_c} = \sin^{-1}\frac{\widetilde{\text{Gr}}_c}{\text{Gr}} = \sin^{-1}\frac{9.55}{15} \approx 40^\circ$$

with the vertical y-direction. The larger the experimental value of Gr is the smaller the allowed modulation angle becomes. This has a straightforward physical explanation: the increase in the value of Grashof number intensifies the vertical basic flow velocity, which in turn results in a stronger vertical alignment of the instability patterns.

The Type III instability boundary is shown in Figure 3.27(a) by the dash-dotted line. The basic flow is stable with respect to this mode below it. As seen from Figure 3.27(b), the corresponding instability patterns have larger wavenumbers (dash-dotted line) than those of the Type I and Type II instabilities (solid and dashed lines, respectively). As follows from Figure 3.27(c), similar to the Type I instability, the Type III instability arises in the form of two waves counter-propagating with speeds that become faster than those of the Type I waves for sufficiently large values of Gr_m. The peculiar feature of the Type III instability seen in Figure 3.27(a) is that its boundary appears to end abruptly at certain values of gravitational and magnetic Grashof numbers. Such an unusual behaviour was discussed in detail in [234]. There it was shown that the Type III instability appears as a result of a sudden qualitative change in the problem's dispersion relation when its branches corresponding to either the Type I or Type II instabilities bifurcate resulting in the appearance of the Type III waves. Experimentally, the appearance of the Type III instability could be detected either by observing a sudden transition from stationary (Type II) to nonstationary (Type III) patterns at relatively small values of $\widetilde{\text{Gr}}$ or from one unsteady pattern (Type I) to another (Type III) with a shorter wavelength.

Another feature distinguishing the Type III instability from its Type I and Type II counterparts, whose patterns are characterised by a fixed spatial orientation, is that the main periodicity direction for the Type III instability depends on the value of the Grashof number. For example, as follows from Figure 3.28(a) for $\text{Gr}_m = 35$, the Type III instability has the largest growth rate at $\widetilde{\text{Gr}}_c = 5.07$ (the left end of the dash-dotted line) where it has the form of vertically propagating waves with $\alpha = \widetilde{\alpha} \approx 3.636$ and $\beta = 0$. However according to the inverse Squire's transformation, for any larger value of the Grashof number, it will be seen as a pair of oblique waves counter-propagating along the direction forming the angle $\cos^{-1}(\widetilde{\text{Gr}}_c/\text{Gr})$ with the vertical y direction. In other words, as the Grashof number (and thus the vertical basic flow velocity) increases, the axes of the Type III instability rolls approach the vertical, and the patterns drift almost horizontally. At the same time, the Type I instability patterns remain horizontal and propagate vertically. For example, at $\mathrm{Gr} = 15$ the Type III instability waves are expected to propagate along the direction forming the angle of $\cos^{-1}(5.07/15) \approx 70°$ with the vertical direction rather than vertically, and their span-wise direction is expected to form the angle of $\sin^{-1}(5.07/15) \approx 20°$ with the vertical.

3.4 Flow Patterns in an Oblique Magnetic Field

While the analysis of convection types arising in a normal magnetic field reveals major physical mechanisms responsible for the formation of various flow patterns, there are a number of motivating factors for considering the situation when magnetic field is applied to the fluid layer under some non-zero inclination angle δ; see Figure 3.2. One of such factors stems from experimental observations reported in [36–38, 238] and also discussed in Section 4.4.1 in this book. It has been observed that when a normal magnetic field is applied to a sufficiently wide and long ferrofluid layer, the convection patterns arising near the edges of the experimental layer differ drastically in both orientation and behaviour from their counterparts seen in the central part of the layer. Namely, while the most prominent pattern in the middle part of the layer is stationary, the propagating structures have been detected near the edges that form some angle with the boundary, see Figures 6.37(d,e,i,j) and 6.39(c,d). The exact reasons for such a different behaviour of a ferrofluid near the edges of the flow domain are still not completely clear to date, but a plausible explanation is that at the boundary between magnetic (ferrofluid) and non-magnetic (container wall) media, the magnetic field lines inevitably refract, which is the consequence of Maxwell's boundary conditions for a magnetic field; see Section 3.2. As a result even if the applied magnetic field is assumed to be normal to the layer, which is the case in the many studies [89, 234, e.g.], the field lines necessarily curve near the layer edges so that they are effectively "sucked in" the magnetic medium; see Figure 4.8. Such a behaviour of magnetic field lines near the borders is very sensitive to the minor details of the border geometry and its defects so that it is virtually impossible to know what the local inclination angle of the magnetic field is. To render the problem tractable to the analysis, here we resort to a compromise: we still consider an infinitely wide and long layer of fluid and assume that the applied field is uniform. However, we allow its arbitrary inclination with respect to the plane of the layer. Effectively, this adds two field direction angles to the problem's parameter list; see Section 2.2.

3.4.1 Linearised Perturbation Equations in Zero Gravity

To reveal effects caused by magnetic field inclination in the most straightforward way, we first focus on pure magnetoconvection regimes and thus set gravity to zero so that the base state is motionless as is the case in classical Rayleigh-Bénard convection. The main governing parameter in this case is magnetic Rayleigh number Ra_m. The linearised nondimensional equations in this case become [202]

$$
\begin{aligned}
&0 = Du_1 + \mathrm{i}(\alpha v_1 + \beta w_1)\,, && (3.45)\\
&\sigma u_1 + \Pr(\alpha^2+\beta^2-D^2)u_1 + DP_1 + e_{10}\mathrm{Ra}_m \Pr DH_{x0}\,\theta_1\\
&\quad +\mathrm{Ra}_m \Pr\theta_0 e_{10} D^2\phi_1 + \mathrm{Ra}_m\Pr\theta_0\left(\mathrm{i}(\alpha e_{20}+\beta e_{30}) + (1-e_{10}^2)\frac{DH_{x0}}{H_0}\right)D\phi_1\\
&\quad -\mathrm{i}\mathrm{Ra}_m\Pr\theta_0 e_{10}(\alpha e_{20}+\beta e_{30})\frac{DH_{x0}}{H_0}\phi_1 = 0\,, && (3.46)\\
&\sigma v_1 + \Pr(\alpha^2+\beta^2-D^2)v_1 + \mathrm{i}\alpha P_1\\
&\quad +\mathrm{i}\alpha\mathrm{Ra}_m\Pr\theta_0 e_{10} D\phi_1 - \alpha\mathrm{Ra}_m\Pr\theta_0(\alpha e_{20}+\beta e_{30})\phi_1 = 0\,, && (3.47)\\
&\sigma w_1 + \Pr(\alpha^2+\beta^2-D^2)w_1 + \mathrm{i}\beta P_1\\
&\quad +\mathrm{i}\beta\mathrm{Ra}_m\Pr\theta_0 e_{10} D\phi_1 - \beta\mathrm{Ra}_m\Pr\theta_0(\alpha e_{20}+\beta e_{30})\phi_1 = 0\,, && (3.48)\\
&\sigma\theta_1 + D\theta_0 u_1 + (\alpha^2+\beta^2-D^2)\theta_1 = 0\,, && (3.49)
\end{aligned}
$$

$$
\begin{aligned}
0 = {} & (D^2-\alpha^2-\beta^2)\phi_1 + (1-e_{10}^2)\left(\frac{\chi_*-\chi}{1+\chi}\mathrm{N}-\theta_0\right)\frac{D^2\phi_1}{H_0}\\
& -\left[e_{10}\left(\frac{\chi_*-\chi}{1+\chi}\mathrm{N}-\theta_0\right)\left(2\mathrm{i}(\alpha e_{20}+\beta e_{30}) + 3(1-e_{10}^2)\frac{DH_{x0}}{H_0}\right)\right.\\
& \left.+(1-e_{10}^2)D\theta_0\right]\frac{D\phi_1}{H_0} - \left[\left(\frac{\chi_*-\chi}{1+\chi}\mathrm{N}-\theta_0\right)\right.\\
& \times\left(\alpha^2+\beta^2-(\alpha e_{20}+\beta e_{30})^2 + \mathrm{i}(\alpha e_{20}+\beta e_{30})(1-3e_{10}^2)\frac{DH_{x0}}{H_0}\right)\\
& \left.-\mathrm{i}(\alpha e_{20}+\beta e_{30})e_{10}D\theta_0\right]\frac{\phi_1}{H_0}\\
& -\left(\mathrm{i}(\alpha e_{20}+\beta e_{30}) + (1-e_{10}^2)\frac{DH_{x0}}{H_0}\right)\theta_1 - e_{10}D\theta_1 \qquad (3.50)
\end{aligned}
$$

with

$$
\begin{aligned}
\mathbf{H}_1 &= [D\phi_1, \mathrm{i}\alpha\phi_1, \mathrm{i}\beta\phi_1]^T\,,\\
H_1 &= \mathbf{H}_1\cdot\mathbf{e}_0 = e_{10}D\phi_1 + \mathrm{i}(\alpha e_{20}+\beta e_{30})\phi_1\,,
\end{aligned}
$$

$$\mathbf{M}_1 = \chi \mathbf{H}_1 - (1+\chi)\theta_1 \mathbf{e}_0 + \frac{(1+\chi)\theta_0 - (\chi_* - \chi)\mathrm{N}}{H_0}(H_1 \mathbf{e}_0 - \mathbf{H}_1)\,,$$
$$M_1 = \mathbf{M}_1 \cdot \mathbf{e}_0 = \chi H_1 - (1+\chi)\theta_1\,.$$

The disturbance velocity and temperature fields are subject to standard homogeneous boundary conditions

$$u_1 = v_1 = w_1 = \theta_1 = 0 \text{ at } x = \pm 1\,. \tag{3.51}$$

As discussed in Section 3.3.1, if there are no induced currents outside the layer and a non-magnetic medium fills the surrounding space, then the perturbation of external magnetic field has potential $\phi_1^e(x)\exp(\sigma t + \mathrm{i}\alpha y + \mathrm{i}\beta z)$, which, as follows from Equations (2.4) and (2.5), satisfies Laplace's equation

$$(D^2 - \alpha^2 - \beta^2)\phi_1^e = 0\,, \tag{3.52}$$

in the regions $x < -1$ and $x > 1$. Upon taking into account (2.5), the linearisation of the magnetic boundary conditions (2.13) in the case of an oblique field leads to

$$D\phi_1^e = \left(1 + \chi + (1 - e_{10}^2)\frac{(\chi_* - \chi)\mathrm{N} \pm (1+\chi)}{H_0}\right) D\phi_1 \tag{3.53}$$
$$-\mathrm{i}e_1(\alpha e_{20} + \beta e_{30})\frac{(\chi_* - \chi)\mathrm{N} \pm (1+\chi)}{H_0}\phi_1\,,$$
$$\phi_1^e = \phi_1 \quad \text{at} \quad x = \pm 1\,, \tag{3.54}$$

where

$$\phi_1^e(x) = \begin{cases} Ae^{\sqrt{\alpha^2+\beta^2}x}\,, & x < -1 \\ Be^{-\sqrt{\alpha^2+\beta^2}x}\,, & x > 1 \end{cases}\,. \tag{3.55}$$

After eliminating A and B from Equations (3.55) and (3.54), we obtain the boundary conditions for ϕ_1 at $x = \pm 1$

$$\left(1 + \chi + (1 - e_{10}^2)\frac{(\chi_* - \chi)\mathrm{N} \pm (1+\chi)}{H_0}\right) D\phi_1 \tag{3.56}$$
$$\pm\sqrt{\alpha^2+\beta^2}\phi_1 - \mathrm{i}e_{10}(\alpha e_{20} + \beta e_{30})\frac{(\chi_* - \chi)\mathrm{N} \pm (1+\chi)}{H_0}\phi_1 = 0\,.$$

Upon using the generalised Squire transformations (3.25), where we also add

$$H_{x0} = \widetilde{H}_{x0}\,,\ H_0 = \widetilde{H}_0\,,\ e_{10} = \widetilde{e}_{10}\,,\ \alpha e_{20} + \beta e_{30} = \widetilde{\alpha}\widetilde{e}_{20}\,,\ \mathrm{Ra}_m = \widetilde{\mathrm{Ra}_m}\,, \tag{3.57}$$

following the procedure similar to that described in Section 3.3.1 equations and keeping tildes only to denote nontrivially transformed quantities (3.45)–(3.50) become

$$0 = Du + \mathrm{i}\widetilde{\alpha}\widetilde{v}\,, \tag{3.58}$$
$$\sigma u + \mathrm{Pr}(\widetilde{\alpha}^2 - D^2)u + DP + e_{10}\mathrm{Ra}_m\mathrm{Pr}DH_{x0}\theta + \mathrm{Ra}_m\mathrm{Pr}\theta_0 e_{10}D^2\phi \tag{3.59}$$

$$+\mathrm{Ra}_m\mathrm{Pr}\theta_0\left[\mathrm{i}\widetilde{\alpha}\widetilde{e}_{20}+(1-e_{10}^2)\frac{DH_{x0}}{H_0}\right]D\phi$$

$$-\mathrm{i}\widetilde{\alpha}\mathrm{Ra}_m\mathrm{Pr}\,\theta_0 e_{10}\widetilde{e}_{20}\frac{DH_{x0}}{H_0}\phi=0\,,$$

$$\sigma\widetilde{v}+\mathrm{Pr}(\widetilde{\alpha}^2-D^2)\widetilde{v}+\mathrm{i}\widetilde{\alpha}P+\widetilde{\alpha}\mathrm{Ra}_m\mathrm{Pr}\,\theta_0(ie_{10}D\phi-\widetilde{\alpha}\widetilde{e}_{20}\phi)=0\,, \tag{3.60}$$

$$\sigma w+\mathrm{Pr}(\widetilde{\alpha}^2-D^2)w+\mathrm{i}\beta P+\beta\mathrm{Ra}_m\mathrm{Pr}\,\theta_0(ie_{10}D\phi-\widetilde{\alpha}\widetilde{e}_{20}\phi)=0\,, \tag{3.61}$$

$$\sigma\theta+D\theta_0 u+(\widetilde{\alpha}^2-D^2)\theta=0\,, \tag{3.62}$$

$$0=(D^2-\widetilde{\alpha}^2)\phi+(1-e_{10}^2)\left(\frac{\chi_*-\chi}{1+\chi}\mathrm{N}-\theta_0\right)\frac{D^2\phi}{H_0}-\left[(1-e_{10}^2)D\theta_0\right.$$

$$+e_{10}\left(\frac{\chi_*-\chi}{1+\chi}\mathrm{N}-\theta_0\right)\left(2\mathrm{i}\widetilde{\alpha}\widetilde{e}_{20}+3(1-e_{10}^2)\frac{DH_{x0}}{H_0}\right)\Bigg]\frac{D\phi}{H_0}$$

$$-\left[\left(\frac{\chi_*-\chi}{1+\chi}\mathrm{N}-\theta_0\right)\left(\widetilde{\alpha}^2(1-\widetilde{e}_{20}^2)+\mathrm{i}\widetilde{\alpha}\widetilde{e}_{20}(1-3e_{10}^2)\frac{DH_{x0}}{H_0}\right)\right.$$

$$\left.-\mathrm{i}\widetilde{\alpha}\widetilde{e}_{20}e_{10}D\theta_0\right]\frac{\phi}{H_0}-\left(\mathrm{i}\widetilde{\alpha}\widetilde{e}_{20}+(1-e_{10}^2)\frac{DH_{x0}}{H_0}\right)\theta-e_{10}D\theta \tag{3.63}$$

with the boundary conditions

$$\left(1+\chi+(1-e_{10}^2)\frac{(\chi_*-\chi)\mathrm{N}\pm(1+\chi)}{H_0}\right)D\phi$$

$$\pm|\widetilde{\alpha}|\phi-\mathrm{i}\widetilde{\alpha}e_{10}\widetilde{e}_{20}\frac{(\chi_*-\chi)\mathrm{N}\pm(1+\chi)}{H_0}\phi=0\,, \tag{3.64}$$

$$u=\widetilde{v}=w=\theta=0\quad\text{at}\quad x=\pm1\,. \tag{3.65}$$

As in Section 3.3.1 the equivalent two-dimensional problem is obtained by splitting Equation (3.61) off the system. However, it is important to note that the notion of such an equivalent two-dimensional problem here is somewhat different. The reason is that after the above transformations, the external magnetic field remains three-dimensional. In general it still has three non-zero components in the x, y and z directions, and thus it needs to be described using two coordinate angles δ and γ that act as independent control parameters of the problem. The above transformations simply mean that we conveniently view the y direction as the periodicity direction of the arising perturbation structures, while the vector of the applied magnetic field can be arbitrarily oriented with respect to it. More specifically, the axes of the instability rolls are considered to be always parallel to the z-axis in Figure 3.2 so that choosing $\gamma=0°$ ($\gamma=90°$) means that the magnetic field has a component in the plane of the fluid layer that is perpendicular (parallel) to the roll axes. The values of $0°<\gamma<90°$ are interpreted accordingly. For the sake of brevity in the following sections, we will conveniently refer to instability patterns computed using the above transformed equations for $\gamma=0°$ and $\gamma=90°$ as transverse and longitudinal rolls, respectively, while patterns obtained for all other values of γ will be referred to as oblique rolls. We will also refer to angle δ as the field inclination angle and angle γ as the angle between the axes of the instability rolls and the in-layer component of the applied magnetic field.

Table 3.4 The critical values of Ra_m, $\widetilde{\alpha}$ and disturbance wave speed $\widetilde{c} = -\sigma^I/\widetilde{\alpha}$ and the corresponding perturbation energy integrals Σ_k and Σ_{m1} for magnetoconvection at $\delta = 0°$, $H^e = 100$ (odd-numbered lines), $H^e = 10$ (even-numbered lines) and various values of χ and χ_*.

χ	χ_*	$\widetilde{\alpha}_c$	Ra_{mc}	$\widetilde{c}_c$	Σ_k	$\mathrm{Re}(\Sigma_{m1})$	$\mathrm{Im}(\Sigma_{m1})$
5	5	1.936	181.7	0	0.0982	1.584	0
		1.920	178.0	0	0.0984	1.559	0
3	5	1.860	159.3	0	0.0997	1.424	0
		1.846	157.0	0	0.0998	1.407	0
3	3	1.915	178.3	0	0.0985	1.565	0
		1.909	176.8	0	0.0986	1.554	0
1.5	2.5	1.847	159.1	0	0.0998	1.427	0
		1.843	158.3	0	0.0998	1.421	0
1	3	1.797	145.9	0	0.1007	1.325	0
		1.792	145.1	0	0.1007	1.320	0
1	2	1.827	155.0	0	0.1001	1.397	0
		1.825	154.5	0	0.1001	1.393	0
0.5	1.5	1.801	150.0	0	0.1005	1.357	0
		1.799	149.2	0	0.1005	1.355	0

3.4.2 Flow Stability Characteristics in Zero Gravity

The numerical values of critical parameters for thermomagnetic convection arising in magnetic fields of various orientations and intensities are given in Tables 3.4, 3.5, and 3.6. The data in the tables warrants a number of general conclusions.

The magnetoconvective instability arising in a normal field remains stationary regardless of the specific magnetic properties of the fluid and the magnitude of the applied magnetic field. This is in agreement with the findings previously reported, for example, in [89, 113] and [234]. However, in contrast to all previous studies, the instability threshold Ra_{mc} is found to depend not only on the values of the magnetic susceptibilities χ and χ_* but also on the magnitude of the applied magnetic field, namely, the decrease of the characteristic nondimensional field parameter N promotes instability and increases the wavelength of the arising patterns. This dependence, however, remains relatively weak: the largest difference between the critical values of magnetic Rayleigh and wavenumbers is found to be under 3.5% and 1.5%, respectively, for a fluid with the highest degree of magnetisation investigated ($\chi = \chi_* = 5$) when the external magnetic field is changed by a factor of 10. The comparison of the current results with our previous study [234] shows that the dependence of the critical parameters on the magnitude of the magnetic field is traced back to the form of the constitutive magnetisation equation (2.8). Its linearisation used in all previous studies cited above eliminates the dependence of the threshold values on the amplitude of the normal magnetic field. However, such an idealisation is only robust for the case of paramagnetic fluids with small magnetic susceptibilities, and it should not be expected to be uniformly valid for realistic ferromagnetic fluids.

If the dependence of the fluid magnetisation on the magnitude of the applied magnetic field remains linear that is if $\chi \approx \chi_*$ (see Figure 2.1), then the magnetoconvection threshold parameters decrease monotonically with the values of magnetic susceptibilities to their limiting values $(\mathrm{Ra}_{mc}, \widetilde{\alpha}_c) \approx (160.5, 1.805)$ that are independent of the magnitude of the applied magnetic field. However, when the fluid magnetisation approaches saturation so that $\chi < \chi_*$, the variation of the differential and integral susceptibilities have opposite influences on the threshold: the decrease of χ at fixed χ_* promotes instability, while the decrease of χ_* at fixed χ delays it. In realistic ferrofluids, however, the values of both χ and χ_* decrease with the increasing magnetic field but at different rates; see Figure 2.1(a). Therefore it is not straightforward to anticipate what the overall effect of a changing magnetic field on the convection onset could be, and one needs to rely on the specific computational results. In particular, the data in Table 3.4 shows that the critical value of magnetic Rayleigh number decreases by more than 10% when progressively stronger magnetic field is applied to a layer of experimental ferrofluid with the initial susceptibilities $\chi = \chi_* \approx 3$ that are reduced to $\chi \approx 1.5$ and $\chi_* \approx 2.5$ during a typical experimental run.

It is remarkable that as seen from Tables 3.5 and 3.6, when an oblique magnetic field is applied to the layer, the trends described above are reversed even for small field inclination angles δ: now the decrease of χ at fixed χ_* delays instability, while the decrease of χ_* at fixed χ promotes it. This indicates the qualitative difference between the instability mechanisms present in normal and oblique fields that we will discuss in more detail in the following sec-

Table 3.5 Same as Table 3.4 but for transverse rolls at $\delta = 10°$ and $\gamma = 0°$.

χ	χ_*	$\widetilde{\alpha}_c$	Ra_{mc}	$\widetilde{c}_c$	Σ_k	$\mathrm{Re}(\Sigma_{m1})$	$\mathrm{Im}(\Sigma_{m1})$
5	5	2.428	2049	0.0256	0.0506	1.9562	−0.0037
		2.468	1935	0.2048	0.0505	1.9993	−0.0354
3	5	2.510	2796	0.0279	0.0472	1.7755	−0.0026
		2.534	2577	0.2332	0.0476	1.8146	−0.0273
3	3	2.116	533.8	−0.0011	0.0822	2.2432	−0.0044
		2.135	548.3	−0.0085	0.0809	2.2420	−0.0421
1.5	2.5	2.051	433.2	−0.0012	0.0873	2.0035	−0.0029
		2.061	439.8	−0.0108	0.0866	2.0058	−0.0287
1	3	2.221	724.9	0.0018	0.0757	1.9704	−0.0033
		2.228	721.9	0.0167	0.0750	1.9575	−0.0321
1	2	1.967	329.4	−0.0014	0.0926	1.8328	−0.0019
		1.973	333.5	−0.0139	0.0923	1.8391	−0.0192
0.5	1.5	1.894	259.3	−0.0012	0.0962	1.6545	−0.0012
		1.898	261.4	−0.0121	0.0960	1.6594	−0.0115

tions. The numerical data given in Tables 3.4, 3.5, and 3.6 also demonstrates a very strong stabilisation effect of the field inclination compared to the normal field situation that is further illustrated in Figures 3.29(a) and 3.31(a). Such a stabilisation is observed regardless of the specific magnetic properties of the fluid for all investigated values of χ and χ_*.

An even more striking effect of the field inclination is evident from the data presented in Table 3.5: the transverse instability rolls computed for $\gamma = 0°$ become oscillatory resulting in waves propagating along the direction of the field component that is tangential to the plane of the fluid layer. This is a somewhat unexpected result given that the unperturbed problem possesses a full planar symmetry with no preferred direction. Moreover [113] even argued that the instability in this problem can only be stationary. The resolution of this apparent paradox is prompted by the comparative computational data presented in Table 3.5 for $H^e = 10$ and $H^e = 100$ and by Figure 3.30. They show that the magnitude of the disturbance wave speed $|\widetilde{c}| = \left|\frac{\sigma^I}{\widetilde{\alpha}_c}\right|$ is approximately inversely proportional to the magnitude of the applied magnetic field H^e, which in turn is proportional to the field parameter N characterising the nonlinearity of the magnetic field distribution within a layer. It is assumed in [113] that $\mathrm{N} \to \infty$ and effectively postulated that the magnetic field within the layer varies linearly. No unsteady patterns were found there. Therefore we conclude that the main reason for the appearance of oscillatory instability in the current problem is the nonlinearity of a magnetic field within the ferrofluid layer.

Table 3.6 Same as Table 3.4 but for longitudinal rolls at $\delta = 10°$ and $\gamma = 90°$.

χ	χ_*	$\widetilde{\alpha}_c$	Ra_{mc}	$\widetilde{c}_c$	Σ_k	$\mathrm{Re}(\Sigma_{m1})$	$\mathrm{Im}(\Sigma_{m1})$
5	5	1.937	385.2	0	0.0982	1.5844	0
		1.943	389.6	0	0.0978	1.6117	0
3	5	1.903	458.9	0	0.0988	1.5105	0
		1.908	457.2	0	0.0984	1.5258	0
3	3	1.916	267.0	0	0.0985	1.5650	0
		1.917	270.3	0	0.0985	1.5747	0
1.5	2.5	1.865	251.6	0	0.0995	1.4640	0
		1.866	253.5	0	0.0994	1.4693	0
1	3	1.842	313.1	0	0.0999	1.4094	0
		1.844	313.7	0	0.0997	1.4146	0
1	2	1.844	226.5	0	0.0998	1.4315	0
		1.845	227.8	0	0.0998	1.4346	0
0.5	1.5	1.816	203.2	0	0.1002	1.3879	0
		1.817	204.1	0	0.1003	1.3893	0

The values of the threshold parameters for longitudinal rolls computed for $\gamma = 90°$ are given in Table 3.6. Remarkably, they remain strictly stationary for all values of the governing parameters. Figure 3.31(a) shows that similar to the critical magnetic Rayleigh number for transverse rolls, the one for longitudinal rolls increases quickly with the field inclination angle δ. However for all non-zero angles, it remains smaller than that of transverse rolls. This is consistent with findings of [113] for paramagnetic fluids and confirms an experimental fact that the axes of thermomagnetic rolls appearing away from the boundaries always align with the tangential component of the magnetic field since this configuration is found to be less stable than a transverse one. Having said this, we emphasise that even though longitudinal rolls are always expected to dominate the observed instability patterns, the possibility of the existence of transverse rolls should not be ignored for at least two reasons. Firstly, unlike in paramagnetic fluids, in ferrofluids they are qualitatively different from their longitudinal counterparts as they are unsteady. They are also characterised by a wavenumber that depends sensitively on the field inclination angle (see Figure 3.29(b)), while the wavenumber of longitudinal rolls remains almost constant as the field inclination is increased; see Figure 3.31(b). Secondly, near the boundaries of a layer, the longitudinal rolls may be suppressed due to the geometry of the boundary or other influences that are not present in unbounded domains so that oscillatory transverse rolls might be preferred. The experimental observations [238] (see Figures 6.37(d,e,i,j), 6.39(c,d) and 6.41) indeed indicate that this might have been the case in the near-boundary regions of a finite experimental enclosure.

Given that the two limiting cases of transverse and longitudinal rolls have qualitatively different characteristics, it is of interest to investigate how and at what value of the intermediate angle the transition between stationary and oscillatory patterns occurs. Thus we have computed the stability characteristics of oblique rolls for various values of magnetic susceptibilities and field inclination angles. These are presented in Figures 3.32, 3.33, and 3.34. They confirm that both the critical magnetic Rayleigh number and wavenumber increase continuously and monotonically from longitudinal to transverse rolls and the rate at which they do grow quickly with the field inclination angle. The only exception is the behaviour of the wavenumber for relatively large field inclination angles when it reaches its maximum value for oblique rolls forming the angle of about 45° with the tangential field component and then starts decreasing. Of particular interest is the behaviour of the disturbance wave speed. It grows continuously from zero for longitudinal rolls to its maximum for transverse rolls; however, the most rapid growth is observed for $\gamma \lesssim 50^\circ$ and $\gamma \gtrsim 130^\circ$. This suggests that if the value of magnetic Rayleigh number is gradually increased in an experiment, then the stationary rolls aligned with the tangential component of the field will appear first. Subsequently, they would be unsteadily modulated by a periodic pattern forming the angle of about 40°–45° with the axes of the stationary rolls. A further increase of magnetic Rayleigh number would lead to the increase of the modulation frequency and wavenumber and to the re-orientation of the modulating pattern so that it would become closer to orthogonal with respect to the original stationary rolls.

It is also noteworthy that the Ra_{mc} and $\widetilde{\alpha}_c$ curves are symmetric with respect to the $\gamma = 90^\circ$ line, while the $\widetilde{c}_c$ line is centrosymmetric with respect to $(\widetilde{c}_c, \gamma) = (0, 90^\circ)$. To shed light on why this is so, refer to Figure 3.5(a) and (c) where the concave south-west/north-east magnetic field lines are shown for $\gamma = 0^\circ$. If γ is changed to 180°, the magnetic field lines re-orient to become north-west/south-east and convex. As has been discussed above, the appearance of oscillatory disturbances is a consequence of the nonlinearity of a magnetic field. Therefore we conclude that it is this change of the curvature of magnetic lines in the plane perpendicular to the roll axes that is responsible for the change of the sign of the disturbance wave speed.

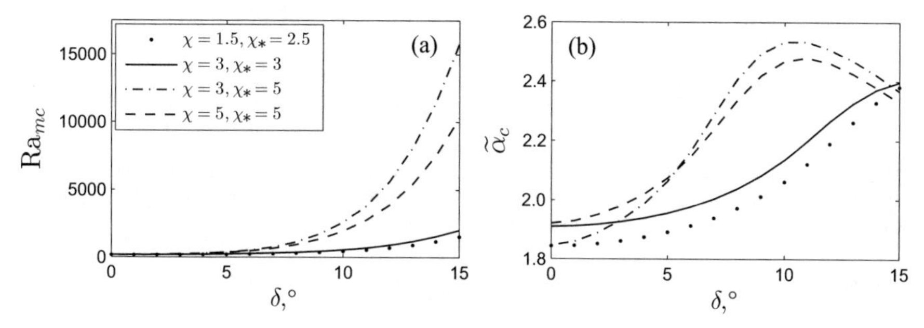

Fig. 3.29 (a) Critical magnetic Rayleigh number Ra_{mc} and (b) wavenumber $\widetilde{\alpha}_c$ as functions of the field inclination angle δ for transverse rolls at $H^e = 10$ and $\gamma = 0°$. The respective plots for $H^e = 100$ are indistinguishable within the figure resolution.

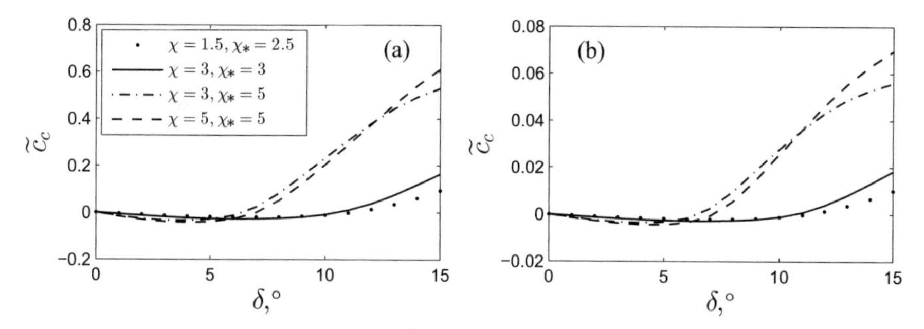

Fig. 3.30 Critical wave speed $\widetilde{c} = -\frac{\sigma^I}{\widetilde{\alpha}_c}$ as the function of the field inclination angle δ for transverse rolls at $\gamma = 0°$ for (a) $H^e = 10$ and (b) $H^e = 100$.

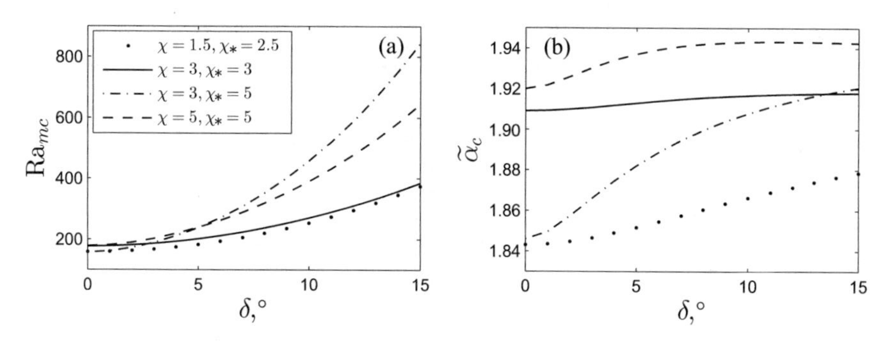

Fig. 3.31 Same as Figure 3.29 but for longitudinal rolls at $\gamma = 90°$.

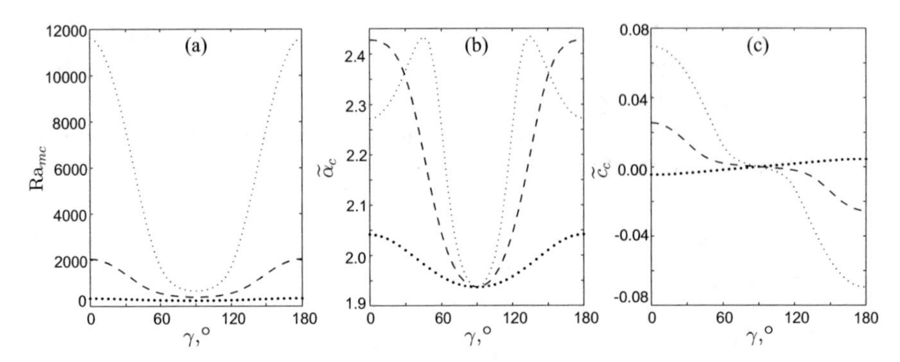

Fig. 3.32 (a) Critical magnetic Rayleigh number Ra_{mc}, (b) wavenumber $\widetilde{\alpha}_c$ and (c) wave speed $\widetilde{c}_c$ as functions of the azimuthal angle γ for $\delta = 5^\circ$ (large dots), 10° (dashed lines) and 15° (small dots) for $H^e = 100$ and $\chi = \chi_* = 5$.

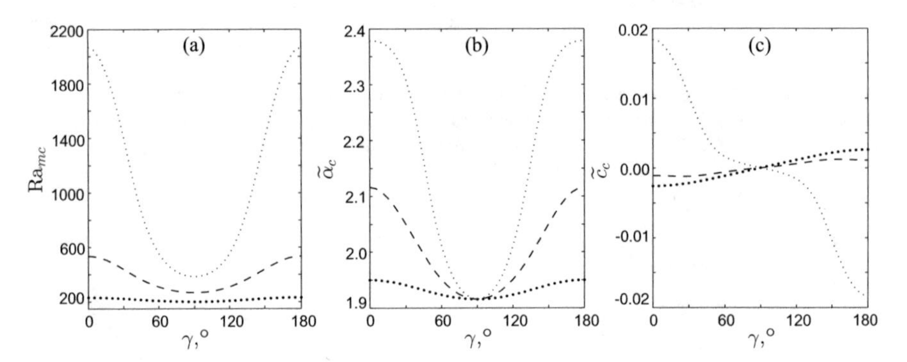

Fig. 3.33 Same as Figure 3.32 but for $\chi = \chi_* = 3$.

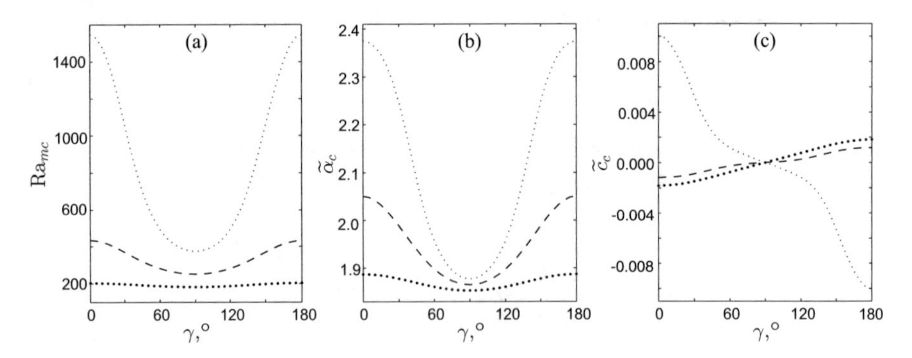

Fig. 3.34 Same as Figure 3.32 but for $\chi = 1.5$ and $\chi_* = 2.5$.

3.4.3 Perturbation Energy Balance in Zero Gravity

To confirm the physical nature of the observed instabilities, we repeat the perturbation energy balance consideration developed in Section 3.3.3 here taking into account the field inclination effects. The energy balance equations in this case become

$$\sigma \Sigma_k = \Sigma_{vis} + \Sigma_{m1} + \Sigma_{m2}\,, \tag{3.66}$$

where

$$\Sigma_k = \int_{-1}^{1} \underbrace{(|u|^2 + |\widetilde{v}|^2)}_{E_k} \mathrm{d}x > 0\,, \tag{3.67}$$

$$\Sigma_{vis} = \int_{-1}^{1} \underbrace{-\mathrm{Pr}(\widetilde{\alpha}^2(|u|^2 + |\widetilde{v}|^2) + |Du|^2 + |D\widetilde{v}|^2)}_{E_{vis}} \mathrm{d}x = -1\,, \tag{3.68}$$

$$\Sigma_{m1} = \int_{-1}^{1} \underbrace{-\mathrm{Ra}_m \mathrm{Pr} D H_{x0} e_{10} \theta \bar{u}}_{E_{m1}} \mathrm{d}x\,, \tag{3.69}$$

$$\Sigma_{m2} = \int_{-1}^{1} E_{m2} \mathrm{d}x \tag{3.70}$$

and

$$\begin{aligned} E_{m2} = & -\mathrm{Ra}_m \mathrm{Pr} D H_{x0} \bar{u} \frac{\theta_0}{H_0}((1 - e_{10}^2) D\phi - \mathrm{i}\widetilde{\alpha} e_{10} \widetilde{e}_{20} \phi) \\ & -\mathrm{Ra}_m \mathrm{Pr} \theta_0 \left(e_{10} \bar{u} D^2 \phi + \mathrm{i}\widetilde{\alpha}(e_{10} \bar{\widetilde{v}} + \widetilde{e}_{20} \bar{u}) D\phi - \widetilde{\alpha}^2 \widetilde{e}_{20} \bar{\widetilde{v}} \phi \right) . \end{aligned}$$

Equation (3.66), which determines the complex growth rate σ of linear instability, does not contain the modified pressure P as it integrates to zero identically. This confirms that the potential component of Kelvin force included in P has no effect on the stability of a layer of a ferromagnetic fluid. Σ_{m1} represents the variation of fluid magnetisation (and thus of the local Kelvin force) due to the thermal perturbations, while Σ_{m2} describes the energy contribution associated with the induced magnetic field variations. Separating the real (Re) and imaginary (Im) parts of (3.66), we obtain at the critical point

$$0 = \mathrm{Re}(\Sigma_{m1} + \Sigma_{m2}) - 1\,, \quad \sigma^I \Sigma_k = \mathrm{Im}(\Sigma_{m1} + \Sigma_{m2})\,. \tag{3.71}$$

The energy terms with positive real parts promote instability, while the ones with negative suppress it. The second of Equation (3.71) demonstrates that the nature of the detected oscillatory instabilities is purely magnetic.

Tables 3.4, 3.5, and 3.6 contain numerical data for various perturbation energy terms that enable us to draw a number of general conclusions. Firstly, the magnitude of the kinetic energy term Σ_k never exceeds the value of about 10% of the viscous dissipation, while the magnitude of the magnetic contribution Σ_{m1} always exceeds the dissipation value. This confirms that the instability is of magnetic rather than hydrodynamic or thermal nature and that the visible fluid motion triggered by the instability is not the main recipient of the energy supplied to the system (in experiments such as those described in [238], the energy is supplied by heat exchangers attached to the layer walls). Secondly, since $\mathrm{Re}(\Sigma_{m1})$ is always positive, we conclude that the specific mechanism triggering the instability is the thermally induced variation of fluid magnetisation. Thirdly, since $\mathrm{Re}(\Sigma_{m1}) > 1$ then, according to the first of Equation (3.71), $\mathrm{Re}(\Sigma_{m2}) < 0$. This means that the variation of the applied magnetic field caused by perturbations always plays a stabilising role. In summary, the analysis of mechanical energy balance shows that the energy received by the system through a thermal exchange with the ambient is mostly spent on varying the local magnetisation of the fluid. In turn the latter triggers fluid motion, which is an observable signature of instability. The remaining part of the received energy is spent on modifying the magnetic field. Since the variation of magnetic field is not limited to the interior of the layer, this energy largely leaves it and thus cannot be used for supporting a mechanical instability within the system.

Typical distributions of the perturbation energy integrands for instability patterns arising in normal and oblique fields are shown in Figures 3.35 and 3.36, respectively. Since the integrand behaviour for longitudinal rolls is found to be qualitatively similar to that for stationary rolls arising in a normal field, only the results for transverse rolls are presented here. As expected, the viscous dissipation E_{vis} occurs mostly near the solid boundaries, and the kinetic energy E_k of perturbations is maximised near the centre of the layer. The middle panels in both figures show that the magnetisation variation effect E_{m1} plays a destabilising role uniformly across the complete width of the layer and with the maximum near its center. On the other hand, the stabilising effect of magnetic field modification E_{m2} is most pronounced near the walls of the layer. This is intuitively expected since the internal magnetic field near the walls defines the external field via the field-matching boundary conditions (3.53) and (3.54). The overall role of the E_{m1} and E_{m2} effects does not change in the oblique field; however the field inclination introduces a noticeable asymmetry. The maximum of the destabilising influence shifts towards the hot wall, comparing the locations of the maxima of the dash-dotted lines in the middle panels of Figures 3.35 and 3.36. This is because the unstable magnetic buoyancy stratification in oblique fields is more pronounced near the hot wall; see Figure 3.5(b) and (d). While the magnitude of $\mathrm{Re}(E_{m1})$ determines whether the instability is present, the right panel in Figure 3.36 shows that it is the magnitude of $\mathrm{Im}(E_{m2})$ that predominantly defines the sign of σ^I and thus the propagation direction of transverse and oblique rolls.

3.4.4 Perturbation Fields in Zero Gravity

We present comparative plots of typical perturbation fields arising in normal and inclined magnetic fields next. The mechanism driving convection is straightforward to see from Figure 3.37 for a normal field. Consider, for

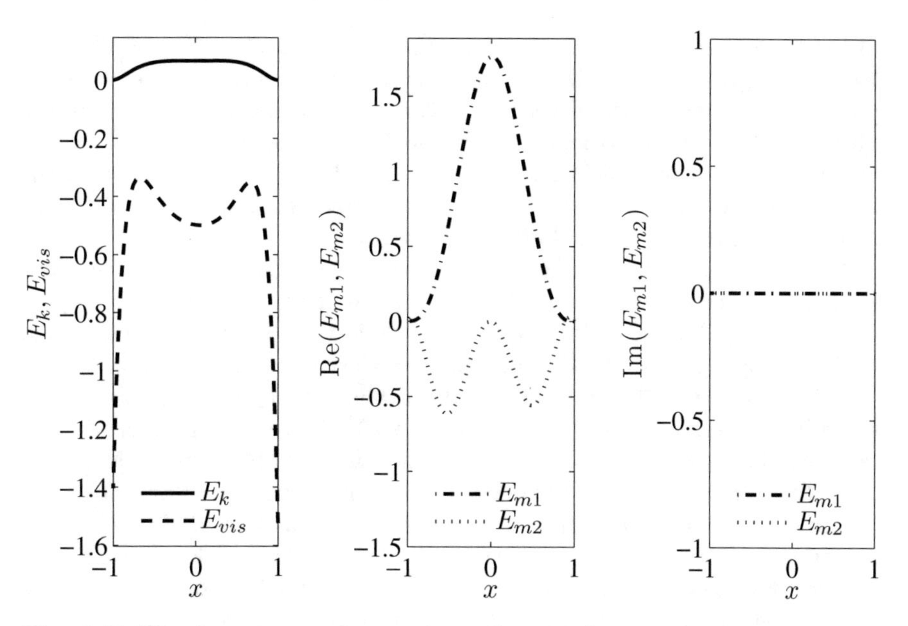

Fig. 3.35 Disturbance energy integrands at the critical point of magnetoconvection threshold $\mathrm{Ra}_{mc} = 176.8$, $\widetilde{\alpha}_c = 1.909$ at $H^e = 10$, $\delta = \gamma = 0°$ and $\chi = \chi_* = 3$.

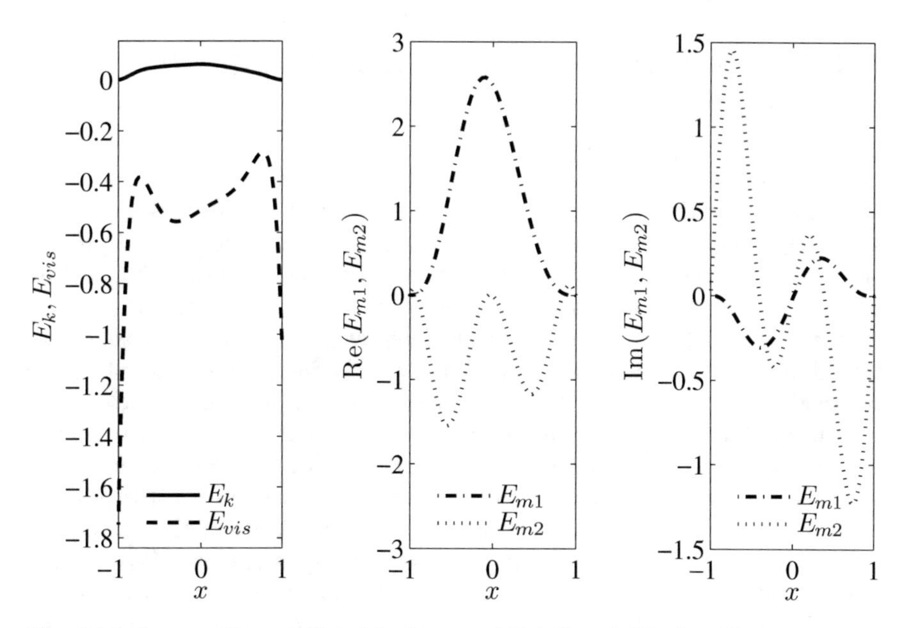

Fig. 3.36 Same as Figure 3.35 but for $\mathrm{Ra}_{mc} = 548.4$, $\widetilde{\alpha}_c = 2.135$, $\delta = 10°$.

example, the region near $\widetilde{y} = 3$. There the thermal perturbation θ_1 leads to local cooling. As a result this region becomes stronger magnetised (see the plot for M_1), and the fluid there is driven towards the hot wall where the basic magnetic field is stronger; see Figure 3.3(a) and (d). This is reflected in the plot of the velocity field showing that indeed cool fluid flows towards the hot wall (from right to left) there. This situation is similar to gravitational convection arising in a fluid heated from below.

When the applied magnetic field is inclined, the mechanism driving convection remains the same even though it is less straightforward to recognise it from Figure 3.38. The thermal and magnetisation perturbation cells align with the applied magnetic field and so does the main fluid flow direction.

It is noteworthy that the perturbation cells for magnetic field H_1 corresponding to transverse rolls do not align with the rest of the perturbation field; see the right panel in Figure 3.38. They also become asymmetric. At the same time, the structure of the perturbation fields for longitudinal rolls in an inclined field (not shown) remains very similar to that seen in Figure 3.37 for a normal field. Therefore it is logical to conclude that the phase shift between the magnetic field H_1 and the rest of the perturbation fields is responsible for the change of the instability character to oscillatory for transverse and oblique rolls.

3.4.5 Linearised Perturbation Equations in Non-zero Gravity

The analysis of pure magnetoconvection undertaken in previous sections has demonstrated a rich variety of physical effects and flow patterns caused by symmetry-breaking effects related to the nonlinear variation of fluid properties across the layer placed in an oblique external magnetic field. Introduction of gravity creates yet another parametric dimension that makes the problem even richer and leads to qualitatively different instabilities than those observed in a gravity-free setting. We analyse these instabilities caused by the interaction of magnetic Kelvin and gravitational buoyancy forces below.

The stability of a non-isothermal flow in a vertical layer of ordinary fluid is one of the classical problems of natural convection [14, 96], and we recollect that for a class of large-Prandtl-number fluids, to which kerosene- and transformer-oil-based ferrofluids belong to, the instability in this configuration occurs in the form of two waves counter-propagating along the direction of the gravity [126]. On the other hand, we have seen in Section 3.3.4 that the most dangerous instability mode detected in a normal magnetic field in the small gravity limit consists of stationary rolls with the axes parallel to the direction of the gravity. At the same time, it has been shown in Section 3.4.2 that the oblique magnetic field tends to align the axes of the rolls with its in-layer component. Therefore, it remains to be seen what exactly pattern orientation will result when an oblique magnetic field and the gravity act simultaneously.

In the presence of gravity Squire-transformed linearised perturbation, Equations (3.58)–(3.65) remain unchanged apart from the addition of the buoyancy term in the momentum equation. It is also traditional to write nondimensional equation using Grashof rather than Rayleigh numbers as main governing parameters:

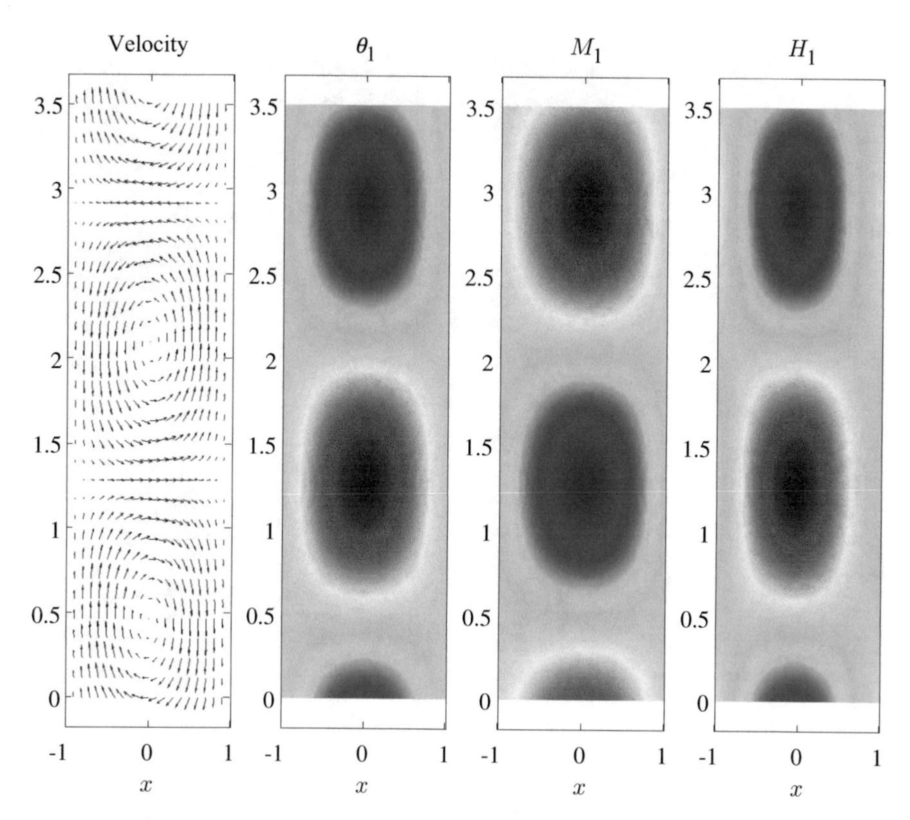

Fig. 3.37 Perturbation eigenfunctions of the fluid velocity $\mathbf{v}_1 = (u, v)$, temperature θ_1, magnetisation M_1 and magnetic field H_1 for magnetoconvection at $H^e = 10$, $\delta = \gamma = 0°$ and $\chi = \chi_* = 3$ at the critical point $\mathrm{Ra}_{mc} = 176.8$, $\widetilde{\alpha}_c = 1.909$. The field values increase from black to white (from blue to red online).

$$\begin{aligned}
&\sigma u + (\widetilde{\alpha}^2 + \mathrm{i}\widetilde{\alpha}\widetilde{v}_0 - D^2)u + DP + e_{10}\mathrm{Gr}_m DH_{x0}\theta + \mathrm{Gr}_m\theta_0 e_{10}D^2\phi \\
&\quad + \mathrm{Gr}_m\theta_0\left[\mathrm{i}\widetilde{\alpha}\widetilde{e}_{20} + (1 - e_{10}^2)\frac{DH_{x0}}{H_0}\right]D\phi \\
&\quad - \mathrm{i}\widetilde{\alpha}\mathrm{Gr}_m\theta_0 e_{10}\widetilde{e}_{20}\frac{DH_{x0}}{H_0}\phi = 0\,, \qquad (3.72)
\end{aligned}$$

$$\begin{aligned}
&\sigma\widetilde{v} + D\widetilde{v}_0 u + (\widetilde{\alpha}^2 + \mathrm{i}\widetilde{\alpha}\widetilde{v}_0 - D^2)\widetilde{v} + \mathrm{i}\widetilde{\alpha}P - \widetilde{\mathrm{Gr}}\theta \\
&\quad + \widetilde{\alpha}\mathrm{Gr}_m\theta_0(\mathrm{i}e_{10}D\phi - \widetilde{\alpha}\widetilde{e}_{20}\phi) = 0\,, \qquad (3.73)
\end{aligned}$$

$$\sigma w + (\widetilde{\alpha}^2 + \mathrm{i}\widetilde{\alpha}\widetilde{v}_0 - D^2)w + \mathrm{i}\beta P + \beta \mathrm{Gr}_m \theta_0(\mathrm{i}e_{10}D\phi - \widetilde{\alpha}\widetilde{e}_{20}\phi) = 0, \quad (3.74)$$

$$\sigma\theta + D\theta_0 u + \left(\frac{\widetilde{\alpha}^2 - D^2}{\mathrm{Pr}} + \mathrm{i}\widetilde{\alpha}\widetilde{v}_0\right)\theta = 0\,. \quad (3.75)$$

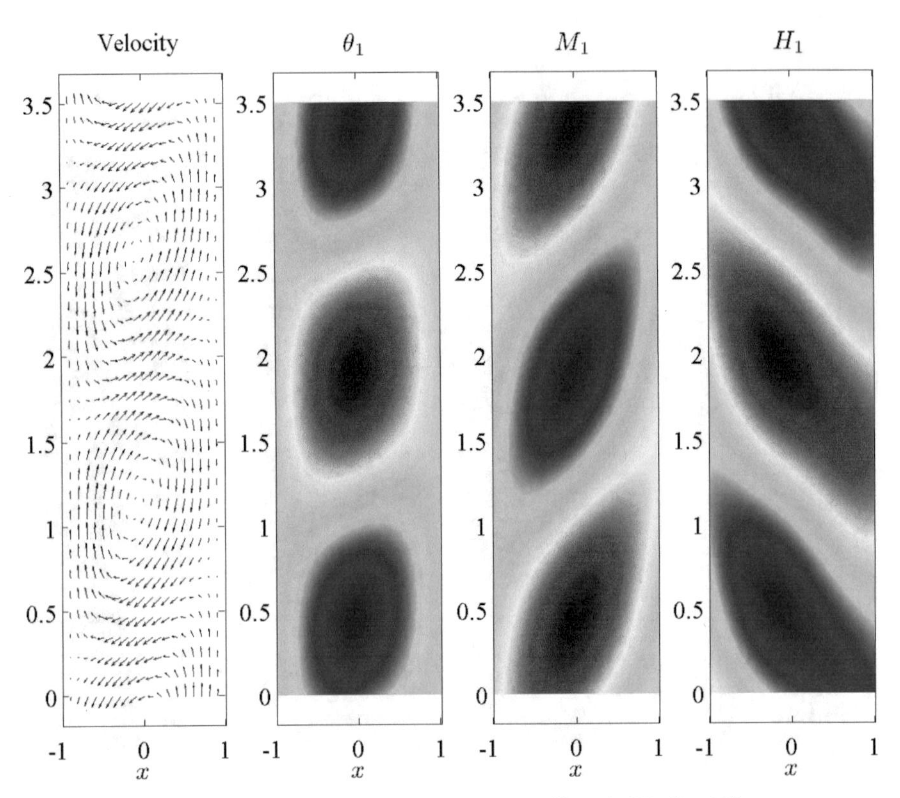

Fig. 3.38 Same as Figure 3.37 but for $\mathrm{Ra}_{mc} = 548.4$, $\widetilde{\alpha}_c = 2.135$, $\delta = 10°$.

Because of the addition of gravity, it is important to realise that now the problem has two characteristic directions in the plane of the layer: the direction of gravity and the direction of the in-layer component of the applied magnetic field. Therefore, the three-dimensional unfolding of results obtained for equivalent two-dimensional problem will need to be treated with care as will be done in the subsequent sections. Here we note that under the adopted transformations

$$\frac{\alpha H_y^e + \beta H_z^e}{H^e} = \alpha \sin\delta\cos\gamma + \beta\sin\delta\sin\gamma = \widetilde{\alpha}\frac{\widetilde{H}_y}{H^e} = \widetilde{\alpha}\sin\delta\cos\widetilde{\gamma}\,,$$

or, for an oblique field with $\delta \neq 0°$, $\alpha\cos\gamma + \beta\sin\gamma = \sqrt{\alpha^2+\beta^2}\cos\widetilde{\gamma}$. Then

$$\gamma = \tan^{-1}\frac{\beta}{\alpha} \pm \widetilde{\gamma}\,. \quad (3.76)$$

In particular, if $\beta = 0$ then $\gamma = \pm\widetilde{\gamma}$. However, when $\alpha = 0$ then $\gamma = 90° \pm \widetilde{\gamma}$. It is convenient to choose $\widetilde{\gamma}$ as an independent problem parameter characterising the magnetic field orientation keeping in mind its meaning given by (3.76).

3.4.6 Wave-Like Instabilities in an Oblique Field in Non-zero Gravity

In this section we will discuss in detail the characteristics of the Type I instability appearing in the form of two counter-propagating thermogravitational waves. The representative critical values similar to those given in Table 3.3 for this instability observed in an oblique magnetic field are presented in Table 3.7 for various field inclination angles. As follows from the data in these tables, the basic flow becomes more stable when the field inclination angle increases beyond $\delta = 5°$. The main (but not the only) reason for this stabilisation is the geometrical reduction of the normal component of the applied magnetic field responsible for the appearance of the cross-layer ponderomotive (Kelvin) force enhancing the instability. The wavenumber of the disturbance waves decreases, and as a result the disturbance wavelength increases. The disturbance waves also propagate quicker with the increase of the field inclination angle.

Table 3.7 The critical values of Grashof number $\widetilde{\mathrm{Gr}}$, wavenumber $\widetilde{\alpha}$ and disturbance wave speed $\widetilde{c} = -\sigma^I/\widetilde{\alpha}$ for magnetogravitational convection waves in oblique magnetic fields for $\mathrm{Gr}_m = 15$, $\widetilde{\gamma} = 0°$, $\mathrm{Pr} = 55$, $H^e = 100$ (odd-numbered lines) and $H^e = 10$ (even-numbered lines).

Wave propagating upward										
		$\delta = 5°$			$\delta = 10°$			$\delta = 15°$		
χ	χ_*	$\widetilde{\alpha}_c$	$\widetilde{\mathrm{Gr}}_c$	$\widetilde{c}_c$	$\widetilde{\alpha}_c$	$\widetilde{\mathrm{Gr}}_c$	$\widetilde{c}_c$	$\widetilde{\alpha}_c$	$\widetilde{\mathrm{Gr}}_c$	$\widetilde{c}_c$
3	3	1.232	57.03	3.643	1.200	61.05	3.921	1.163	65.14	4.201
		1.232	56.88	3.632	1.215	59.85	3.841	1.180	63.73	4.106
1.5	2.5	1.244	55.21	3.521	1.206	60.02	3.852	1.168	64.42	4.152
		1.246	54.94	3.502	1.221	58.81	3.771	1.183	63.22	4.071
Wave propagating downward										
3	3	1.231	57.12	−3.649	1.195	61.39	−3.944	1.159	65.47	−4.223
		1.223	57.91	−3.703	1.174	63.18	−4.065	1.142	66.94	−4.321
1.5	2.5	1.243	55.31	−3.528	1.202	60.33	−3.874	1.164	64.68	−4.170
		1.234	56.13	−3.584	1.183	61.90	−3.979	1.150	65.85	−4.249

The inclination of magnetic field in the vertical plane ($\delta \geq 5°$, $\widetilde{\gamma} = 0°$) changes the asymmetry in the behaviour of the Type I waves. In contrast to the normal field case, in an oblique field the upward propagating waves become unstable at the slightly smaller values of $\widetilde{\mathrm{Gr}}$ than those for downward waves. The waves propagating upwards are characterised by somewhat larger wavenumbers than those of their counterparts moving downwards, and their wave speeds are always smaller than those of the downward waves. These trends remain when the fluid approaches magnetic saturation and its magnetic susceptibilities are reduced.

As follows from Table 3.7, regardless of whether the fluid is close to magnetic saturation ($\chi < \chi_*$) or not ($\chi = \chi_*$), the wave propagating upwards in a thermomagnetically more sensitive[2] ($H^e = 10$) fluid is characterised by a larger wavenumber and smaller wave speed than those of its less sensitive counterpart ($H^e = 100$). The basic flow becomes less stable with respect to this wave when H^e decreases. The trends detected for the downward waves are exactly opposite: they become more stable, longer and propagate faster as the fluid's thermomagnetic sensitivity increases. These observations lead us to a qualitative conclusion that in an oblique field, the waves propagating upwards are expected to be observed experimentally first, and the faster they grow the shorter they become and the slower they propagate.

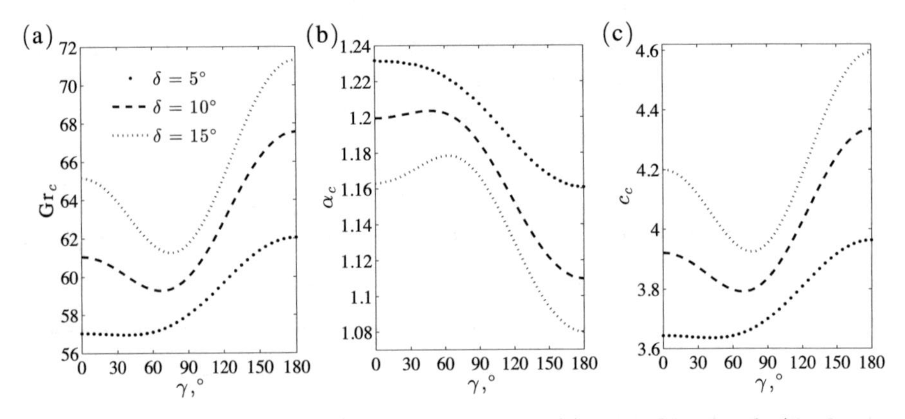

Fig. 3.39 Comparison of the critical parameter values: (a) Grashof number Gr (the flow is stable under the respective curves), (b) wavenumber α and (c) wave speeds c as functions of the field inclination and orientation angles δ and γ for $\mathrm{Gr}_m = 15$, $H^e = 100$, $\mathrm{Pr} = 55$ and $\chi = \chi_* = 3$.

To this point the dependence of flow stability characteristics on the values of χ, χ_* and the field inclination angle δ has been investigated for the zero azimuthal field orientation angle $\widetilde{\gamma}$. To investigate the influence of the field orientation angle $\widetilde{\gamma}$, the stability results are computed for a representative value of $\mathrm{Gr}_m = 15$. The critical parameter values for the case of a linear magnetisation law $\chi = \chi_* = 3$ are shown in Figure 3.39 as functions of the magnetic field inclination and orientation angles. The flow is stable in the regions below the respective curves in Figure 3.39(a). Therefore this type of instability occurs in the form of two-dimensional patterns that are periodic in the vertical y direction with a wavenumber $\widetilde{\alpha} = \alpha$. In this case $\widetilde{\gamma} = \gamma$ and $\widetilde{\mathrm{Gr}} = \mathrm{Gr}$, and thus in the rest of this section, the tildes are omitted. Regardless of the field orientation, the basic flow becomes more stable at larger field inclination angles δ. This is primarily due to the geometric reduction of the active normal component of the applied magnetic field, which is proportional to $\cos\delta$ (see discussion in [202]). With the increase of the field inclination angle δ, the wavenumber decreases (see Figure 3.39(b)) so that the distance

[2] As defined in Section 3.2.

between the instability rolls increases. It follows from Figure 3.39(c) that as the field inclination angle increases, the wave speed also increases. The similar numerical results for a stronger magnetisable fluid with $\chi = \chi_* = 5$ remain qualitatively the same. Thus they are not presented here. However, we note that the basic convection flow of a stronger magnetisable fluid generally becomes more stable for all field orientation angles γ, and its instability patterns are characterised by a smaller wavenumber and a faster wave speed.

As seen from Figure 3.39(a), the instability detected for $\gamma = 180°$ occurs at noticeably higher values of Gr_c than those for $\gamma = 0°$. This indicates that the up-down symmetry of the field influence is broken. This can be traced back to the curvature of magnetic field lines within the layer of ferrofluid discussed in Section 3.2 (see also [202]). Specifically, as follows from Figures 3.32, 3.33, and 3.34, in the absence of the gravitational field, changing the field orientation angle γ from 0 to 180 degrees reverses the sign of the curvature of magnetic field lines. This leads to the reversal of the sign of the wave speed of thermomagnetically driven disturbances. At the same time when the gravity is taken into account, the computational data reported so far indicates that the wave propagating upwards near the hot wall remains most dangerous, at least for $\delta \geq 5°$. Thus changing the field orientation angle γ by 180 degrees leads to the change from the arrangement when gravitationally and thermomagnetically induced disturbances propagate in the same direction to that when they counter-propagate, and the overall instability is suppressed in the latter case.

The most prominent feature of Figure 3.39(a) is the existence of the minima of the $\text{Gr}_c(\gamma)$ curves. Such minima are more pronounced in stronger magnetisable fluids characterised by the larger value of χ (not shown in the figure). Their existence demonstrates that for each field inclination angle δ, there exists a preferred field orientation angle γ that promotes the onset of magnetogravitational instability the most.

As we have seen in Section 3.4.2, in zero gravity environment, the most dangerous instability patterns are aligned with the in-layer component of the applied magnetic field. It is also known [126] that in the absence of magnetic field (i.e. when $\text{Gr}_m = 0$), the thermogravitational waves arising in a large-Prandtl-number fluid consist of the horizontally uniform structures. Therefore intuitively one might expect that when both Gr and Gr_m are non-zero, the least stable situation would occur when the direction of the in-layer component of the applied oblique magnetic field is horizontal, that is, when $\gamma = 90°$ and $H_y^e = 0$. Yet the computational results presented in Figure 3.40(a) show that the field orientation angle $\gamma_{\min}$ for which the instability first occurs tends to 90° only for sufficiently large field inclination angles δ, that is, when the in-layer component of the magnetic field becomes sufficiently large. When such a component is small ($\delta \lesssim 3.5°$), the instability depends on the field orientation only weakly, and the most unstable situation corresponds to $\gamma = 0°$. However, for larger field inclination angles, the behaviour of $\gamma_{\min}$ becomes a sensitive function of δ. The likely reason for such a peculiar behaviour is due to the fact reported in Section 3.4.2 for pure magnetic convection. It was demonstrated there that, similar to the gravity, an inclined magnetic field breaks a planar

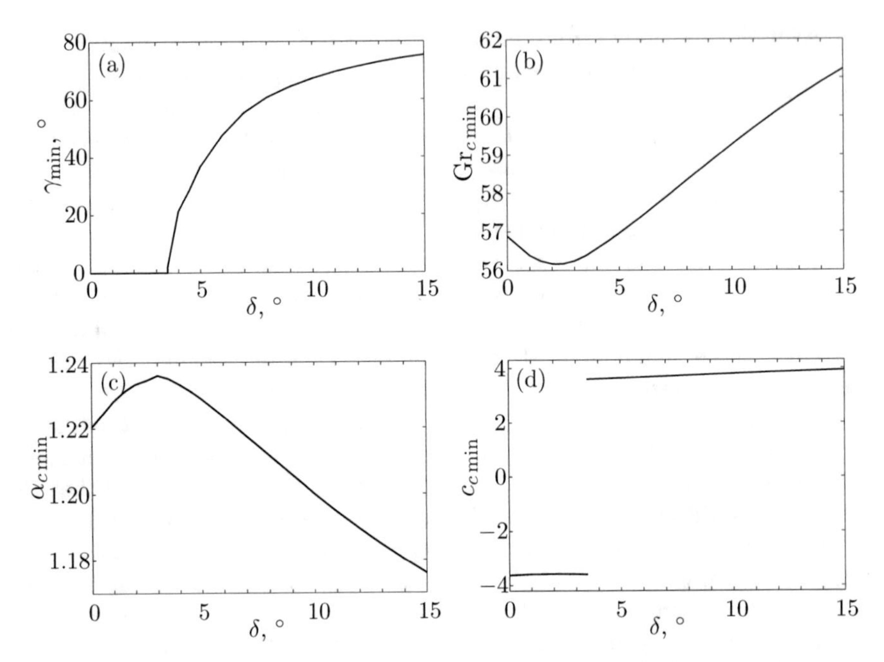

Fig. 3.40 (a) The value of the field orientation angle $\gamma_{\min}$ at which the instability first occurs and (b–d) the corresponding critical parameters as functions of the field inclination angle δ for $\mathrm{Gr}_m = 15$, $H^e = 100$, $\mathrm{Pr} = 55$ and $\chi = \chi_* = 3$.

symmetry of the arising convection flows so that the most amplified instability patterns align with the in-layer component of the applied magnetic field (see also [106], where a similar effect of an oblique magnetic field is discussed in the context of free surface phenomena in magnetic fluids). Such preferentially aligned patterns remain stationary in the absence of gravity. Therefore even though the geometrical optimality might favour the alignment of the two instability patterns described above, the competition between travelling thermogravitational waves and stationary magnetoconvection rolls would, to some degree, hinder the development of the overall instability. On the other hand, it was also shown in Section 3.4.2 that the thermomagnetic instability patterns that are not aligned with the in-layer component of the magnetic field (i.e. observed for $\gamma \neq 90°$ in the current context) while characterised by a smaller growth rate have a non-zero wave speed. Such nonstationary wave-like patterns therefore could be favoured when the magnetic instability overlaps with the vertically propagating horizontally uniform thermogravitational waves. Therefore the choice of $\gamma_{\min} \neq 90°$ appears to be due to the competition between the two optimality criteria: maximising the amplification rate of combined thermogravitational and thermomagnetic instabilities and matching their propagation speeds.

It follows from Figure 3.40(b) that there always exists the overall optimal orientation of magnetic field $(\delta_{\min}, \gamma_{\min})$, which minimises the value of the critical Grashof number $\mathrm{Gr}_{c,\min}$. In particular, for $\mathrm{Gr}_m = 15$, $\chi = \chi_* = 3$ and

$\mathrm{Pr} = 55$, $\mathrm{Gr}_{c,\min} \approx 56.16$, $\delta_{\min} \approx 2°$ and $\gamma_{\min} = 0°$. Such a global minimum corresponds to a disturbance waves with a wavenumber $\alpha_{c,\min} \approx 1.234$; see Figure 3.40(c). Figure 3.40(d) indicates another noteworthy feature of the wave-like instabilities detected in an oblique field: the upward propagating wave becomes the most dangerous for all optimal field orientations at $\delta \gtrsim 3.5°$. For smaller field inclination angles, the most unstable wave propagates downwards (consistently with findings reported in Table 3.3 for normal field) although the stability characteristics of the wave propagating upwards remain very close. At larger field inclination angles, the symmetry-breaking effect of a magnetic field becomes more pronounced, and the switch of the dominant instability mode to the upward propagating wave occurs; see Figure 3.40(d). This switch is accompanied by the appearance of a well-defined non-zero optimal field orientation angle as seen in Figure 3.40(a). The comparison of the critical parameters for the waves propagating upwards and downwards is presented in Figure 3.41 for a representative field inclination angle $\delta = 5°$ in a linear magnetisation regime $\chi = \chi_* = 3$. The difference between the characteristics of the two waves becomes finite but remains relatively small so that the waves are expected to co-exist in realistic experiments. Therefore the experimental ability to observe both waves is important.

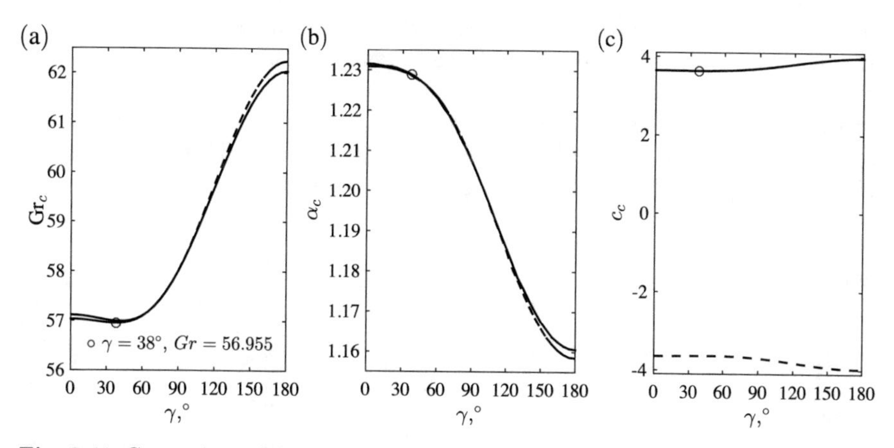

Fig. 3.41 Comparison of the critical parameter values for the waves propagating upwards (solid line) and downwards (dashed line): (a) Grashof number Gr (the flow is stable under the respective curves), (b) wavenumber α and (c) wave speeds c as functions of the azimuthal angle γ for $\mathrm{Gr}_m = 15$, $H^e = 100$, $\mathrm{Pr} = 55$, $\delta = 5°$ and $\chi = \chi_* = 3$.

Figure 3.41 demonstrates that the critical parameter curves for both waves have qualitatively similar shapes. The basic flow becomes unstable with respect to the upward wave for somewhat smaller values of Gr. Quantitatively, the differences between the critical parameters for the two waves are more evident for $\gamma \to 0°$ or $\gamma \to 180°$, that is, when the applied magnetic field belongs to a vertical plane perpendicular to the fluid layer walls. For such a field orientation, the wavelength of the upward propagating waves is slightly shorter than that of the downward waves.

Our computations (not shown here) confirm that when the fluid approaches magnetic saturation (i.e. both χ and χ_* decrease and become unequal), qualitatively the critical parameter curves for the wave-like disturbances remain similar to those seen in Figure 3.41. However, the values of both the critical Grashof number and the optimal field orientation angle γ decrease (e.g. from $\mathrm{Gr}_c \approx 56.96$ and $\gamma \approx 38°$ at $\chi = \chi_* = 3$ to $\mathrm{Gr}_c \approx 55.15$ $\gamma \approx 35°$ at $\chi = 1.5$ and $\chi_* = 2.5$).

In conclusion of this section, we compare the stability characteristics of the basic flow with respect to wave-like disturbances of Type I for thermomagnetically less ($H^e = 100$) and more ($H^e = 10$) sensitive fluids; see Figure 3.42. The waves propagating upwards remain the most dangerous in both types of fluids, so only the critical parameters corresponding to them are shown. There are a number of general trends that are evident from Figure 3.42. Firstly, the flows of thermomagnetically more sensitive fluids placed in a magnetic field with a predominantly vertical in-layer component (γ close to 0 or 180 degrees) are generally less stable than those of their less sensitive counterparts. When the applied oblique magnetic field is mostly horizontal ($\gamma \sim 90°$), that is, when the curvature of the magnetic field lines within the fluid layer is in the plane perpendicular to the direction of the gravity, the magnetic sensitivity of a fluid does not appear to play a significant role in defining the flow stability parameters. Secondly, the wave-like instability patterns arising in a more thermomagnetically sensitive fluid are characterised by a larger wavenumber and thus by convection structures that are closer packed in the direction of the gravity when $\gamma > 90°$, once again demonstrating the symmetry-breaking effect of an oblique magnetic field. Thirdly, the instability waves arising in a thermomagnetically more sensitive fluid generally

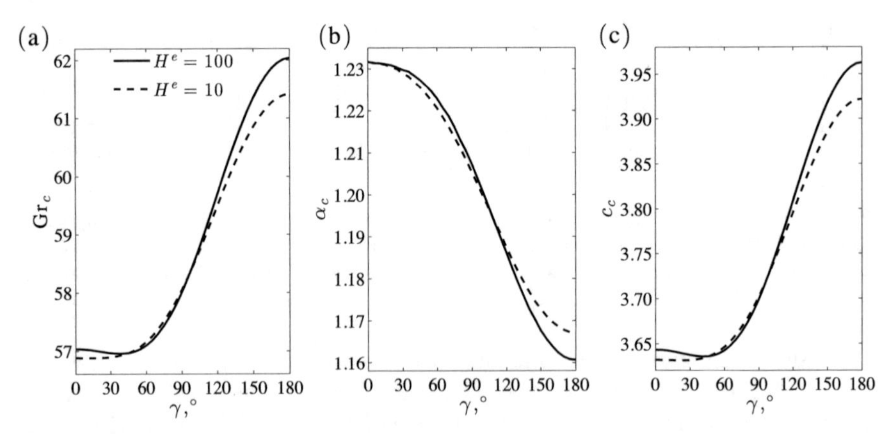

Fig. 3.42 Comparison of the critical parameter values for thermomagnetically less ($H^e = 100$, solid line) and more ($H^e = 10$, dashed line) sensitive fluids: (a) Grashof number Gr (the flow is stable under the respective curves), (b) wavenumber α and (c) wave speeds c as functions of the azimuthal angle γ for $\mathrm{Gr}_m = 15$, $\mathrm{Pr} = 55$, $\delta = 5°$ and $\chi = \chi_* = 3$. Type I instability.

have a somewhat smaller wave speed. Therefore increasing the fluid's thermomagnetic sensitivity quenches the propagation of disturbance waves. This is consistent with the findings discussed in Section 3.4.2 that the most amplified thermomagnetically driven instability patterns remain stationary in the absence of gravity. The computational data (not shown) also demonstrates that these trends are not affected when the fluid approaches its magnetic saturation with $\chi < \chi_*$.

3.4.7 Stability Diagrams for an Equivalent Two-Dimensional Problem in an Oblique Field and Non-zero Gravity

In Section 3.4.6 the flow instability properties associated with the wave-like Type I disturbances have been discussed in detail. Here we will identify parametric regions where different physical mechanisms lead to the onset of instability in the considered geometry. To do that, we consider the representative complete stability diagrams for an equivalent two-dimensional problem given in Figure 3.43. Note that plots (a)–(c) in Figure 3.43 are identical to those in Figure 3.27 discussed in detail in Section 3.3.6. They are repeated here only to enable an easy comparison with various inclined field cases.

3.4.7.1 Field Inclined in the Plane Containing the Main Periodicity Direction ($\widetilde{\gamma} = 0°$)

The magnetic field inclination adds further complexity to the already quite complicated instability picture in the presence of both gravitational and magnetic effects. The stability diagram for an oblique magnetic field ($\delta = 5°$, $\widetilde{\gamma} = 0°$) is shown in Figure 3.43(d). The comparison with Figure 3.43(a) for a normal field shows that the flow stability region becomes larger in an oblique magnetic field. This is consistent with the numerical results given in Tables 3.3 and 3.7. As follows from Figure 3.43(e), similar to the normal field case, the Type I instability[3] is characterised by a smaller wavenumber (the solid line) compared to that of the Type III instability (the dash-dotted line). However the symmetry of the disturbance thermal waves propagation is broken in an oblique field, and the upward wave becomes more dangerous. Therefore in Figure 3.43(f), only the critical wave speed for this wave is shown. It increases monotonically with Gr_m.

It is remarkable that the qualitative change in stability diagram occurs even for such small field inclination angles. The solid and dashed stability boundary lines distinguished in Figure 3.43(a) merge in Figure 3.43(d)

[3] Refer to Section 3.3.2 for definitions of instability types.

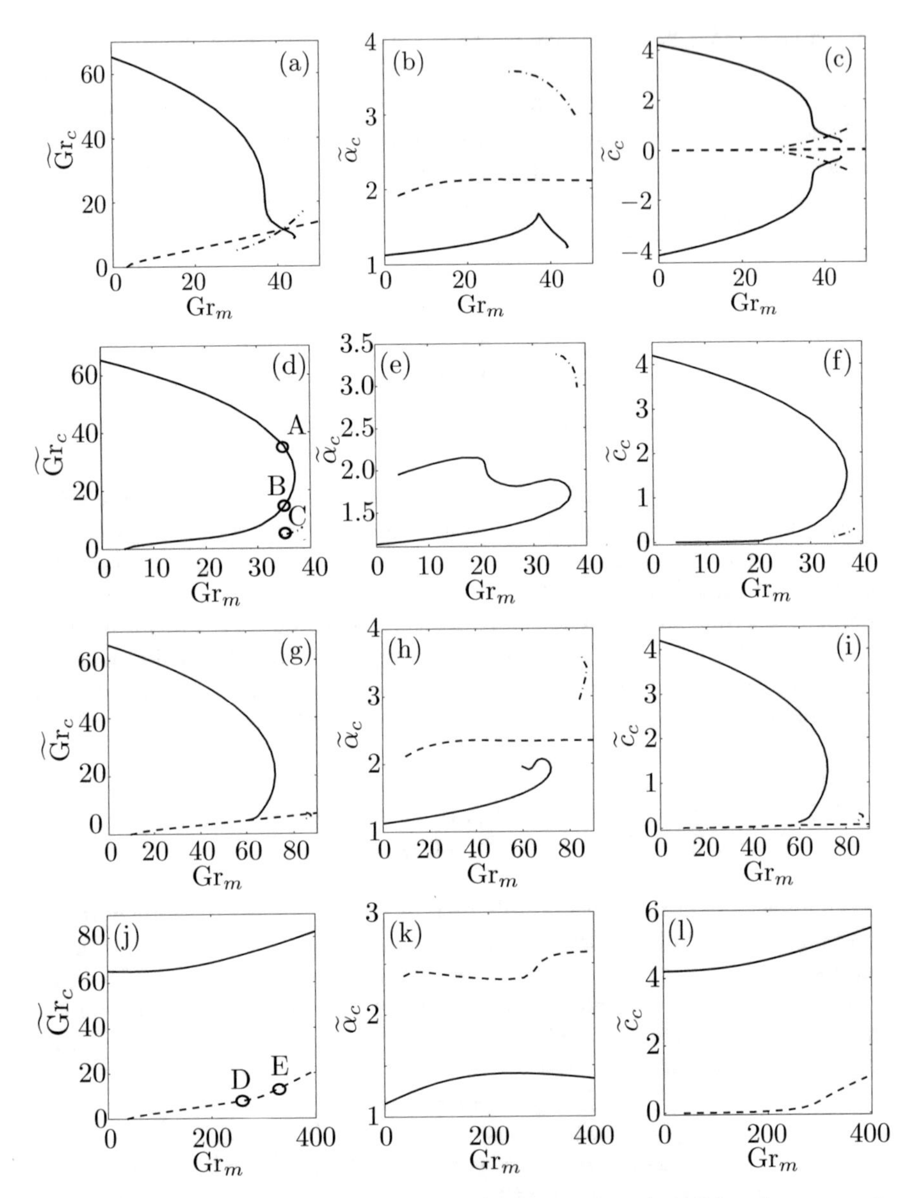

Fig. 3.43 Variation of stability diagrams (left), wavenumbers (middle) and wave speeds (right) with the magnetic field inclination for $H^e = 100$, $\mathrm{Pr} = 55$, $\chi = \chi_* = 3$, $\widetilde{\gamma} = 0°$ and (a–c) $\delta = 0°$, (d–f) $\delta = 5°$, (g–i) $\delta = 10°$ and (j–l) $\delta = 15°$.

indicating that the distinction between the Type I and Type II instabilities becomes blurred when the applied magnetic field is inclined in a plane containing the periodicity direction. The dash-dotted line in the lower right corner in Figure 3.43(a) becomes much shorter meaning that the Type III instability could be hard to detect experimentally in an oblique field. Even though the solid line originates from $\widetilde{\mathrm{Gr}} = 0$ in Figure 3.43(d), it corresponds to nonstationary magnetoconvection; see the lower part of the solid curve in Figure 3.43(f). This is consistent with the results of Section 3.4.2 where it has been shown that the thermomagnetic instability patterns that are not aligned with the in-layer component of the applied magnetic field are always nonstationary.

While the orientation of the Types II and III instability patterns remains qualitatively unaffected by the small field inclination, the qualitative changes occur in the orientation of the Type I instability patterns at sufficiently large values of Gr_m. As seen from Figure 3.44(a), the $\sigma^R(\widetilde{\mathrm{Gr}})$ curve crosses the

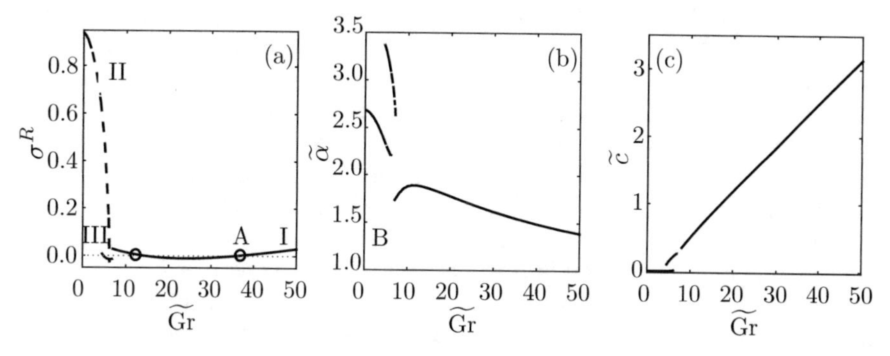

Fig. 3.44 (a) Maximum amplification rate for an equivalent two-dimensional problem, (b) the corresponding wavenumbers $\widetilde{\alpha}$ and (c) wave speeds for $\mathrm{Gr}_m = 35$, $H^e = 100$, $\mathrm{Pr} = 55$ and $\chi = \chi_* = 3$ in an oblique magnetic field at $(\delta, \widetilde{\gamma}) = (5°, 0°)$.

zero level twice at points A and B (see also the corresponding points in Figure 3.43(d)). According to the inverse Squire's transformation, the decreasing segment of the $\sigma^R(\widetilde{\mathrm{Gr}})$ curve to the left of point B in Figure 3.44(a) indicates the existence of oblique instability structures with the orientation depending on the value of Gr. For the relatively small values of $\mathrm{Gr} < \mathrm{Gr}_B$, the fastest-growing Type I instability pattern is almost horizontal with $\beta = 0$, but as Gr increases, the most unstable patterns turn and approach vertical (β increases at the expense of α). This continues until Gr reaches the value of Gr_A. At this point another pair of the Type I waves appears that are horizontally uniform ($\beta = 0$) and propagate vertically. Figure 3.44(c) also confirms the conclusion made earlier that due to the symmetry-breaking effect of nonlinear fluid magnetisation, the most dangerous Type I pattern switches from the wave propagating downwards in a normal field to the one propagating upward in an oblique field. However the growth rates of both waves remain close so that a counter-propagating wave pair is likely to be seen in experiments. Figure 3.44(a) also demonstrates that for the sufficiently large values of Gr_m, the Type II instability has a much larger growth rate than those of the Type I and Type III patterns. Therefore stationary vertical thermomagnetic rolls are expected to dominate the overall disturbed flow. Yet the presence of the Type I and III instabilities should be visible experimentally as nonstationary three-dimensional modulations of vertical rolls, which has indeed been detected in experiments reported in [237, 238].

The magnetic field inclination continues to play a very important role in shaping the parametric stability boundaries of the considered flow when δ is increased further. The stability diagram for $\delta = 10^\circ$ and $\widetilde{\gamma} = 0^\circ$ shown in Figure 3.43(g) demonstrates that a significant stabilisation of the flow with respect to the Type I disturbances is observed (the area bounded by the solid line increases). The Type I and Type II instabilities are again easily distinguished. The waves propagating upwards still remain the most dangerous for the non-zero values of the magnetic Grashof number.

At even larger field inclination angles, another qualitative change occurs. As seen from Figure 3.43(j), for $\delta = 15^\circ$, the Type III instability is not detected over the investigated range of the governing parameters. The critical values of $\widetilde{\mathrm{Gr}}$ for the Type I instability now increase monotonically with Gr_m. This is traced back to the aligning influence of the applied magnetic field. With the increasing field inclination angle δ and $\widetilde{\gamma} = \gamma = 0^\circ$, the vertical in-layer field component increases as well and so does its "pattern aligning" effect. Thus the vertically propagating and horizontally uniform Type I waves are suppressed by the inclined field applied in a vertical plane and require a much stronger gravitational buoyancy characterised by $\widetilde{\mathrm{Gr}}$ to arise. The Type II instability characteristics presented for the field inclined at $\delta = 15^\circ$ and shown in Figure 3.43(j)–(l) change in a peculiar manner. For small values of the magnetic Grashof number, the Type II instability boundary (the basic flow is unstable below the dashed line) rises almost linearly, and the critical wavenumber remains almost constant at $\widetilde{\alpha} \approx 2.5$. This instability remains nearly stationary up to $\mathrm{Gr}_m \sim 300$ (e.g. point D in Figure 3.43(j)). However for larger values of Gr_m, the slope of the stability boundary changes rapidly (even though in a continuous manner) to a larger value (e.g. point E in Figure 3.43(j)) and so does the value of $\widetilde{\alpha}$. The critical wave speed becomes non-zero and starts growing.

In conclusion of this section, we note that our computations not discussed here in detail show that the main effect of the approaching magnetisation saturation regime $\chi < \chi_*$ is the disappearance of the Type III instability even at the field inclination angles as small as $\delta = 5^\circ$. However, the Types I and II instabilities remain qualitatively unchanged. Thus no detailed discussion of magnetic saturation regimes will be given here. We have also explored the influence of the fluid's thermomagnetic sensitivity on the stability characteristics of the flow by computing the results for $H^e = 10$. Again it was found that the only qualitative effect the variation of this parameter leads to is the disappearance of the Type III instability in a more magnetically sensitive fluid.

Therefore the specific results presented in this section appear to be robust. They provide a sufficiently complete view of the instability processes taking place in the considered geometry for a wide range of physical conditions.

3.4.7.2 Arbitrary Field Orientation

All stability diagrams discussed so far have been computed for $\widetilde{\gamma} = 0°$. It is of interest now to compare the flow stability characteristics in the applied fields of different orientations. This is done in Figure 3.45 for $\delta = 10°$. The figure demonstrates a sensitive dependence on the choice of $\widetilde{\gamma}$. However this variation is not monotonic. As $\widetilde{\gamma}$ increases from $0°$, the parametric stability region in Figure 3.45(a) initially shrinks and then starts growing. Therefore the general conclusion is that there exists an optimal field orientation angle $\widetilde{\gamma}$ for which the basic flow becomes most unstable. The existence of the optimal field orientation angle $\gamma_{\min}$ for the Type I instability has been discussed in detail in Section 3.4.6, and here we will focus on the Type II instability.

The most unstable pattern of the Type II instability corresponds to vertical rolls with $\alpha = 0$ and $\beta = \widetilde{\alpha}$. However rolls of all other orientations can also

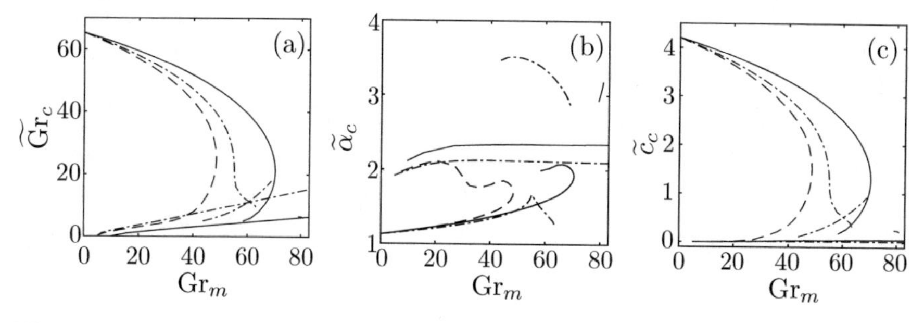

Fig. 3.45 Comparison of the critical values for $H^e = 100$, $\mathrm{Pr} = 55$, $\chi = \chi_* = 3$, $\delta = 10°$ and the field orientation angles $\widetilde{\gamma} = 0°$ (solid lines), $67°$ (dashed lines) and $90°$ (dash-dotted lines): (a) stability diagram for an equivalent two-dimensional problem, (b) the critical wavenumbers and (c) the corresponding wave speeds along the stability boundaries shown in plot (a).

exist up to $\mathrm{Gr} = \widetilde{\mathrm{Gr}}_c$ corresponding to the values shown in Figure 3.46(a) (computed for $\mathrm{Gr}_m = 15$ as an example). The vertical alignment of instability patterns occurs for the larger values of Gr. The value of $\widetilde{\mathrm{Gr}}_c$ depends on the field orientation angle $\widetilde{\gamma}$. As follows from Figure 3.46(a), such an alignment for $\mathrm{Gr}_m = 15$ and $\delta = 5°$ is most delayed when the field is oriented at the angle $\gamma = 90° \pm \widetilde{\gamma} \approx 237°$ or $-57°$ to the vertical y axis. Figure 3.46(b) demonstrates that the orientation of the applied magnetic field also affects the wavelength of the Type II instability patterns: it slightly increases as the field orientation angle approaches the optimal value. As follows from Figure 3.46(c) the Type II instability in this example remains essentially stationary for all field orientation angles (this, however, is not the case for larger values of Gr_m, as seen from Figure 3.43(f)).

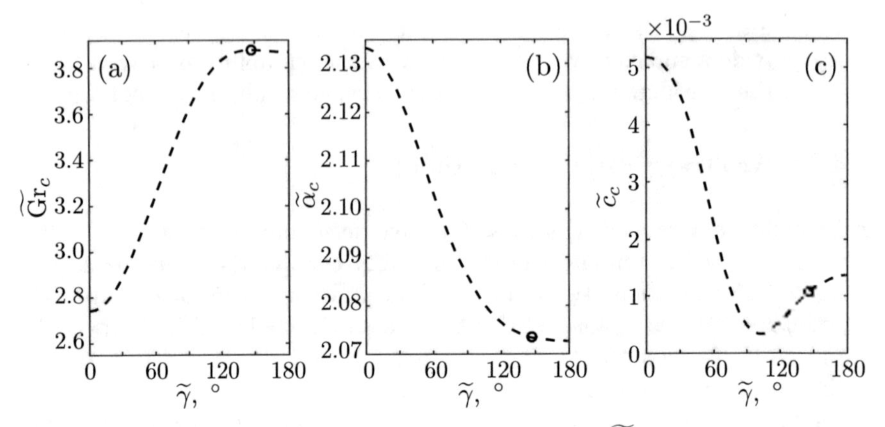

Fig. 3.46 The critical parameter values: (a) Grashof number $\widetilde{\mathrm{Gr}}$ (the flow is stable above the curve), (b) wavenumber $\widetilde{\alpha}$ and (c) wave speeds $\widetilde{c}$ as functions of the field orientation angle $\widetilde{\gamma}$ for $\delta = 5°$, $\mathrm{Gr}_m = 15$, $H^e = 100$, $\mathrm{Pr} = 55$ and $\chi = \chi_* = 3$. The circles denote values corresponding to $\widetilde{\mathrm{Gr}} \approx 3.878$ computed for $\widetilde{\gamma} = 147°$.

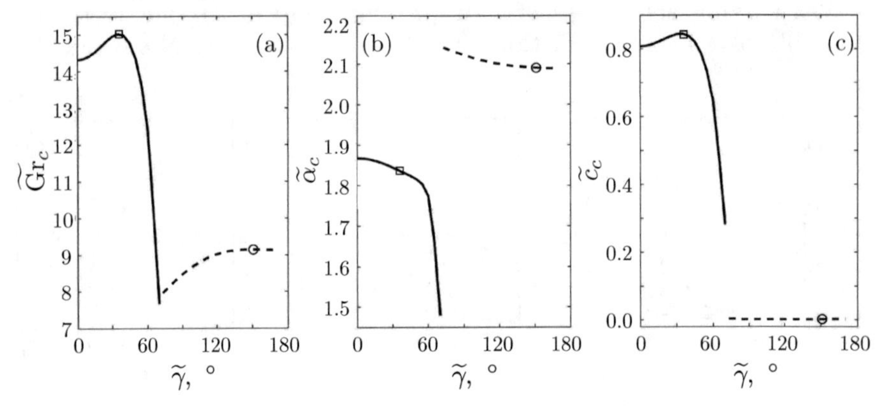

Fig. 3.47 Same as Figure 3.46 but for $\mathrm{Gr}_m = 35$. The circles and squares denote values corresponding to $\widetilde{\mathrm{Gr}}_c \approx 9.142$ computed for $\widetilde{\gamma} = 151°$ and to $\widetilde{\mathrm{Gr}}_c \approx 15.023$ computed for $\widetilde{\gamma} = 36°$, respectively.

For the larger values of Gr_m, the dependence of the critical values on the field orientation angle $\widetilde{\gamma}$ becomes more complicated. As seen from Figure 3.47 unlike for smaller values of Gr_m, both the Type II (dashed line) and Type I (solid line) instabilities can be detected when varying $\widetilde{\gamma}$. The two types of instabilities here are distinguished by the values of their wavenumbers (the Type I instability has a longer wavelength; see Figure 3.47(b)) and wave speeds (the Type I instability is wave-like, while the Type II patterns are virtually stationary; see Figure 3.47(c)). Both instabilities have their own optimal field orientation angles (shown by symbols). Namely, as follows from the inverse Squire's transformation discussed earlier, for the representative value of $\mathrm{Gr}_m = 35$, the vertical alignment of the Type II instability patterns is delayed the most (up to $\mathrm{Gr} \approx 9.142$) if the field is oriented at $\gamma = 90° \pm \widetilde{\gamma} \approx 241°$

or $-61°$. At the same time, such a delay (up to Gr ≈ 15.023) in the vertical alignment of the Type I instability patterns is most evident for the field orientation angles in the range $0° < \tilde{\gamma} < 90°$ (see the maximum of the solid curve marked by the square in Figure 3.47(a)). Therefore, we conclude that the orientation of the applied magnetic field can have a profound influence on the type of the observed convection patterns [237, 238], which we will demonstrate in the subsequent chapters.

3.5 Weakly Nonlinear Consideration of Thermomagnetic Convection

3.5.1 Amplitude Expansion

The theoretical and computational results reported in the previous sections show that in flat vertical ferrofluid layers, thermomagnetic convection is the only instability mode arising in a wide range of governing parameters (regions [3] and [10] in Figure 3.21(a)). It sets in the form of vertical rolls that are periodic in the horizontal z direction. The linear analysis also shows that its growth rate does not depend on whether the gravity is taken into account, and thus it can be determined by formally setting Gr $=$ Ra $= 0$ in the governing equations. Two important questions, however, cannot be answered within the framework of linear analysis. Namely, it remains unknown whether the critical values of the governing parameters for the onset of thermomagnetic instability determined assuming idealised infinitesimal disturbances remain unchanged in realistic conditions when disturbances have a finite amplitude. In other words, it is not known whether the flow bifurcation leading to the appearance of convection structures is supercritical or subcritical (in the former case, linear analysis gives the unambiguous critical values of parameters regardless of the strength of the initial perturbation, while in the second, the threshold depends on the type and amplitude of the initial perturbation). The second question is regarding the strength of the resulting convection pattern. Linear stability consideration states that once set convection patterns grow exponentially in time. Clearly, this cannot happen in reality, and flows must reach some finite saturation when the growth of instability is balanced by the dissipation present in the system. The characteristics of such a saturated state are of major practical interest.

To investigate the temporal evolution of thermomagnetic roll structure detected in regions [3] and [10] of the parametric space shown in Figure 3.21, we assume that to the leading order, the perturbed field remains periodic in the horizontal z direction with the corresponding wavenumber β. Our goal is to determine the time-dependent amplitude of such a periodic spatial structure that characterises the strength of the arising magnetoconvection.

This amplitude is assumed to be finite but sufficiently small so that it can be used in an asymptotic series the low-order truncation of which gives an approximate solution to the governing equations. General algebraic details of the version of a weakly nonlinear analysis that employ below are quite involved and will not be discussed here. An interested reader is referred to [241–243] and references therein for an introductory reading on the procedure. Here we follow [78], where the application of the amplitude expansion in the context of thermomagnetic convection was first developed and look for the solution of a system of governing equations formulated in Section 2.2 in the form

$$\mathbf{w} = \mathbf{w}_{00} + \varepsilon A_{11}\mathbf{w}_{11}E + \varepsilon^2 A_{20}\mathbf{w}_{20} + \varepsilon^2 A_{22}\mathbf{w}_{22}E^2 + \varepsilon^3 A_{31}\mathbf{w}_{31}E + \cdots + \text{c.c.}\,, \tag{3.77}$$

where $A_{kl} = A_{kl}(t)$, $k, l = 1, 2, \ldots$ are the amplitudes of perturbations, k is the order of the small amplitude, l is the order of the Fourier component E, $\mathbf{w} = (u, v, w, \theta, P, \phi)^T$, $\mathbf{H} = \nabla\phi$, $E = \exp(\mathrm{i}\beta z)$ and c.c. denotes the complex conjugate of the terms containing E. Parameter ε is introduced formally for book-keeping purposes so that the terms of different orders can be easily distinguished. Once the derivation is completed, this parameter will be removed so that only the amplitude itself will remain as a small parameter. The higher-order terms are induced by the fundamental harmonic $\varepsilon A_{11}\mathbf{w}_{11}E$ through quadratic nonlinearity of the governing equations. We introduce multiple timescales $t_0 = t$, $t_1 = \varepsilon t$, $t_2 = \varepsilon^2 t$, $\ldots$ so that $A_{kl} = A_{kl}(t_0, t_1, t_2, \ldots)$ and thus the time derivative takes the form of

$$\frac{d}{dt} = \frac{\partial}{\partial t_0} + \varepsilon\frac{\partial}{\partial t_1} + \varepsilon^2\frac{\partial}{\partial t_2} + \cdots\,. \tag{3.78}$$

In the limit $A_{kl} \to 0$ expansion (3.77) reduces to the steady basic flow solutions $\mathbf{w}_{00} = (0, v_0, 0, \theta_0, P_0, \phi_0)^T$.

3.5.2 Linearised Disturbances

Without loss of generality, we identify $\varepsilon A_{11} \equiv A$. Then terms of the order $\varepsilon^1 E^1$ contribute to equations linearised about the basic state

$$\frac{\partial A}{\partial t_0}u_{11} + A[(\beta^2 - D^2)u_{11} + Dp_{11} + \mathrm{Gr}_m DH_0\,\theta_{11} + \mathrm{Gr}_m\theta_0 D^2\phi_{11}] = 0\,, \tag{3.79}$$

$$\frac{\partial A}{\partial t_0}v_{11} + A(\beta^2 - D^2)v_{11} = 0\,, \tag{3.80}$$

$$\frac{\partial A}{\partial t_0}w_{11} + A[(\beta^2 - D^2)w_{11} + \mathrm{i}\beta p_{11} + \mathrm{i}\beta\mathrm{Gr}_m\theta_0 D\phi_{11}] = 0\,, \tag{3.81}$$

$$\frac{\partial A}{\partial t_0}\theta_{11} + A\left[D\theta_0 u_{11} + \frac{\beta^2 - D^2}{\mathrm{Pr}}\theta_{11}\right] = 0\,, \tag{3.82}$$

$$A\left[Du_{11} + i\beta w_{11}\right] = 0\,, \tag{3.83}$$

$$A\left[\left(D^2 - \beta^2\left(1 + \frac{\chi_* - \chi}{\chi + 1}\frac{\mathrm{N}}{H_0} - \frac{\theta_0}{H_0}\right)\right)\phi_{11} - D\theta_{11}\right]\,, \tag{3.84}$$

with boundary conditions

$$u_{11} = v_{11} = w_{11} = \theta_{11} = 0\,, \quad D\phi_{11} \mp \frac{|\beta|}{1+\chi}\phi_{11} = 0 \quad \text{at } x = \mp 1\,. \tag{3.85}$$

Equations (3.79)–(3.85) can be written in a matrix form as

$$\left(A\mathscr{A}_\beta + \frac{\partial A}{\partial t_0}\mathscr{B}\right)\mathbf{w}_{11} = \mathbf{0}\,, \tag{3.86}$$

where operators $\mathscr{A}_\beta$ and $\mathscr{B}$ are defined as

$$\mathscr{A}_\beta = \begin{bmatrix} \beta^2 - D^2 & 0 & 0 & \mathrm{Gr}_m DH_0 & D & \mathrm{Gr}_m\theta_0 D^2 \\ 0 & \beta^2 - D^2 & 0 & 0 & 0 & 0 \\ 0 & 0 & \beta^2 - D^2 & 0 & \mathrm{i}\beta & \mathrm{i}\beta\mathrm{Gr}_m\theta_0 D \\ D\theta_0 & 0 & 0 & \dfrac{\beta^2 - D^2}{\mathrm{Pr}} & 0 & 0 \\ D & 0 & \mathrm{i}\beta & 0 & 0 & 0 \\ 0 & 0 & 0 & -D & 0 & D^2 - \beta^2(1+r) \end{bmatrix},$$

$$\mathscr{B} = \begin{bmatrix} 1&0&0&0&0&0 \\ 0&1&0&0&0&0 \\ 0&0&1&0&0&0 \\ 0&0&0&1&0&0 \\ 0&0&0&0&0&0 \\ 0&0&0&0&0&0 \end{bmatrix}, \quad r = \frac{\chi_* - \chi}{\chi + 1}\frac{\mathrm{N}}{H_0} - \frac{\theta_0}{H_0}\,.$$

The system of linear differential equations (3.86) has a solution in the form $A(t)\mathbf{w}_{11}$, where

$$A(t) = \widetilde{A}(t_1, t_2, \ldots)\exp(\sigma t_0) \text{ and } \frac{\partial A}{\partial t_0} = \sigma A\,. \tag{3.87}$$

Then Equation (3.86) becomes

$$\mathscr{L}_{\beta,\sigma}\mathbf{w}_{11} \equiv (\mathscr{A}_\beta + \sigma\mathscr{B})\mathbf{w}_{11} = \mathbf{0}\,. \tag{3.88}$$

In order for the homogeneous Equation (3.88) to have a nontrivial solution, the operator $\mathscr{L}_{\beta,\sigma}$ in its left-hand side must be singular. Then the complex disturbance amplification rate $\sigma(\beta;\Pi) = \sigma^R(\beta;\Pi) + i\sigma^I(\beta;\Pi)$ and the nontrivial solution vector $\mathbf{w}_{11} = \mathbf{w}_{11}(x;\beta,\Pi)$, where Π represents a set of physical governing parameters listed in Section 2.2, must be the eigenvalue and the eigenvector of the generalised eigenvalue problem (3.88). Given that eigenvectors are defined up to a multiplicative constant, without loss of generality, here we choose to scale them in such a way that $\max(|\theta_{11}|) = 1$ so that the perturbation amplitude becomes a direct measure of the size of thermal perturbations relative to the basic flow temperature variation. The critical point is defined by the condition $\max_\beta(\sigma^R(\Pi_c)) = 0$. The wavenumber value β_c corresponding to such a zero maximum defines the spectral position of the parametric bifurcation point Π_c that signifies the transition to the basic flow instability. The goal of the current section is to develop a weakly nonlinear model of the flow in a close, but not necessarily asymptotically close, vicinity of the critical point Π_c. As such, the eigenmode $\mathbf{w}_{11}$ that represents the fastest-growing ($\sigma^R_{\max} > 0$) periodic perturbation component and that is referred to as the fundamental harmonic below is computed for the actual physical parameters of interest, that is, for $\Pi \neq \Pi_c$ in general.

3.5.3 Mean Flow Correction and Second Harmonic

(i) Collecting terms of the order $\varepsilon^2 E^0$, we obtain the equations that describe the mean perturbation to the flow. They can be written in the operator form as

$$\varepsilon^2 \left(A_{20}\mathscr{A}_0 + \frac{\partial A_{20}}{\partial t_0}\mathscr{B} \right) \mathbf{w}_{20} = |A|^2 \mathbf{f}_{20} \,, \tag{3.89}$$

where $\mathbf{f}_{20}$ is given by

$$\mathbf{f}_{20} = - \begin{bmatrix} \mathrm{i}\beta(u_{11}\bar{w}_{11} - \bar{u}_{11}w_{11}) + D|u_{11}|^2 + \mathrm{Gr}_m\beta^2\theta_0 D\dfrac{|\phi_{11}|^2}{H_0} \\ \quad +\mathrm{Gr}_m(\theta_{11}D^2\bar{\phi}_{11} + \bar{\theta}_{11}D^2\phi_{11}) \\ \mathrm{i}\beta(v_{11}\bar{w}_{11} - \bar{v}_{11}w_{11}) + u_{11}D\bar{v}_{11} + \bar{u}_{11}Dv_{11} \\ \mathrm{i}\beta\mathrm{Gr}_m(\bar{\theta}_{11}D\phi_{11} - \theta_{11}D\bar{\phi}_{11}) + u_{11}D\bar{w}_{11} + \bar{u}_{11}Dw_{11} \\ \mathrm{i}\beta(\bar{w}_{11}\theta_{11} - w_{11}\bar{\theta}_{11}) + u_{11}D\bar{\theta}_{11} + \bar{u}_{11}D\theta_{11} \\ 0 \\ \dfrac{\beta^2}{H_0^2}\left[|\phi_{11}|^2(D\theta_0 + 2rDH_0) - rH_0D|\phi_{11}|^2\right] \end{bmatrix} ,$$

with the boundary conditions

$$u_{20} = v_{20} = w_{20} = \theta_{20} = 0 \,, \tag{3.90}$$

$$\varepsilon^2 A_{20} D\phi_{20} = |A|^2 \frac{\beta^2}{H_0^2} |\phi_{11}|^2 \left(\mp 1 - \frac{\chi - \chi_*}{1+\chi} \mathrm{N} \right) \quad \text{at } x = \mp 1 \,. \quad (3.91)$$

The compatibility of the left- and right-hand sides of the above equations requires that $\varepsilon^2 A_{20} = |A|^2$. Therefore $\varepsilon^2 \frac{\partial A_{20}}{\partial t_0} = \frac{\partial |A|^2}{\partial t_0} = 2\sigma^R |A|^2$, and Equation (3.89) becomes

$$\mathscr{L}_{0,2\sigma^R} \mathbf{w}_{20} = \mathbf{f}_{20} \,. \quad (3.92)$$

(ii) Terms of the order $\varepsilon^2 E^1$ result in

$$\frac{\partial A}{\partial t_1} \mathscr{B} \mathbf{w}_{11} = \mathbf{0} \text{ or } \frac{\partial A}{\partial t_1} = 0 \quad (3.93)$$

since $\mathscr{B}\mathbf{w}_{11} \neq \mathbf{0}$. Thus the disturbance evolution is independent of the slow time t_1, and therefore $A(t) = A(t_0, t_2, \ldots)$.

(iii) Collecting terms of the order $\varepsilon^2 E^2$, we obtain

$$\varepsilon^2 \left(A_{22} \mathscr{A}_{2\beta} + \frac{\partial A_{22}}{\partial t_0} \mathscr{B} \right) \mathbf{w}_{22} = A^2 \mathbf{f}_{22} \,, \quad (3.94)$$

where $\mathbf{f}_{22}$ is given by

$$\mathbf{f}_{22} = - \begin{bmatrix} \mathrm{i}\beta u_{11} w_{11} + \frac{1}{2} D u_{11}^2 + \mathrm{Gr}_m \theta_{11} D^2 \phi_{11} - \frac{1}{2} \mathrm{Gr}_m \beta^2 \theta_0 D \frac{\phi_{11}^2}{H_0} \\ \mathrm{i}\beta v_{11} w_{11} + u_{11} D v_{11} \\ \mathrm{i}\beta w_{11}^2 + u_{11} D w_{11} + \mathrm{i}\beta \mathrm{Gr}_m \theta_{11} D\phi_{11} - \mathrm{i}\beta^3 \mathrm{Gr}_m \theta_0 \frac{\phi_{11}^2}{H_0} \\ \mathrm{i}\beta w_{11} \theta_{11} + u_{11} D\theta_{11} \\ 0 \\ \frac{\beta^2}{H_0^2} \left[2 H_0 \phi_{11} \theta_{11} - \frac{1}{2} \phi_{11}^2 D\theta_0 + \frac{3}{2} r H_0 D\phi_{11}^2 - r\phi_{11}^2 D H_0 \right] \end{bmatrix} ,$$

with the boundary conditions

$$u_{22} = v_{22} = w_{22} = \theta_{22} = 0 \,, \quad (3.95)$$

$$\varepsilon^2 A_{22} \left(D \mp \frac{2\beta}{1+\chi} \right) \phi_{22} = A^2 \frac{\beta^2}{2H_0^2} \phi_{11}^2 \left(\frac{\chi - \chi_*}{1+\chi} \mathrm{N} \pm 1 \right) \quad (3.96)$$

at $x = \mp 1$. Given that $A \sim \exp(\sigma t_0)$, the compatibility of the left- and right-hand sides of (3.94) requires that $\varepsilon^2 A_{22} = A^2$ so that

$$\varepsilon^2 \frac{\partial A_{22}}{\partial t_0} = 2\sigma A^2 .$$

Hence Equation (3.94) becomes

$$\mathscr{L}_{2\beta,2\sigma} \mathbf{w}_{22} = \mathbf{f}_{22} \,. \tag{3.97}$$

In the above, the implicit assumption is made that the operators in the left-hand sides of Equations (3.92) and (3.97) are non-singular so that the unique solutions can be found. This is, indeed, so in the considered physical problem. The reader is referred to [241], [242] and [235] for the description of the procedure required when this is not the case and resonant interactions have to be carefully accounted for.

3.5.4 Fundamental Harmonic Distortion and Landau Equation

Terms of the order $\varepsilon^3 E^1$ result in

$$\varepsilon^3 \left(A_{31} \mathscr{A}_\beta + \frac{\partial A_{31}}{\partial t_0} \mathscr{B} \right) \mathbf{w}_{31} = A|A|^2 \mathbf{f}_{31} - \varepsilon^2 \frac{\partial A}{\partial t_2} \mathscr{B} \mathbf{w}_{11} \,, \tag{3.98}$$

where elements of $\mathbf{f}_{31} = \left[f_{31}^{(1)}, f_{31}^{(2)}, f_{31}^{(3)}, f_{31}^{(4)}, 0, f_{31}^{(6)} \right]^T$ are given by

$$f_{31}^{(1)} = - \left[\begin{array}{l} \mathrm{i}\beta(u_{11}w_{20} - \bar{u}_{11}w_{22} + 2u_{22}\bar{w}_{11}) + D(u_{11}u_{20}) + D(\bar{u}_{11}u_{22}) \\ \quad + \mathrm{Gr}_m \left(\theta_{20} D^2 \phi_{11} + \theta_{11} D^2 \phi_{20} + \bar{\theta}_{11} D^2 \phi_{22} + \theta_{22} D^2 \bar{\phi}_{11} \right) \\ \quad + \beta^2 \mathrm{Gr}_m \left[\theta_{11} D \frac{|\phi_{11}|^2}{H_0} - \frac{1}{2} \bar{\theta}_{11} D \frac{\phi_{11}^2}{H_0} + 2\theta_0 D \frac{\bar{\phi}_{11}\phi_{22}}{H_0} \right. \\ \qquad \left. + \frac{1}{2} \theta_0 \phi_{11}^2 D \frac{D\bar{\phi}_{11}}{H_0^2} - \frac{1}{2} \theta_0 \bar{\phi}_{11} D \frac{D\phi_{11}^2}{H_0^2} \right] \end{array} \right] ,$$

$$f_{31}^{(2)} = - \left[\begin{array}{l} \mathrm{i}\beta(v_{11}w_{20} - \bar{v}_{11}w_{22} + 2v_{22}\bar{w}_{11}) \\ \quad + u_{11} D v_{20} + \bar{u}_{11} D v_{22} + u_{20} D v_{11} + u_{22} D \bar{v}_{11} \end{array} \right] ,$$

$$f_{31}^{(3)} = - \left[\begin{array}{l} \mathrm{i}\beta(w_{11}w_{20} + \bar{w}_{11}w_{22}) + u_{20} D w_{11} + u_{22} D \bar{w}_{11} + u_{11} D w_{20} + \bar{u}_{11} D w_{22} \\ \quad + \mathrm{i}\beta \mathrm{Gr}_m (\theta_{20} D \phi_{11} - \theta_{22} D \bar{\phi}_{11} + 2\bar{\theta}_{11} D \phi_{22}) \\ \quad + \frac{\mathrm{i}\beta^3}{H_0} \mathrm{Gr}_m (2\theta_0 \bar{\phi}_{11} \phi_{22} - \bar{\theta}_{11} \phi_{11}^2) \\ \quad + \frac{\mathrm{i}\beta^3}{2H_0^2} \mathrm{Gr}_m \theta_0 (\phi_{11}^2 D \bar{\phi}_{11} - \bar{\phi}_{11} D \phi_{11}^2) \end{array} \right] ,$$

$$f_{31}^{(4)} = - \left[\begin{array}{l} \mathrm{i}\beta(w_{20}\theta_{11} - w_{22}\bar{\theta}_{11} + 2\bar{w}_{11}\theta_{22}) \\ \quad + u_{20} D \theta_{11} + u_{22} D \bar{\theta}_{11} + u_{11} D \theta_{20} + \bar{u}_{11} D \theta_{22} \end{array} \right] ,$$

$$f_{31}^{(6)} = -\left[\begin{array}{l} \frac{\beta^2}{H_0}\left[\theta_{20}\phi_{11} - \theta_{22}\bar{\phi}_{11} + 2\bar{\theta}_{11}\phi_{22} + r\left(\phi_{11}D\phi_{20} - 3\bar{\phi}_{11}D\phi_{22}\right)\right] \\ \quad + \frac{\beta^2}{H_0^2}\left[2D\theta_0\bar{\phi}_{11}\phi_{22} + D\theta_{11}|\phi_{11}|^2 - \frac{1}{2}D\bar{\theta}_{11}\phi_{11}^2 \right. \\ \qquad -2(\bar{\theta}_{11}\phi_{11} - \theta_{11}\bar{\phi}_{11})D\phi_{11} \\ \qquad +r\left(\frac{3}{2}\beta^2\phi_{11}|\phi_{11}|^2 + 3\bar{\phi}_{11}(D\phi_{11})^2 - 2\phi_{11}|D\phi_{11}|^2\right. \\ \qquad\quad \left.\left. +4\bar{\phi}_{11}\phi_{22}DH_0 + 2|\phi_{11}|^2D^2\phi_{11} - \phi_{11}^2D^2\bar{\phi}_{11}\right)\right] \\ + \frac{\beta^2}{H_0^3}[D\theta_0\left(\phi_{11}^2D\bar{\phi}_{11} - 2|\phi_{11}|^2D\phi_{11}\right) + \left(\bar{\theta}_{11}\phi_{11}^2 - 2\theta_{11}|\phi_{11}|^2\right)DH_0 \\ \qquad -3r\left(2|\phi_{11}|^2D\phi_{11} - \phi_{11}^2D\bar{\phi}_{11}\right)DH_0] \end{array}\right],$$

with the boundary conditions

$$u_{31} = v_{31} = w_{31} = \theta_{31} = 0\,, \tag{3.99}$$

$$\begin{aligned} \varepsilon^3 A_{31}\left(D\phi_{31} \mp \frac{\beta}{1+\chi}\phi_{31}\right) &= A|A|^2\frac{\beta^2}{H_0^3} \\ &\times(2|\phi_{11}|^2D\phi_{11} - 2H_0\bar{\phi}_{11}\phi_{22} - \phi_{11}^2D\bar{\phi}_{11})\left(\frac{\chi-\chi_*}{1+\chi}\mathrm{N} \pm 1\right) \end{aligned} \tag{3.100}$$

at $x = \mp 1$. The compatibility of the left- and right-hand sides of the above equations requires that $\varepsilon^3 A_{31} = A|A|^2$, so that $\varepsilon^3\frac{\partial A_{31}}{\partial t_0} = (\sigma + 2\sigma^R)A|A|^2$, and $\varepsilon^2\frac{\partial A}{\partial t_2} = KA|A|^2$, where $K = K^R + iK^I$ is a generally complex coefficient traditionally referred to as the Landau constant. Equation (3.98) now takes the form

$$\mathscr{L}_{\beta,\sigma+2\sigma^R}\mathbf{w}_{31} = \mathbf{f}_{31} - K\mathscr{B}\mathbf{w}_{11}\,. \tag{3.101}$$

In the limit of $\sigma^R \to 0$, that is, in the asymptotic vicinity of a critical point, the operator in the left-hand side of (3.101) becomes identical to that in (3.88) and thus singular. Therefore, in the limit $\Pi \to \Pi_c$ (3.101) is solvable only if its right-hand side is in the range of $\mathscr{L}_{\beta,\sigma}$, that is, if

$$\langle \mathbf{w}_{11}^\dagger, \mathbf{f}_{31} - K\mathscr{B}\mathbf{w}_{11}\rangle = 0\,, \text{ or } K = \langle \mathbf{w}_{11}^\dagger, \mathbf{f}_{31}\rangle\,, \tag{3.102}$$

where $\mathbf{w}_{11}^\dagger$ is the eigenfunction of the adjoint problem defined as

$$\mathscr{L}_{\beta,\sigma}^\dagger\mathbf{w}_{11}^\dagger \equiv (\mathscr{A}_\beta^{*T} + \sigma^*\mathscr{B}^{*T})\mathbf{w}_{11}^\dagger = \mathbf{0} \tag{3.103}$$

and normalised as $\langle \mathbf{w}_{11}^\dagger, \mathscr{B}\mathbf{w}_{11}\rangle = 1$. Here the superscript $*T$ denotes conjugate transpose, and the angle brackets stand for a suitably chosen inner product. The numerical results for a discretised problem are obtained here using a standard inner product for column complex-valued vectors $\mathbf{a}$ and $\mathbf{b}$: $\langle \mathbf{a}, \mathbf{b} \rangle \equiv \mathbf{a}^{*T} \cdot \mathbf{b}$.

However, a finite distance away from the critical point $\sigma^R \neq 0$. Then the operator $\mathscr{L}_{\beta,\sigma+2\sigma^R}$ is non-singular so that Equation (3.101) is solvable for any value of K. To proceed from here, we choose to follow the approach suggested in [241]. Using the linearity of problem (3.101), when $\sigma^R \neq 0$, that is, away from the critical point, we look for the solution in the form

$$\mathbf{w}_{31} = \boldsymbol{\chi}_1 + \boldsymbol{\chi}_2 \,, \tag{3.104}$$

where $\boldsymbol{\chi}_k = (u_{\chi k}, v_{\chi k}, w_{\chi k}, \theta_{\chi k}, P_{\chi k}, \phi_{\chi k})$, $k = 1, 2$ are the solutions of the following compound problems:

$$\mathscr{L}_{\beta,\sigma+2\sigma^R} \boldsymbol{\chi}_1 = \mathbf{f}_{31} \,, \tag{3.105}$$

with boundary conditions

$$u_{\chi 1} = v_{\chi 1} = w_{\chi 1} = \theta_{\chi 1} = 0 \,, \tag{3.106}$$

$$D\phi_{\chi 1} \mp \frac{\beta}{1+\chi} \phi_{\chi 1} = \frac{\beta^2}{H_0^3} \times (2|\phi_{11}|^2 D\phi_{11} - 2H_0 \bar{\phi}_{11} \phi_{22} - \phi_{11}^2 D\bar{\phi}_{11}) \left(\frac{\chi - \chi_*}{1+\chi} \mathrm{N} \pm 1 \right) \tag{3.107}$$

at $x = \mp 1$, and

$$\mathscr{L}_{\beta,\sigma+2\sigma^R} \boldsymbol{\chi}_2 = -K \mathscr{B} \mathbf{w}_{11} \tag{3.108}$$

with boundary conditions

$$u_{\chi 2} = v_{\chi 2} = w_{\chi 2} = \theta_{\chi 2} = 0 \,, \quad D\phi_{\chi 2} \mp \frac{\beta}{1+\chi} \phi_{\chi 2} = 0 \,. \tag{3.109}$$

If $\sigma^R > 0$, that is, in supercritical regimes, the operator $\mathscr{L}_{\beta,\sigma+2\sigma^R}$ is guaranteed to be non-singular given that $\mathrm{Re}(\sigma + 2\sigma^R) = 3\sigma^R > \sigma^R$, where by our choice σ^R is the maximum linear amplification rate observed in the system. Therefore in this case, system (3.105) has a unique solution. Given (3.88) we find that system (3.108) also has a unique solution

$$\boldsymbol{\chi}_2 = -\frac{K}{2\sigma^R} \mathbf{w}_{11} \,. \tag{3.110}$$

for any value of K.

Note that the solution of (3.105) can be generally written in a projection form as

$$\boldsymbol{\chi}_1 = r \mathbf{w}_{11} + \hat{\boldsymbol{\chi}}_1 \,, \tag{3.111}$$

where

$$r = \frac{\langle \mathbf{w}_{11}, \boldsymbol{\chi}_1 \rangle}{\langle \mathbf{w}_{11}, \mathbf{w}_{11} \rangle} \text{ and } \langle \hat{\boldsymbol{\chi}}_1, \mathbf{w}_{11} \rangle = 0 \,. \tag{3.112}$$

Combining the terms in (3.88) corresponding to mode E^1, we obtain

$$A\left\{\left[1+|A|^2\left(r-\frac{K}{2\sigma^R}\right)\right]\mathbf{w}_{11}+|A|^2\hat{\chi}_1+\cdots\right\}E^1, \tag{3.113}$$

that is, the term proportional to $\mathbf{w}_{11}$ appears at different orders of amplitude. In order to remove this redundancy, we choose

$$K = 2\sigma^R r. \tag{3.114}$$

Any other choice of the Landau coefficient destroys the uniform validity of the expansion (3.77) in the limit $\sigma^R \to 0$. As shown in [241], Definition (3.114) of the Landau coefficient in the limit $\sigma^R \to 0$ is identical to that obtained from a conventional solvability condition:

$$\lim_{\sigma^R\to 0} K = \lim_{\sigma^R\to 0}(2\sigma^R r) = \langle \mathbf{w}_{11}^{\dagger}, \mathbf{f}_{31}\rangle .$$

Reconstructing the time derivative of the amplitude, we now have

$$\frac{dA}{dt} = \frac{\partial A}{\partial t_0} + \varepsilon^2\frac{\partial A}{\partial t_2} + \cdots = \sigma A + KA|A|^2 + \cdots . \tag{3.115}$$

Neglecting the higher-order terms, we obtain Landau equation

$$\frac{dA}{dt} = \sigma A + KA|A|^2 \tag{3.116}$$

describing the temporal evolution of the amplitude of the linearly most amplified instability mode. Using the polar representation of a complex amplitude $A = |A|\exp(i\varphi)$, this equation is equivalently rewritten as two real equations for the modulus and the phase:

$$\frac{d|A|}{dt} = \sigma^R|A| + K^R|A|^3, \tag{3.117}$$

$$\frac{d\varphi}{dt} = \sigma^I + K^I|A|^2. \tag{3.118}$$

Equation (3.117) can have two equilibrium solutions:

- $|A_e| = 0$, which always exists but is stable only if $\sigma^R < 0$;
- and

$$|A_e| = \sqrt{-\frac{\sigma^R}{K^R}}, \tag{3.119}$$

  which only exists if $\sigma^R/K^R < 0$ and is stable if $\sigma^R > 0$. If $K^R < 0$, then this equilibrium solution exists only for $\sigma^R > 0$, i.e. in linearly unstable regimes. In this case the bifurcation is supercritical; otherwise it is subcritical.

3.5.5 Numerical Results and Their Physical Interpretation

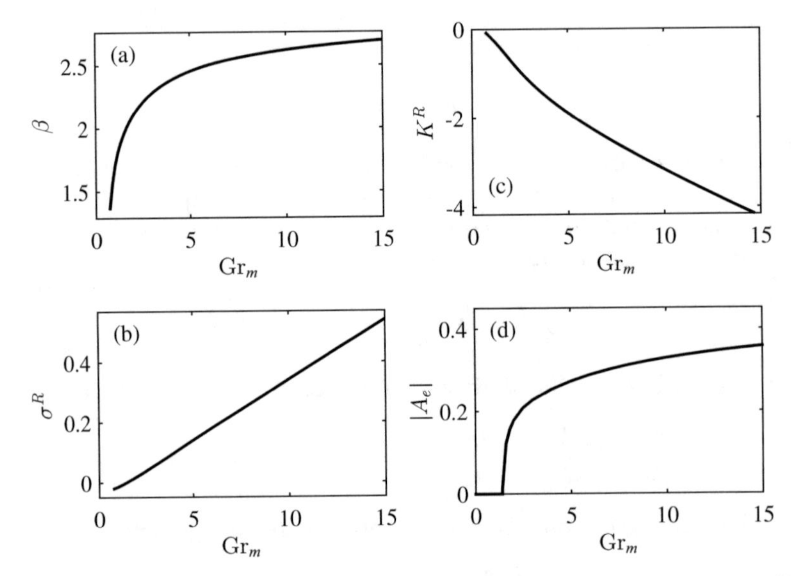

Fig. 3.48 (a) Wavenumber β, (b) amplification rate σ^R, (c) Landau constant $K = K^R$ and (d) equilibrium disturbance amplitude $|A_e|$ as functions of the magnetic Grashof number Gr_m for pure thermomagnetic convection in a flat infinite layer of ferrofluid.

The numerically evaluated flow parameters for thermomagnetic convection regimes are summarised in Figure 3.48. As seen from Figure 3.48(a), the wavenumber of the fastest-growing infinitesimal disturbance changes rapidly in the vicinity of the critical point. As the value of the magnetic Grashof number increases beyond the critical value determined from a linear stability analysis (see point D in Table 3.7), so does the wavenumber. This means that the strengthening of magnetic effects results in the appearance of additional more compactly spaced convection rolls. The disturbance amplification rate σ^R grows almost linearly with Gr_m. The Landau constant remains real and negative, and its magnitude also increases with Gr_m.

The main physical conclusion that we draw based on these facts about the instability leading to the appearance of thermomagnetic rolls is that it is a result of a supercritical pitchfork bifurcation: the real part of the Landau constant remains negative, $K^R < 0$; see Figure 3.48(c). Experimentally such a bifurcation would correspond to the appearance of convection patterns at the well-defined values of control parameters. The equilibrium amplitude of convection rolls given by (3.119) increases very sharply near the critical point; see Figure 3.48(d). Decreasing the control parameter values from supercritical to critical would lead to a gradual decay of convection at its onset point.

In practice this enables one to reasonably accurately detect the convection threshold via visual observations or heat transfer measurements as will be discussed in the subsequent chapters of the book. As the parameter values are increased beyond critical, the typical for a pitchfork bifurcation square-root-like growth of the disturbance amplitude given by (3.119) is hindered by the increase of $|K^R|$; see Figure 3.48(c) and (d). Thus after a quick onset of thermomagnetic convection, its strength and qualitative spatial structure are expected to remain unchanged for a wide range of supercritical parameter values. Computations also show that $\sigma^I = K^I = 0$ for pure thermomagnetic convection meaning that the arising roll instability patterns remain stationary as the disturbance phase variation equation (3.118) is trivial in this case. Both above conclusions are consistent with the experimental observations reported in e.g. [238] for a homogeneous ferrofluid.

In Figure 3.49 the cross-layer view is shown of typical fields that exist in supercritical regimes after convection sets in the form of stationary rolls, and their amplitude reaches the saturation value. The fields are reconstructed to the second order in amplitude as

$$\mathbf{w} = \mathbf{w}_{00} + |A_e|^2 \mathbf{w}_{20} + [A_e \mathbf{w}_{11} E + A_e^2 \mathbf{w}_{22} E^2 + c.c.] \,.$$

They illustrate the thermomagnetic nature of the instability. Thermal perturbations that are shown in the left panel lead to a nonuniform magnetisation of a fluid as seen in the third panel (cool regions are stronger magnetised). The variation of magnetisation is pronounced much stronger than that of other flow quantities. In particular, the distribution of the magnitude of magnetic field presented in the right panel remains much closer to that of the parallel basic flow. The magnetic field magnitude is larger near the hot left wall, while strongly magnetised fluid is located preferentially near the cold wall. Therefore the ponderomotive Kelvin force occurs driving cooler and stronger magnetised fluid from the right wall towards the hot left wall. The warm and less magnetised fluid is then displaced from the hot wall so that convection rolls are formed as demonstrated by the fluid velocity vectors shown in the left panel. The second panel from the left shows the distribution of magnetic pressure in the disturbed flow. It has a maximum near the centre of the layer so that the arising magnetic pressure gradient leads to the appearance of a magnetostatic force acting in the outward directions. This force can be sufficiently large to lead to bulging of convection chamber walls as, indeed, was observed in experiments reported in [238]. Therefore perspective thermomagnetic heat management systems have to be designed so that the fluid enclosure has a sufficient mechanical strength to withstand such a deformation.

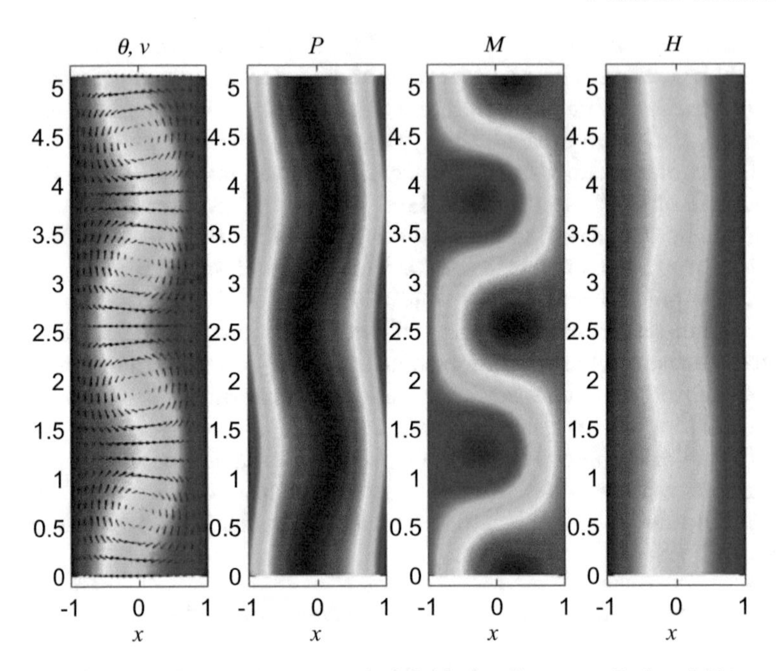

Fig. 3.49 Cross-layer view of the disturbed fields for $\mathrm{Gr}_m = 5$. Red and blue regions correspond to large and small field values, respectively.

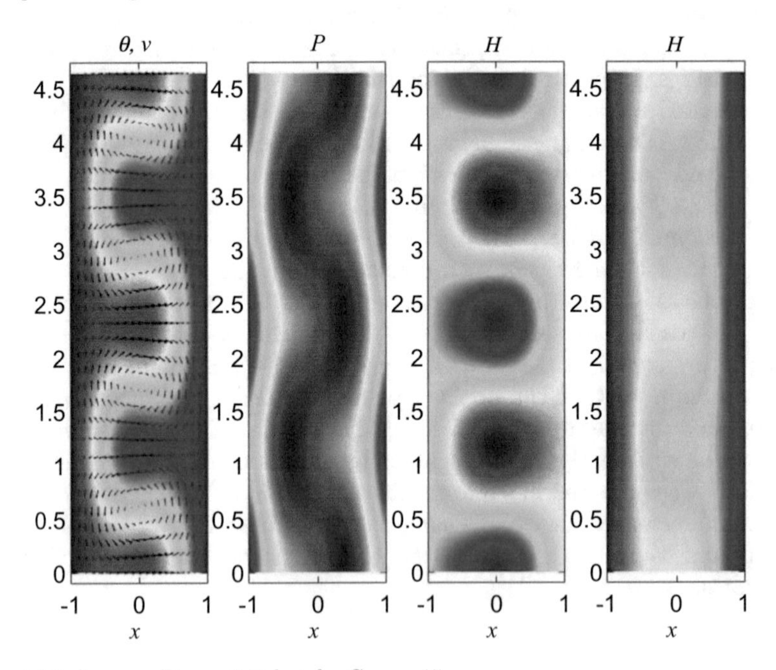

Fig. 3.50 Same as Figure 3.49 but for $\mathrm{Gr}_m = 15$.

When the value of the magnetic Grashof number increases, the qualitative structure of the perturbed flow remains unchanged; see Figure 3.50. However, the second harmonic becomes more pronounced, e.g. see the panel depicting the magnetic pressure field in Figure 3.50. The separation between the regions of strong and weak magnetisation becomes more obvious, and the shape of these regions becomes almost rectangular in the cross-layer plane. The strength of magnetoconvection vortices increases, and as a result the mushroom-like thermal plumes develop; see the leftmost panel. The distribution of the internal magnetic field follows that of the temperature; compare the left- and rightmost panels in Figure 3.50.

3.5.6 Conclusions

To summarise, the theoretical analysis of this chapter shows that thermomagnetic convection in an infinite layer of homogeneous ferrofluid sets in the form of stationary rolls as a result of supercritical bifurcation from a basic motionless or parallel-flow state. Thermomagnetic rolls are arbitrarily oriented in a horizontal layer if the applied magnetic field is normal to it. If the layer is placed in an oblique magnetic field, the magnetoconvection rolls preferably align with the in-layer component of the field. However, if the layer is placed in the gravitational field, which causes buoyancy-driven parallel basic flow, the preferred orientation of the thermomagnetic rolls is determined by the balance between the competing influences of magnetic and basic flow roll alignments. When the gravitational effects are negligible, thermomagnetic convection arising in a normal field is stationary. However, if the field is inclined with respect to the layer, the symmetry-breaking effects of nonlinear field strength variation across the layer can lead to a drift of thermomagnetic rolls.

It is important to remember that all theoretical results reported and discussed in this chapter are obtained by adopting two major assumptions: that the fluid layer is laterally infinite, so that the influence of its boundaries can be neglected, and that the fluid magnetisation depends only on its temperature. In reality, ferrofluids always occupy a finite domain, and the presence of boundaries may be felt over a large part of the fluid volume. Moreover, unless the fluid is previously homogenised via mechanical stirring, shaking or by applying convective mixing, its properties may depend on its local composition, which can vary from point to point depending on the history of fluid's storage and use. Currently, very sophisticated and sometimes ambiguous influences of container boundaries and compositional nonuniformity of the fluid are beyond the reach of a rigorous analytical treatment, and an experimental investigation to which the rest of this book is devoted remains the only way of shedding light onto an intricate and multifaceted behaviour of realistic non-isothermal ferrofluids flowing in finite domains.

Chapter 4
Experimental Methodology

Abstract The methodology of experimental investigation of thermogravitational and thermomagnetic convection in ferrofluids is discussed. The necessary physical conditions for the observation of thermomagnetic convection in external uniform magnetic field are identified. The methods of registering heat fluxes and visualising convection flow patterns in nontransparent ferrofluids are discussed. Various designs of experimental chambers, sensors and measuring devices are presented, and the main features distinguishing the behaviour of magneto-polarisable fluids from that of their non-magnetic counterparts are highlighted. Specifically, the influence of the working chamber geometries, sizes and boundaries on the distribution of a magnetic field inside cavities and thus on the characteristics of the arising convective flows and heat transfer is emphasised.

4.1 Properties of Ferrofluids

The design of experiments involving flows of ferrofluids depends crucially on the knowledge of their physical properties. Generally, these have to be determined in dedicated experiments conducted either by the manufacturer [13, 19, 23, 209, 248] or in-house. Among other factors the properties of magnetic colloids are defined by the way they are manufactured, the size distribution of solid phase particles, the interaction among these particles, the presence of free molecules of stabilising (surfactant) agent and the composition of the carrier fluid [9, 53, 85, 134, 193]. For this reason, to obtain conclusive results, it is essential that experiments are conducted using samples of

See Appendix B for the list of previously published materials re-used in this chapter with permission.

A. A. Bozhko, S. A. Suslov, *Convection in Ferro-Nanofluids: Experiments and Theory*, Advances in Mechanics and Mathematics 40,
https://doi.org/10.1007/978-3-319-94427-2_4

stable fluid (synthesised under patent 2517704 of the Russian Federation [6]) from the same manufacturer's batch [152]. To be specific here, we will discuss fluids containing particles with the average size of 10 nm. The typical average physical properties of several such ferrofluids are summarised in Table 4.1. However, Figures 4.1 and 4.2 demonstrate that the properties of ferrofluids provided by different manufacturers could vary significantly even though the solid phase concentration and the average particle size remain the same.

The density ρ of ferrofluids is mostly determined by the concentration of heavy solid phase. It can be reasonably assumed that they are linearly related. However, the mass density of ferrofluid is easier to measure in in-house experiments when, for example, the concentration of ferrofluids needs to be changed from manufacturer's specifications. For this reason below we characterise various ferrofluid properties by their dependence on the fluid density.

One of the most important characteristics of a ferrofluid is its magnetisation of saturation M_s that defines the intensity of interaction of ferrofluid with the applied magnetic field. Figure 4.1 demonstrates that fluid magnetisation grows approximately linear with the density of the fluid.

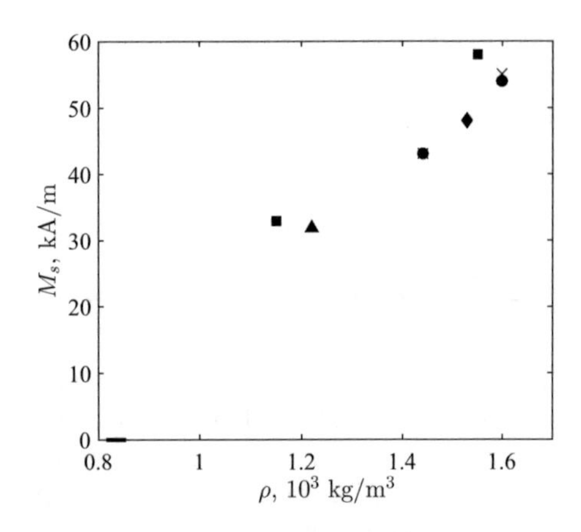

Fig. 4.1 Magnetisation of saturation of ferrofluids with different densities reported by various manufacturers: line, kerosene; circles, [13, 88, 208]; squares, data provided by the Laboratory of Dispersed Systems of the Institute of Continuous Media Mechanics of the Ural Branch of the Russian Academy of Sciences, Perm, Russian Federation; triangles, [219]; crosses, [152]; diamonds, Ferrotec (at the time of writing available from `http://www.ferrofluidics.de/en/htmls/fluid.data.php`).

The dynamic viscosity η of ferrofluids also increases with their density as seen from Figure 4.2. One of the reasons for this is the appearance of the so-called rotational viscosity component that is associated with the tendency of solid particles to resist their rotation in a shear flow. While the theoretical description of rotational viscosity is well known from the seminal work of Einstein [82], the experimental data presented in Figure 4.2 shows that in reality the total viscosity of ferrofluids can vary by the factor of two or more for the same fluid density (e.g. see the values for $\rho \approx 1.25 \times 10^3$ kg/m^3 in

Table 4.1 Physical properties of common magnetic fluids.

Carrier	M_s, kA/m	ρ, kg/m^3	χ	η, Pa s	λ, W/(m K)	α, m^2/s	D m^2/s	β 1/K
Kerosine 1	43.0	1.44×10^3	2.88	7.66×10^{-3}	0.19	1.0×10^{-7}	1.9×10^{-11}	7.7×10^{-4}
Kerosine 2	55.0	1.55×10^3	5.72	0.009	0.21	1.1×10^{-7}	1.9×10^{-11}	7.5×10^{-4}
Transfor-mer oil	44.9	1.37×10^3	4.30	0.069	0.20	1.0×10^{-7}	6.2×10^{-13}	6.1×10^{-4}
Polyethyl-siloxane	48.0	1.49×10^3	4.62	0.376	0.23	1.2×10^{-7}	1.1×10^{-13}	6.2×10^{-4}

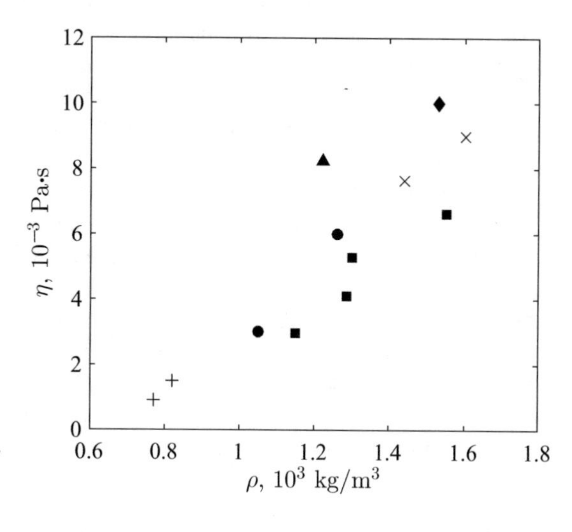

Fig. 4.2 Dynamic viscosity of ferrofluids with different densities reported by various manufacturers in the absence of magnetic field: pluses, kerosene; circles, [13, 88, 208]; squares, data provided by the Laboratory of Dispersed Systems of the Institute of Continuous Media Mechanics of the Ural Branch of the Russian Academy of Sciences, Perm, Russian Federation; triangles, [219]; crosses, [152]; diamonds, Ferrotec (at the time of writing available from `http://www.ferrofluidics.de/en/htmls/fluid.data.php`).

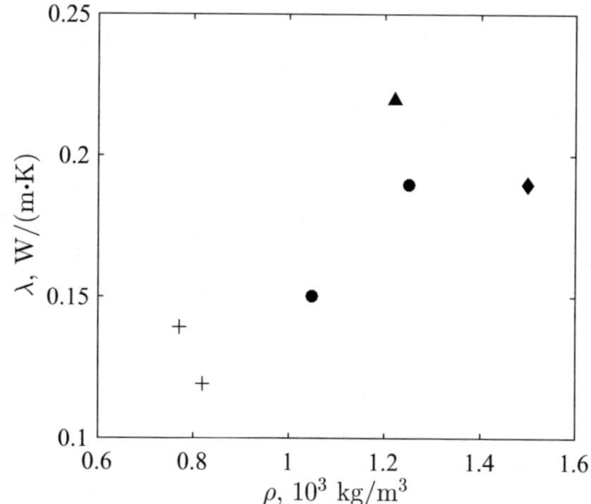

Fig. 4.3 Thermal conductivity of ferrofluids with different densities reported by various manufacturers in the absence of magnetic field: pluses, kerosene; circles, [13, 88, 208]; triangle, [219]; diamond—[23].

Figure 4.2). A similar variation of values of thermal conductivity λ is also evident in Figure 4.3.

When ferrofluids are placed in an external magnetic field, their transport properties change yet again and become dependent not only on the local structure of the shear flow but also on the local configuration of the magnetic field that is influenced by the ferrofluid itself. Such a strong coupling between ferrofluid properties and its flow and magnetic environment can be traced back to the presence of particle aggregates containing from several to several dozens particles [9, 53, 60, 187, 193]. While the addition of surfactants significantly reduces the probability of forming such aggregates, the possibility of this cannot be excluded completely. Due to the magnetic dipole interaction, the number of aggregates increases when magnetic field is applied, but their exact number and sizes remain unknown because they depend on the history and specifics of experiment. As a result, the gradients associated with the gravitational sedimentation and thermo- and magnetophoresis of aggregates of such fluid properties as rotational and magnetic viscosities and magnetic susceptibility become virtually impossible to quantify. The qualitative behaviour of the observed ferrofluid flows does not seem to be affected by such an ambiguity. However, presenting experimental results in a criterial form becomes rather problematic as the values of the governing nondimensional parameters can only be estimated using the assumed average but not actual local fluid properties. Moreover, the knowledge of the dimensional magnetic field and temperature difference rather of their scaled counterparts is preferred for technical applications. For these reasons the discussion of experimental data has to be given in terms of directly measurable quantities, while the comparison with analytical results obtained in terms of nondimensional groups such as Rayleigh, Grashof and Prandtl numbers involving unknown transport properties has to be limited to qualitative and/or averaged values.

4.2 Requirements for Experimental Setup

While natural gravitational convection inevitably occurs in ferrofluids in ground-based laboratory experiments with non-uniformly heated ferrofluid, it is a magnetically driven convection, which is of primary interest here. Thus the design of experimental setup should serve the purpose of ensuring that such convection is the preferred mode of heat transfer across the experimental chamber. The main conditions for that can be derived from the analysis of the governing parameters defined in Chapter 2. Namely, the geometry of the experimental chamber has to be such that the value of magnetic Grashof number Gr_m is larger than, or at least comparable with, the value of gravitational Grashof number Gr.

Given that

$$\frac{\mathrm{Gr}_m}{\mathrm{Gr}} \sim \frac{K^2 \Delta T}{d} \tag{4.1}$$

it is clear that magnetoconvection can be dominant in highly thermosensitive magnetic fluids that is in fluids characterised by large values of the pyromagnetic coefficient K. The thermo-sensitivity of magnetic properties of ferrofluids can be improved by increasing the concentration of magnetic phase in the colloid. However, this has its limitations because at large concentrations a fluid becomes non-Newtonian with noticeable memory effects, which is undesirable from the application point of view. An alternative way of promoting convection in ferrofluids is by reducing the viscosity of its liquid base. That is why kerosene-based fluids with the concentration of a magnetic phase not exceeding 10% by volume are frequently used in studies of thermomagnetic effects in ferrofluids.

As follows from Equation (4.1), the relative importance of magnetic effects in ferrofluids increases proportionally to the applied characteristic temperature difference ΔT. However, the maximum value of ΔT is limited by the thermal stability of materials (such as Plexiglas) used to build an experimental setup and of magnetic fluid itself (the highest temperature cannot exceed the boiling point of a carrier fluid). Moreover, at large ΔT the transport properties of a ferrocolloid, e.g. its viscosity, cannot be assumed constant. This would make the comparison of experimental results with theoretical predictions currently available in literature and estimated in the limit of constant viscosity and linear variation of fluid density and magnetisation with temperature less straightforward. For these reasons in experiments described below, the maximum temperature difference in the flow domain was limited to 70 K.

Another way of enhancing a relative role of thermomagnetic effects that follows from Equation (4.1) would be decreasing the characteristic thickness d of the fluid layer in the direction of the temperature gradient. Yet this has its own limitations as well. The convection onset that is of primary interest corresponds to a sufficiently large value of the governing parameters such as the corresponding Grashof or Rayleigh numbers that are proportional to the square or cube of the fluid layer thickness [97]. This imposes the limit of a few millimetres for typical experimental layers using kerosene or transformer-oil-based ferrofluids.

The geometry of the experimental cavity that is to be filled with ferrofluid has a strong influence on the distribution of magnetic field there. A uniform external magnetic field only remains uniform inside ellipsoidal cavities (in particular, inside a sphere) [144]. A limiting case of an ellipsoid with an infinite aspect ratio is a flat layer. Thus when designing an experimental setup, enclosures of such simple geometric shapes are preferred because the interpretation and theoretical modelling of the observed phenomena are much more straightforward and less ambiguous in such enclosures. However, despite the use of simple geometric configurations, the distortion of the magnetic field lines near the flat layer edges is unavoidable due to the refraction of magnetic field lines at the boundary separating the media with significantly different magnetic susceptibilities. To some degree the influence of edge effets can be reduced if flat layers of large aspect ratio are used in experiments, but this inevitably increases the cost of investigation requiring a large quantity of ferrofluid to fill the layer. Moreover, wide and long layers with transparent Plexiglas walls enabling visual observation of fluid thermal fields (see below) tend to bulge due to magnetic pressure developing inside the ferrofluid layer placed in magnetic field. This unpredictable deformation in turn influences strongly the orientation of the internal magnetic field yet again making the interpretation of the observed flows and thermal distributions in such deformed layers more complicated.

Further complications in experimental studies of ferrofluids are associated with the fact that common ferrofluids are nontransparent. Therefore optical or particle tracer methods of flow observations cannot be applied to ferrofluid layers thicker than a few tenth of a millimetre. The range of experimental methods therefore narrows to that of indirect flow visualisation based on registering thermal fluxes and spatial temperature distributions caused by ferrofluid flows [37, 38, 41, 169, 217, 219, 259].

4.3 Experimental Chamber Design

The existence of multiple physical limitations and conflicting physical trends makes the design of experimental chambers a challenging task. Several basic configurations have been chosen that we describe in detail next emphasising the difficulties that had to be resolved before a successful experimental setup could be built.

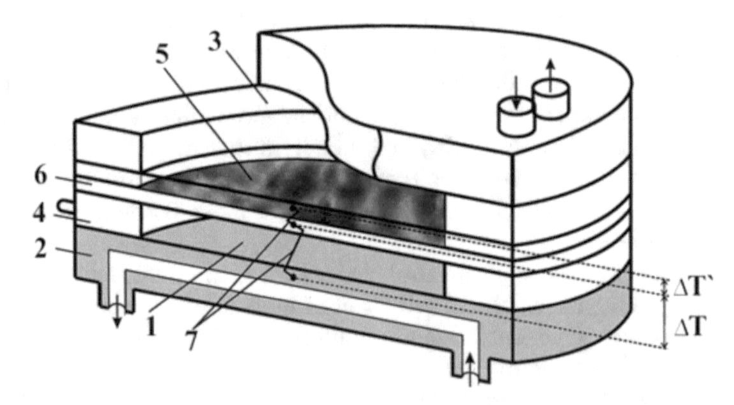

Fig. 4.4 Disk-shaped experimental chamber: 1, cavity filled with ferrofluid; 2 and 3, copper and Plexiglas heat exchangers; 4, ring-shaped frame; 5, thermosensitive film; 6, protective plate. Temperature differences ΔT and $\Delta T'$ across the ferrofluid layer and between the surface of a protecting plate facing the fluid and the Plexiglas heat exchanger, respectively, were measured using thermocouples 7 installed in the centre of the cavity walls [38].

The disk-shaped experimental chamber 1 is shown schematically in Figure 4.4. It had the height of 3.5 ± 0.1 mm and the diameter of 75 mm. The chamber was formed by the copper heat exchanger plate of thickness 10 mm and diameter 98 mm on one wide side and transparent heat exchanger 3 on the other. The set temperature of the copper heat exchanger was maintained by pumping water through parallel channels drilled in the body of the plate. As the fluid flows along the channel, it cools down, and thus the part of a heat exchanger adjacent to the channel inlet is inevitably hotter than that near its outlet. To avoid such a thermal non-uniformity, the channels were arranged in such a way that the water was flowing in the opposite directions through the adjacent channels. Such arrangement of channels ensured the most uniform heating of the copper plate. The water temperature was controlled by jet thermostats with the accuracy of ± 0.02 K.

The transparent heat exchanger 3 consisted of two 2 mm thick Plexiglas plates separated by the 10 mm gap through which cooling fluid was pumped. The perimeter of the cavity was formed by a ring-shaped Plexiglas frame. In the absence of convection, a uniform temperature gradient existed across the working chamber between the heat exchangers.

The coefficients of thermal diffusivity for heat exchanger materials were 4.0×10^2 W/(m·K) (copper) and 0.19 W/(m·K) (Plexiglas), while the thermal diffusivity of experimental fluids varied in the range 0.12–0.21 W/(m·K). Therefore, the ratios of the thermal diffusivities of the fluids and cavity walls were $(3.0\text{–}5.3) \times 10^{-4}$ for copper and 0.6–1.1 for Plexiglas heat exchangers. Thus the uniform temperature equal to that of the thermostat water could be safely expected on the copper wall, while the temperature of the Plexiglas wall remained close to the local temperature of ferrofluid. Therefore, the temperature readings along the Plexiglas wall provided a sufficiently accurate information about the actual temperature distribution in ferrofluid. To visualise the thermal field, a thermally sensitive liquid crystal film 5 (see Figure 4.4) of thickness 0.1 mm was glued on the surface of the Plexiglas heat exchanger. Such a film changed its colour from dark brown to green to bright blue in the temperature interval 24–27 °C. The colour reading accuracy was ± 0.5 degrees [264]. The colour of the film was not affected by the application of a magnetic field [33].

The colour distribution over the surface of a thermosensitive film is the signature of flow structures developing in the layer of a ferrofluid. In the absence of convection, the film had a uniform colour indicating that the surface of a transparent heat exchanger was isothermal. When convection sets the working fluid flows forming convection cells or rolls with nonzero cross-layer velocity components in the bulk of the fluid. The fluid elements moving towards the liquid crystal film from the opposite wall have the temperature different to that of the initial temperature of the film. Once they reach the film, its local temperature changes so that the colour pattern becomes a footprint of the cross-layer velocity field [233]. This in turn enables one to get an idea of the geometrical shape of the arising flow patterns. For example, elongated colour stripes indicate the presence of convection rolls, while closed colour contours evidence the presence of localised convection cells. Numerous experiments and numerical calculations [97, 99, e.g.] show that the temporal variation of convection patterns is sufficiently slow (compared to the thermal response time of a liquid crystal film) so that the observed colour pattern uniquely corresponds to the underlying flow field. Moreover, when convective velocities are small, the amplitude of the surface temperature variation detected by the film is directly proportional to the magnitude of the transverse velocity component. Thus the colour contrast can be used to judge the local intensity of convective fluid motions. It is important to note, however, that the surface visualisation of a thermal field cannot be used to register convective motions that occur primarily in the planes parallel to that of thermosensitive film. For example, the presence of a primary gravitational convection flow with a velocity field parallel to the cavity walls as occurs in differentially heated vertical or inclined with respect to the gravity vector flat layers leaves no thermal signature. However, the existence of such a flow in experimental layers of finite extent still can be detected via the thermal field visualisation near the edges of the layer where the fluid has to turn and thus acquires a transverse velocity component resulting in the variation of film colour from background.

While the idea of using a thermosensitive film for the visualisation of non-isothermal fluid flows might appear natural and logical working with it in practice is not easy. Since the typical liquid crystal films are manufactured on a polymer base, they are characterised by a strong anisotropy of mechanical properties and thus deform strongly when exposed to spatially variable thermal fields. Its non-uniform thermal expansion can lead to its corrugation that would adversely affect the accuracy of its thermoindicator properties. To avoid this and to ensure the maximum possible accuracy of the temperature readings, the film had to be glued using a transparent epoxy resin to the side of a Plexiglas heat exchanger facing ferrofluid. Consequently, to protect the film from the chemical action of a kerosene-based ferrofluid, it was covered by the 1 mm thick protective Plexiglas plate 6 (see Figure 4.4). Unfortunately, the presence of this plate reduced the effective sensitivity and spatial resolution of the film. Therefore, to accurately detect the onset of convection in the chamber, thermocouples were installed in the centre of cavity walls as schematically shown in Figure 4.4. This enabled measuring the local heat flux using the Schmidt-Milverton method [216] by comparing the temperature difference ΔT across the fluid layer with the temperature variation $\Delta T'$ across the protective plate. The details of this method will be discussed in Section 4.5.

To eliminate the influence of the edge curvature, experiments have also been conducted in rectangular convection cells of the length $l = 250\,\mathrm{mm}$, width $w = 70\,\mathrm{mm}$ and the layer thickness $d = 4.0$ and $6.0\,\mathrm{mm}$, see Figure 4.5.

Apart from the shape difference, the overall design of a rectangular convection chamber was similar to that of a disk-shaped chamber shown in Figure 4.4. Because of a larger heat exchanger face area, the thermosensitive film was not glued to the surface of the transparent heat exchanger but rather laminated between two transparent 0.075 mm thick plastic sheets. Since no protective plate separating film 5 from ferrofluid layer 1 was necessary in this design, the spatial resolution of such a thermosensor was improved compared to the design shown in Figure 4.4. For even more accurate temperature measurements, two differential thermocouples 6 were fitted to measure the temperature difference between the surface of heat exchanger 2 and the temperature inside the symmetry plane of the fluid 43 and 46 mm away from the short edges of the layer and protruding 0.5 and 0.8 mm into the fluid, respectively.

A variant of an experimental chamber where the transparent heat exchanger was replaced with a thin textolite plate that served as one of the cavity walls was also used [41, 225]. In this case the temperature distribution on the textolite plate was registered using an infrared camera with the manufacturer-specified resolution of 640×512 pixels and the temperature range from −20 to 3000 °C. The visualised thermal fields were recorded using the computer-controlled digital photo or video camera.

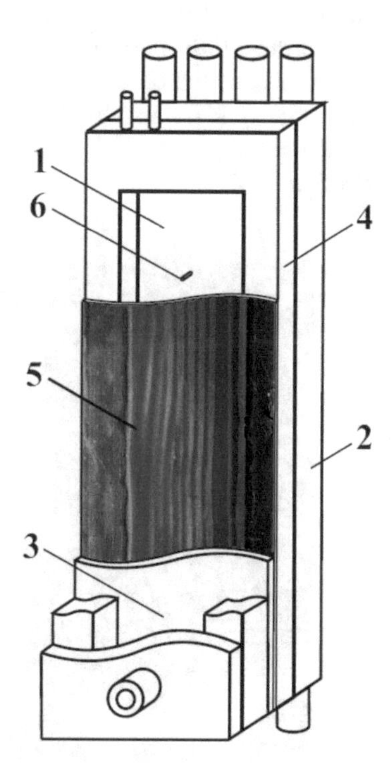

Fig. 4.5 Rectangular experimental chamber: 1, cavity filled with ferrofluid; 2 and 3, brass and Plexiglas heat exchangers; 4, Plexiglas frame; 5, thermosensitive film; 6, thermocouple [238].

4.4 Interpretation of Typical Thermal Field Visualisations

The visualisation of temperature fields in flat ferrofluid layers can be accomplished using thermosensitive film. The examples of such a visualisation are shown in Figures 4.6 and 4.7. The images show photographs of differentially heated horizontal ferrofluid layers. The warm and cool regions correspond to blue (white) and brown (black) areas in the images, respectively. They result from either cell or roll structures with horizontal axes located approximately at the border between different colours. Each blue (white) stripe surrounded by a brown (black) stripes corresponds to a pair of convection rolls with opposite vorticity. In both figures a magnetic field is perpendicular to the view plane.

The left panel in Figure 4.6 corresponds to Rayleigh-Bénard configuration when fluid is heated from below, but the applied temperature difference is below the critical value for the onset of thermogravitational convection. In ordinary non-magnetic fluids, the temperature of the fluid layer facing the observer is expected to be uniform due to the absence of fluid motion. However, a bright circular region along the perimeter of the cavity is clearly seen in the left panel in Figure 4.6 indicating the presence of noticeable toroidal motion there bringing warm fluid from the heated bottom wall to the top. Such a ferrofluid motion in the boundary regions of experimental chambers is the characteristic feature of non-isothermal ferrofluid flows occurring in magnetic fields that should not be confused with convection flows arising in a non-magnetic fluid. The nature of such near-edge flows is discussed in Section 4.4.1.

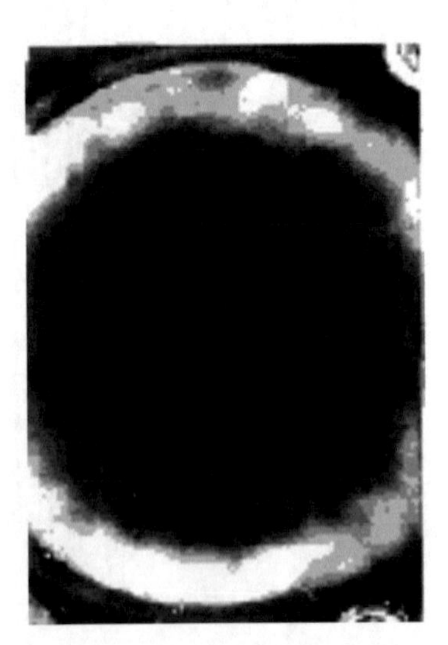

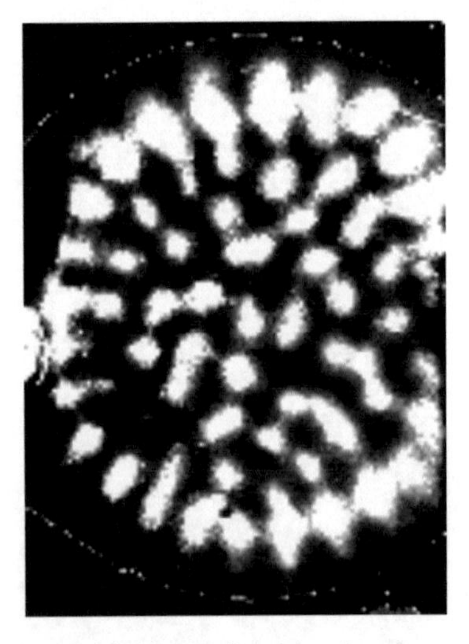

Fig. 4.6 Black-and-white photographs of the thermal field visualisation using thermosensitive film (temperature range 24–27°C) on the upper surface of a disk-shaped cavity filled with ferrofluid ($M_s = 55\,\text{kA/m}$) heated from below and placed in a vertical magnetic field $H = 13\,\text{kA/m}$ for the temperature difference between the heat exchangers $\Delta T < \Delta T_c$ (left) and $\Delta T > \Delta T_c$ (right), where $\Delta T_c = 4.2\,\text{K}$ is the critical temperature difference for the onset of convection at $H = 13\,\text{kA/m}$. Warm and cool regions correspond to white and dark areas, respectively [38].

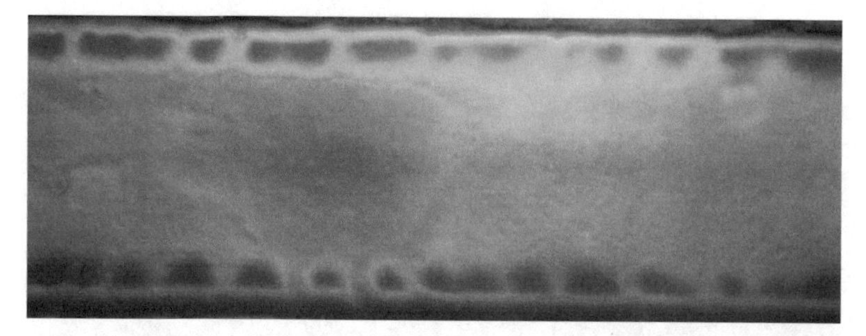

Fig. 4.7 Thermal field visualisation using thermosensitive film (temperature range 17–22°C) on the upper surface of a rectangular cavity filled with ferrofluid ($M_s = 43\,\mathrm{kA/m}$) heated from above and placed in a vertical magnetic field $H = 21\ \mathrm{kA/m}$ for the temperature difference between the heat exchangers $\Delta T = 23\,\mathrm{K}$. Warm and cool regions correspond to blue and brown areas, respectively.

4.4.1 Edge Effects in Magnetic Fluid Flows

The near-edge flows in ferrofluids occur almost immediately after the magnetic field is switched on. The physical reason behind such flows is the strong non-uniformity of the applied magnetic field that inevitably arises at the boundary between media with different magnetic susceptibilities. As follows from Maxwell boundary conditions for a magnetic field, its lines that are not strictly parallel or normal to a boundary separating two media refract in such a way that their density increases inside the medium with a higher magnetic susceptibility. A frequently used analogy is that magnetic field lines are "sucked in" the stronger magnetised medium. Even if the uniform external magnetic field is applied normally to the main observation face of the chamber, near the corners, where the normal to the cavity surface changes its direction, the magnetic field lines necessarily deform in such a way that the magnitude of the magnetic field increases there compared to that in the regions located further away from the edges. This is illustrated in Figure 4.8. Such a focussing of a magnetic field near the corners leads to the appearance of the magnetic field gradient ∇H_b directed towards the nearest edge of the cavity. If the two opposite surfaces of the cavity have different temperatures, ferrofluid adjacent to them will have different magnetisation M so that the ponderomotive force $M\nabla H_b$ directed to the edge will be stronger near the cold surface (the observation surface in Figure 4.6 and the bottom surface in Figure 4.9) than that near the hot surface (the observation surface in Figure 4.7 and the top surface in Figure 4.9). This force misbalance leads to the appearance of the observed near-edge toroidal flow structure, where a non-isothermal ferrofluid flows towards the edge along the colder surface and away from it along the hot one. This is illustrated in Figure 4.7 where the temperature distribution along the heated top surface of a ferrofluid is visualised. Cool fluid is drawn towards the edges along the bottom (invisi-

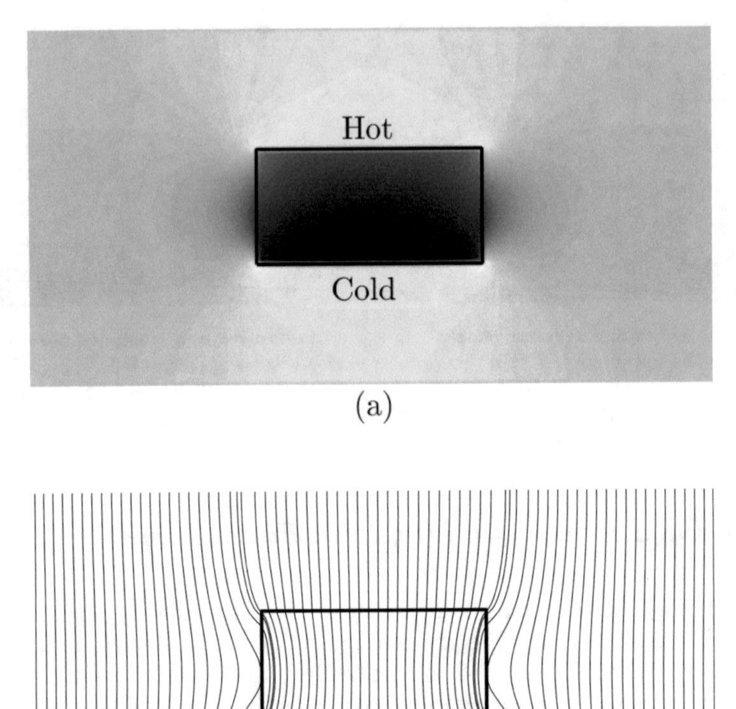

Fig. 4.8 Distortion of a uniform external magnetic field by a rectangular layer of non-uniformly heated ferrofluid: (a) the intensity of magnetic field H (light regions correspond to a stronger field), (b) the structure of magnetic field lines.

ble) surface then rises along the edge wall. This cold stream impinging the top warm surface is visualised as a narrow brown stripe close to the top and bottom edges of the image. The warm fluid displaced by the cold near-edge jet sinks forming a roll (brown-blue stripe in the image). While such a fluid motion is detected for all values of the applied magnetic field regardless of its orientation and the orientation of the temperature gradient with respect to the gravity, the width of the boundary region affected by it depends on both the magnitude of the magnetic field and on the temperature difference between the differentially heated faces of the experimental chamber and could reach up to 2–3 cm in experiments depicted in Figures 4.6 and 4.7.

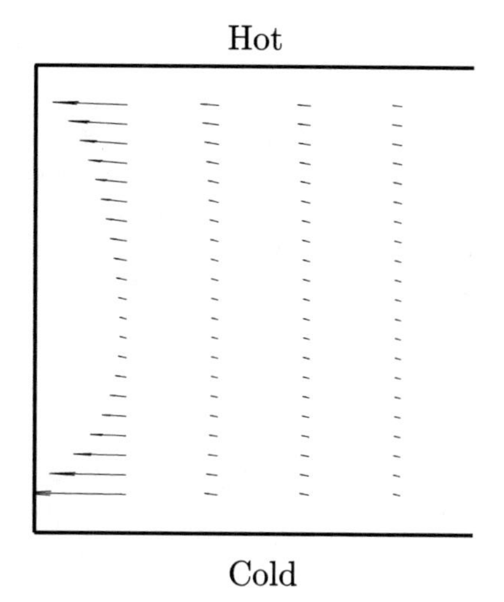

Fig. 4.9 Schematic view of the Kelvin force field near the edge of a non-isothermal ferrofluid layer placed in a uniform vertical external magnetic field.

4.4.2 Convection Patterns

When the fluid is heated from below and the applied temperature difference exceeds the critical value, the Rayleigh-Bénard type of convection flow establishes throughout the horizontal fluid layer with the difference that gravitational buoyancy force in magnetic fluids is now assisted by the magnetic buoyancy that results from the non-uniformity of Kelvin force $\mathbf{F}_K = M\nabla H$ acting on the non-isothermal and thus non-uniformly magnetised ferrofluid. While away from the edges convection takes the form of Rayleigh-Bénard type cells, near the boundaries roll structures perpendicular to them seen in the right panel in Figure 4.6 dominate. Such an orientation of rolls is dictated by two main reasons. Firstly, similar to Rayleigh-Bénard-Poiseuille problem, where convection rolls tend to align with the throughflow, near the boundaries convection patterns align with the velocity of toroidal vortex that mostly belong to the plane perpendicular to the boundary. Secondly, in magnetic fluids convection structures tend to align with the in-layer component of a magnetic field [13, 202, 203]. While away from the boundaries the applied magnetic field is normal to the observation surfaces due to the refraction of magnetic field lines at the cavity edges, magnetic field there inevitably has an in-layer component that is perpendicular to the boundaries. We also note that the near-edge toroidal flow structures are subject to an instability that breaks them into a shorter cells as seen in Figure 4.7. The nature of this instability has not been investigated to date, but it is likely to be of magnetic nature mentioned above: the flow patterns tend to align with the inward component of the distorted magnetic field near the edges. The thermogravitational (Rayleigh-Bénard) mechanism of the toroidal vortex breakup can be excluded because this instability is observed even when the bulk of the fluid remains in stable mechanical equilibrium as indeed is the case in Figure 4.7 (the cavity is heated from above).

4.5 Heat Flux Measurements

While the visualisation of thermal fields discussed in Section 4.4 offers a qualitative insight into the structure of the flows developing in a ferrofluid, it is not capable of providing an accurate quantitative information about such flows. The measurements of a heat flux across the fluid layer can fill this gap. However, given the specifics of ferrofluid flows in magnetic field discussed in Section 4.4.1, a care has to be exercised to make sure that such measurements are not influenced by the edge effects. This necessitates a special design of the experimental chamber for heat flux measurements that is shown schematically in Figure 4.10.

The disk-shaped enclosure is an appropriate choice of the experimental geometry that insures the symmetry of the physical setup and thus enables the quantitative characterisation of heat transfer in ferrofluids via measuring the integral rather than local heat flux. The setup shown in Figure 4.10 had a cavity 1 of height 2.00 ± 0.05 or 5.00 ± 0.05 mm and diameter 75 mm. It was sandwiched between two copper heat exchangers 2 and 3 with the thickness of 10 and diameter of 98 mm each. Ring-shaped Plexiglas frame 4 was used as a spacer defining the height of the chamber. The integral heat flux sensor consisted of a fluoroplastic (thermal diffusivity 0.25 W/(m·K)) plate 6 of thickness $h = 0.2$ or 2.0 mm for a $d = 2$ or 5 mm high cavities, respectively, and a copper plate 5. The fluoroplastic insert was required to enable the heat flux measurements and the determination of the onset of convection using Schmidt-Milverton method [216]. They use the continuity of the heat flux across the adjacent materials with distinct (known) thermal

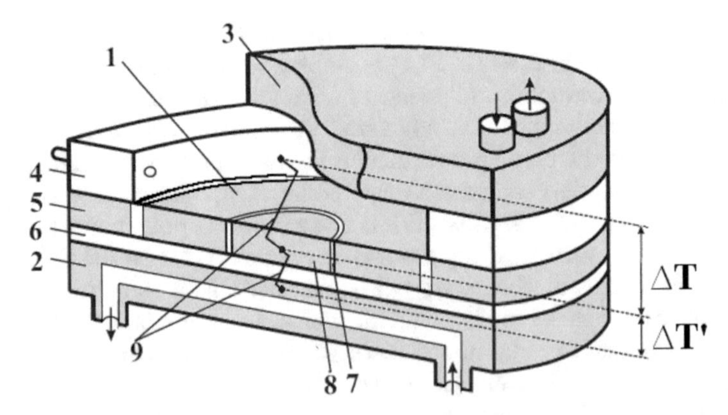

Fig. 4.10 Experimental chamber for heat flux measurements: 1, cavity filled with ferrofluid; 2 and 3, copper heat exchangers; 4, ring-shaped frame; 5, copper plate; 6, fluoroplastic insert; 7, thermo-insulating ring; 8, copper cylinder; 9, thermocouples. Temperature differences ΔT and $\Delta T'$ across the ferrofluid layer and the fluoroplastic insert, respectively, were measured using thermocouples installed in 1 mm channels drilled radially in the copper plates towards their centres.

conductivities. In particular, for the setup depicted in Figure 4.10, one can write the equation for the conduction heat flux balance as

$$\kappa_s \frac{\Delta T'}{h} = \kappa_e \frac{\Delta T}{d} , \tag{4.2}$$

where κ_s is the thermal conductivity of solid (Plexiglas) insert and κ_e is the effective thermal conductivity of the fluid in the cavity. In a conduction regime when the fluid is at rest, $\kappa_e = \kappa$ is the actual (tabulated) thermal conductivity of a fluid, while $\kappa_e > \kappa$ when convection is present. From this expression it follows that the measured temperature differences ΔT (across the fluid layer) and $\Delta T'$ (across the insert) are related linearly $\Delta T' = k_e \Delta T$, where the slope $k_e = \kappa_e h/(\kappa_s d)$ remains constant in the conduction regime. In convection regime the effective thermal conductivity κ_e of the fluid increases and so does the slope k_e. Therefore the sudden change of the slope of the experimental $\Delta T'(\Delta T)$ curve signals the onset of convection and determines the corresponding critical temperature drop ΔT_c across the fluid layer. The temperature differences were measured using copper-constantan thermocouples with the sensitivity of $40\,\mu\text{V/K}$. The thermocouple reading error did not exceed $\pm 0.05\,\text{K}$.

It is easy to show that Equation (4.2) can be conveniently rearranged to define the experimental Nusselt number characterising the ratio of the total heat flux across the fluid layer to its conduction component

$$\text{Nu} = \frac{1}{k} \frac{\Delta T'}{\Delta T} , \tag{4.3}$$

where $k = \kappa h/(\kappa_s d)$. It is equal to 1 in the conduction regime but increases once convection sets in. Note that while insert 6 is required for heat flux measurements, it has to be sufficiently thin to ensure that the as-large-as-possible portion ΔT of the total temperature drop between the heat exchangers (up to 85 K in the reported experiments) would occur across the fluid layer so that the thermomagnetic effects would be maximised.

The temperature readings across the integral sensor were recorded using differential thermocouples 9 that were installed as shown in Figure 4.10 in the radial channels drilled in copper towards the centre of the sensor. The temperature distribution in the system and the accuracy of temperature readings are extremely sensitive to various experimental design factors and can be strongly distorted if special care is not consistently taken. To reduce the reading errors associated with the unaccounted heat exchange with the ambient through thermocouple wires, they were thermo- and electrically insulated by fluoroplastic pipes that were inserted into the drilled channels. The complete assembly was secured by 5 mm (non-magnetic) brass bolts. To avoid a parasitic heat flux across the assembly via the bolts, they were placed into fluoroplastic sleeves and thermo-insulated by textolite washers from direct contact with the copper plate and heat exchangers.

When designing an experimental setup for ferrofluid experiments, it is important to make sure that only non-magnetic materials such as brass, copper, plastic and Plexiglas are used. Even then a special care has to be taken to prevent uncontrollable distortion of the magnetic field or at least to reduce its influence on experimental observations. To make quantitative measurements more predictable and to enable their unambiguous interpretation, the use of uniform external magnetic field is required. This is especially so in experiments aimed at the study of thermomagnetic convection, which is of the fluid motion caused by the variation of fluid magnetisation with temperature rather than by the non-uniformity of the applied field. The strength of thermomagnetic effects is proportional to the variation of the local fluid magnetisation $\Delta M = K\Delta T = \beta_m M \Delta T$, where thermomagnetic coefficient $K = \beta_m M$ of the fluid is defined in terms of the relative pyromagnetic coefficient β_m (which is constant to the leading order when the relative variations of the local temperature and magnetic field remain small) and the local fluid magnetisation M. For typical ferrofluids away from magnetic saturation, $\beta_m \sim 10^{-3}\,1/\mathrm{K}$ [13]. Therefore the relative variation of ferrofluid magnetisation due to thermomagnetic effects $\beta_m \Delta T$ does not exceed few per cent. At the same time, the variation of the applied magnetic field and thus of fluid magnetisation near the edges of the experimental cavity can be much stronger due to the refraction of magnetic field lines there. As seen in Figures 4.6 and 4.7, such field non-uniformity can cause ferrofluid motion and distort thermal readings even when the conditions for mechanical equilibrium of the fluid (pure conduction regime) are well satisfied away from the edges. To reduce the influence of such uncontrollable edge flows, the heat flux measurements have to be performed sufficiently far away from the boundaries of the experimental enclosure. Therefore in the setup shown in Figure 4.10, the thermocouples were installed in the central part 8 of the copper wall that was separated from the rest of the plate by the thermo-insulating ring 7. Such a design guaranteed that heat transfer characteristics were recorded for the section of the layer exposed to a well-controlled uniform (within 1%) external magnetic field not influenced by the edge effects.

4.6 Spherical Configuration

An alternative way of avoiding dealing with the edge effects is to use a spherical enclosure. While visualisation of a thermal field in this configuration becomes impossible, the unpredictable distortion of the applied magnetic field in this case can be avoided because the external uniform magnetic field penetrating a uniformly magnetised spherical ball remains so inside it [144, 254]. A further benefit of using a spherical geometry stems out from the fact that inside a spherical cavity carved in an infinite solid, a constant temperature gradient corresponding to the temperature distribution

$$T = T_0 + \left(1 + \frac{1 - \kappa/\kappa_s}{2 + \kappa/\kappa_s}\right) Az\,, \tag{4.4}$$

where T_0 is the temperature at the centre of the sphere (at $z = 0$), κ_s and κ are the thermal conductivities of the material of a solid and of the fluid in the cavity, respectively, and A is the temperature gradient far away from the cavity, establishes when a constant temperature gradient is created in a solid away from a sphere that leads to the temperature distribution in the solid given by [180]

$$T = T_0 + \left(1 + \frac{1 - \kappa/\kappa_s}{2 + \kappa/\kappa_s}\left(\frac{R}{r}\right)^3\right) Az\,, \tag{4.5}$$

where R is the radius of a spherical cavity and r is the distance from its centre. The temperature distributions in the cavities cut in Plexiglas and filled with transformer oil and ferrofluid based on transformer oil are shown in Figure 4.11. If the direction of such a gradient is antiparallel to the direction of the gravity and the cavity is filled with a homogeneous fluid, then a mechanical equilibrium or pure conduction regime establishes in the cavity with horizontal isothermal planes [97, 180, 260]. This offers a convenient reference state that could be used for further comparative studies of fluid flows caused by various mechanisms in such a geometry.

In practice, the solid block inside which the cavity is carved cannot be infinite so that its sides facing the isothermal surfaces of heat exchangers are finite distance away from the poles of the cavity. In this case the isotherms inside the cavity curve. However, as can be seen from (4.5), the deviation of the temperature distribution (which is maximum in polar regions closest to heat exchangers) inside the sphere from that given by (4.4) decays inversely proportionally to the cube of the distance from the centre of the cavity to the thermostat surfaces [180]. Moreover, if the thermal conductivities κ_s and κ

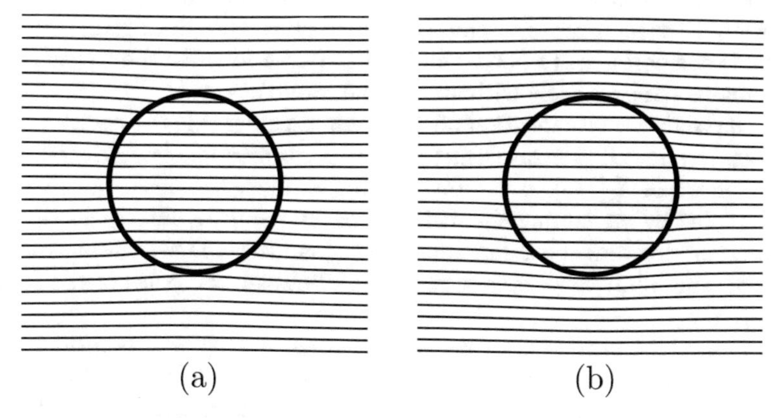

Fig. 4.11 Isotherms in a Plexiglas block with a spherical cavity filled with (a) transformer oil ($\kappa/\kappa_s = 0.8$) and (b) ferrofluid based on transformer oil ($\kappa/\kappa_s = 1.3$).

of the solid and fluid are close as, for example, when a transformer-oil-based ferrofluid fills a cavity carved inside a Plexiglas block, then such a deviation becomes negligible even if the block thickness is close to the diameter of the cavity.

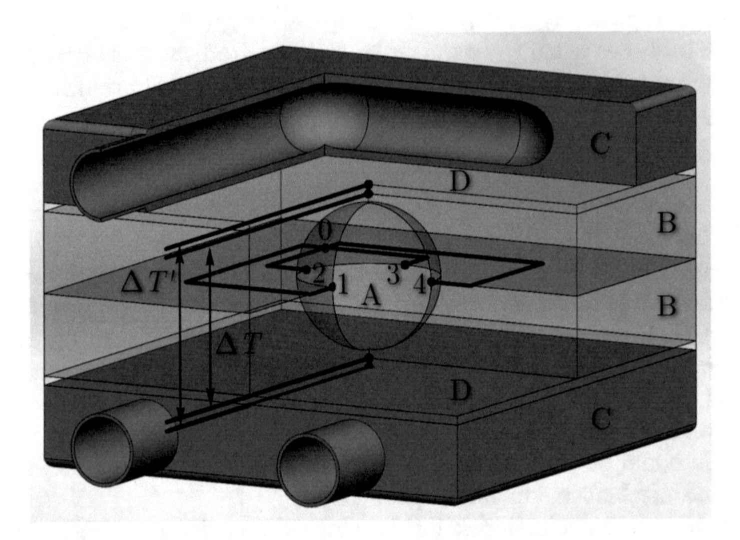

Fig. 4.12 Spherical experimental chamber: A, spherical cavity filled with a ferromagnetic nanofluid; B, Plexiglas plates; C, aluminium heat exchangers; D, Plexiglas inserts; 1, 2, 3, 4, thermocouples; ΔT and $\Delta T'$ are the temperature differences between the poles of the sphere and the heat exchangers, respectively [134].

In view of these facts, a setup schematically shown in Figure 4.12 was built and used in experiments. A spherical cavity A of diameter 16.0 ± 0.1 mm was formed by two hemispheres carved inside two Plexiglas plates B with dimensions $53 \times 53 \times 8.0\,\mathrm{mm}^3$ each glued together. The cavity was filled with a ferrofluid through pipes of the 2 mm diameter. The pipes remained open during the experiment to allow the fluid to exit and enter the cavity during a thermal expansion cycle without damaging the assembly.

Since the direct visualisation of the thermal field inside the cavity was impossible, the heat flux measurements were performed following the Schmidt-Milverton method [216]. According to this method, the onset of convection was detected by registering the variation of temperature differences $\Delta T'$ and ΔT measured across the 1 mm thick Plexiglas inserts D separating the cavity from aluminium heat exchangers C and between the poles of the sphere, respectively.

To detect the structure of convective flows in a sphere, additional four thermocouples were placed equidistantly in the equatorial plane of the cavity as shown in Figure 4.12. Each thermocouple protruded 3 mm from the wall towards the centre of a sphere and had 1 mm soldered joint. The thermocouple stems were made of 0.1 mm wires, which were sufficiently thin, not

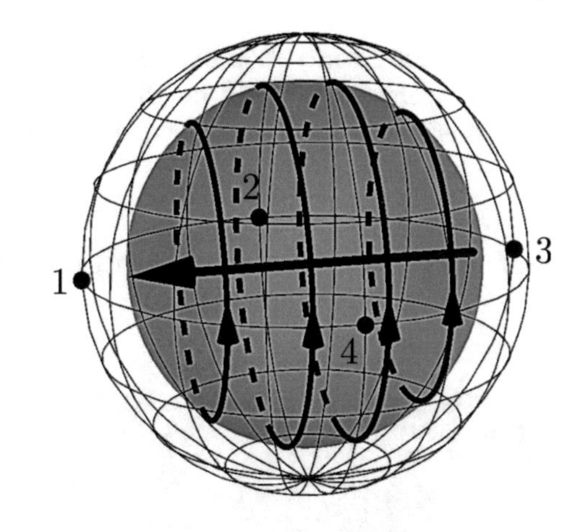

Fig. 4.13 Schematic view of a base vortex: 1, 2, 3 and 4 are thermocouples located in the equatorial plane (see Figure 4.12); the arrow represents a vector of angular velocity of the first base convection vortex [136].

to have strong influence on convective flows arising inside the non-uniformly heated cavity. The first flow instability mode in a sphere heated from below corresponds to a single convection vortex with an arbitrarily oriented horizontal axis parallel to the angular velocity vector ω [97, 180]. Such a motion can be considered as a superposition of two base vortices with the orthogonal axes containing the pairs of thermocouples 1 and 3 (see Figure 4.13) and 2 and 4 (not shown). The vortices are characterised by the angular velocity vectors ω_{I} and ω_{II} such that $|\omega|^2 = |\omega_{\mathrm{I}}|^2 + |\omega_{\mathrm{II}}|^2$. These vortices break the planar symmetry of the temperature distribution within a spherical cavity. For example, vortex ω_{I} leads to the appearance of the temperature difference detected in the equatorial plane by the diametrically opposite thermocouples 2 and 4 (see Figure 4.13). In the case of a nearly linear vertical temperature profile in the central part of the sphere, the magnitude $|\omega|$ of the angular velocity is approximately proportional to the total convective perturbation

$$\theta = \sqrt{\theta_{\mathrm{I}}^2 + \theta_{\mathrm{II}}^2}\,, \tag{4.6}$$

where the component thermal perturbations are $\theta_{\mathrm{I}} = \theta_1 - \theta_3$ and $\theta_{\mathrm{II}} = \theta_2 - \theta_4$ and θ_1, θ_2, θ_3 and θ_4 are the readings of the corresponding thermocouples relative to their common juncture. Such thermocouple signals were used to establish the existence of various convective motions that have components in the form of vortices with horizontal axes [97]. However, the thermocouple arrangement used in experiments enabled detecting and distinguishing between not only the first single-vortex convection mode but also subsequent modes, e.g. toroidal mode, if they appear [39, 97]. In the absence of magnetic field, such modes would arise if the ratio of thermal conductivities of the fluid and the block material is in the range 1–3 [97, 267]. For ferrofluids based on polymethylsiloxane and transformer oil, this ratio is 1 and 0.83, respectively; for pure transformer oil, it is 0.61. Thus, the onset of higher modes in the natural convection experiments was not expected, yet its possibility in magnetoconvection could not be ruled out a priory, which necessitated the four-thermocouple design discussed above.

4.7 Experiments in Magnetic Field

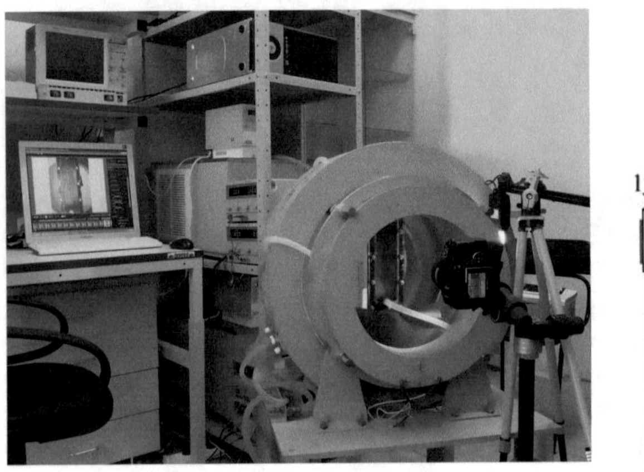

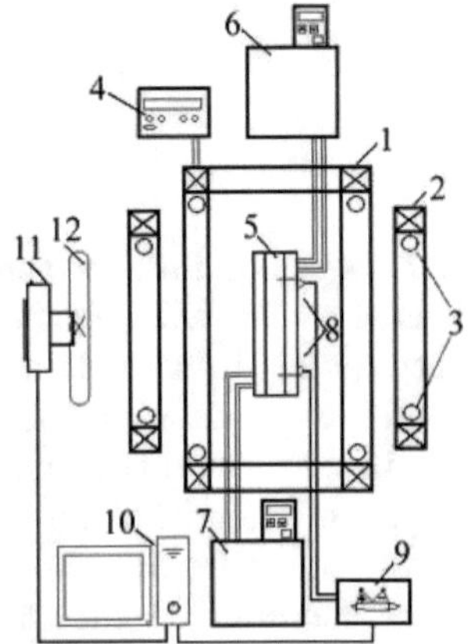

Fig. 4.14 Overall view of an experimental setup: 1, Helmholtz coils; 2, correcting coils; 3, cooling system; 4, DC power supply; 5, experimental chamber; 6 and 7, thermostats; 8, copper-constantan thermocouples; 9, thermocouple signal recorder; 10, computer; 11, photocamera; 12, fluorescent lamp [238].

The ability to create and maintain a uniform magnetic field over a sufficiently large area is crucial for conclusive experimental studies of thermomagnetic convection. To achieve this two methods were used for creating a magnetic field: Helmholtz coils and electromagnets. The advantage of using Helmholtz coils is that they offer a much more precise control of the strength of the created magnetic field and that they are capable of creating a uniform field over a much larger area than electromagnets. At the same time, they require excessively high electric currents to create field of high intensity and thus are prone to overheating and mechanical damage caused by strong electromagnetic repulsion of wiring coils. For this reason the use of electromagnets is preferred when strong magnetic field created over a relatively small area is required. Because of these factors, experiments with relatively small disk-shaped or spherical cavities shown in Figures 4.4, 4.10 and 4.12 were performed in magnetic fields created either by Helmholtz coils or by electromagnets, while large rectangular chambers such as the one depicted in Figure 4.5 had to be placed between Helmholtz coils.

Three sets of Helmholtz coils with the diameters of the working area of 200, 320 and 450 mm were used in experiments. The maximum strength of the uniform magnetic field created over the working area was 35, 20 and 35 kA/m, respectively. The choice of the geometrical parameters of the coils followed the design characteristics formulated in [148, 254].

The photograph and a schematic view of an experimental setup using Helmholtz coils 1 of the largest diameter are shown in Figure 4.14. The uniformity of the magnetic field over the complete working area of the ferrofluid layer was insured using the correcting secondary coils 2. To avoid overheating of the coils that carried the electric current of up to 5 A, the wires were cooled by cold water pumped through fluoroplastic pipes 3 woven over the coils. The direct current feeding the coils was supplied by a power source 4. The experimental chamber 5 was placed in the centre between the coils, where the uniformity of the created magnetic field could be guaranteed within 2%.

The magnetic field created by an electromagnet in experiments with smaller disk-shaped and spherical cavities could reach much higher values of up to 220 kA/m with the degree of the field non-uniformity over the working area of 80 mm in diameter not exceeding 1.5%.

4.8 A Note on Evaluating Nondimensional Governing Parameters

Prior to discussing the detailed experimental results, it is important to note that it is preferable to present quantitative experimental data in terms of nondimensional groups and governing parameters such as Rayleigh or Grashof numbers. This enables one to take advantage of the geometric and dynamic similarities of the problem and make the comparison with theoretical and computational results more straightforward. Unfortunately, in ferrofluid experiments converting the data to conventional dimensionless form is not always possible. This is because transport properties of a ferrofluid such as its viscosity may depend strongly on the experimental conditions. In particular, the local characteristics of the applied magnetic field influence magnetoviscosity of a fluid [171], and the history of a ferrofluid use determines the number and the size of solid particle aggregates forming in it, which in turn leads to an unpredictable variation of the fluid's viscosity and thermal diffusivity. The variation of the values of these coefficients in experimental conditions from the manufacturer-specified values can range from several percent to several times. Thus the accurate evaluation of transport coefficients of ferrofluid would require dedicated in situ measurements that are either impractical or technically impossible. For these reasons presenting experimental results frequently has to be done in terms of directly measurable quantities, that is, in terms of the recorded temperature differences and the magnetic field intensity, which also are the variable parameters in physical and industrial applications. The values of Rayleigh, Grashof and Prandtl numbers can then be estimated using the assumed values of fluid viscosity, but such an estimation can only be approximate.

Chapter 5
Experimental Investigation of Thermogravitational Convection in Ferrofluids

Abstract Specific features of thermogravitational instability and thermogravitational flows arising in ferrofluids are discussed. It is shown that in contrast to ordinary fluids, the characteristics of convection setting in ferrofluids depend on the history of its storage and the conditions of experiment. This is found to be due to a complex composition of ferrocolloids containing carrier fluid, solid particles, their aggregates and surfactant that make them essentially multiphase systems. The results of a comprehensive experimental study of convective heat transfer and flow patterns arising in such fluids when they are non-uniformly heated are presented. Spatially and temporally chaotic ferrofluid flows similar to those previously found in gases and binary mixtures are detected in close proximity of the convection onset. In particular, regimes are detected where convection sets and decays spontaneously drastically changing heat transfer across the domain occupied by the fluid. It is noted that the possibility of such a behavior of nanofluids must be taken into account to avoid malfunction of advanced heat exchangers make use of nanofluids as working media.

5.1 Introduction

Flows of non-isothermal fluids driven by gravitational buoyancy have been a subject of numerous investigations over many decades. Comprehensive reviews of the field can be found, for example, in [50, 95, 97–99, 132]. Here we only mention studies focussing on oscillatory and irregular flows observed in the close vicinity of the convection onset because such regimes frequently

See Appendix B for the list of previously published materials reused in this chapter with permission.

A. A. Bozhko, S. A. Suslov, *Convection in Ferro-Nanofluids: Experiments and Theory*, Advances in Mechanics and Mathematics 40,
https://doi.org/10.1007/978-3-319-94427-2_5

occur in magnetic fluids. The existence of such irregular near-onset regimes is a prominent feature of convective ferrofluid flows distinguishing them from their single-component fluid counterparts.

Irregular spatio-temporal convection in the vicinity of the convection onset in liquid and gas systems was first discovered in the beginning of the 1980s. Initially it was detected in liquid helium [3] and mercury [87]. Subsequently, shadowgraph visualisation confirmed the existence of such regimes in compressed air [65, 66]. The phenomenon has attracted attention of many researchers since; see [4, 7, 10, 16, 21, 27, 28, 50, 67, 68, 73, 76, 95, 98, 128, 129, 163, 164, 189, 207, 265] to name a few. In contrast to fully developed turbulence, flow structures observed in spatio-temporal chaos regimes preserve the value of the characteristic wavenumber.

One of the irregular structures frequently observed in gas mixtures consists of spiral rolls and defects appearing in layers with large ($\sim 10^2$) aspect ratios [10, 28, 163, 189, 207, 265]. Such a "spiral defect chaos" can develop as the degree of supercriticality increases [163] or when the layer rotates [265]. At the same time, transverse vibration of a horizontal layer has been found to suppress the formation of spiral rolls [207]. The competition between spiral defects, ideal rolls and convection cells was studied experimentally in [10, 163]. Numerical modelling using Boussinesq approximation of the governing equations revealed the existence of a strange attractor corresponding to such a spiral chaos [73].

Other types of irregular structures include large multi-arm spirals filling a complete layer as well as foci at the layer boundaries ("PanAm") and in its centre ("targets") [28, 189]. In particular, "target" chaos was observed in the vicinity of a thermodynamic critical point in sulphur fluoride when its Prandtl number changed rapidly from 2 to 115 [7].

By now irregular spatio-temporal behaviour near the onset of convection has been observed in binary mixture experiments [4, 21, 128, 129, 164], liquid crystals [76] and nanofluids [79]. In binary mixtures convection sets in the form of propagating waves consisting of rolls that chaotically break into cells and then restore. This regime is referred to as a zipper state. The amplitude of convection rolls in binary mixtures can also change irregularly in time. This is known as a blinking state [21, 129]. Spatio-temporal modulation of convection patterns in binary mixtures can also result in a very intricate state where zones with developed convection alternate with regions where convection is absent. Such convection "spots" are termed as confined [21, 128, 164] or localised states [4, 70, 71, 76]. Evolution of such states depends on the concentration of binary species [4]. In particular, localised states observed in concentrated solution of ethanol in water eventually gave way to permanent planform convection. In weak solutions convection spots had intermittent character: they would disappear completely leading to pure conduction state and then emerge again in different location. Localised convection states were also found in experiments with nanofluids containing 44 nm tetrafluoroethylene particles suspended in water [79]. After existing for about a week, they were observed to decay giving rise to pure conduction state. Thus, spatio-temporal chaos in the vicinity of convection onset can exist in the form of various irregular structures and defects. Such flow structures arise due to the density gradients caused by several competing mechanisms of a thermal and concentrational nature. In particular, convection patterns could result from double diffusion in binary mixtures. In this case, if the thermodiffusion coefficient is negative [4, 79] and a mixture is heated from below, the

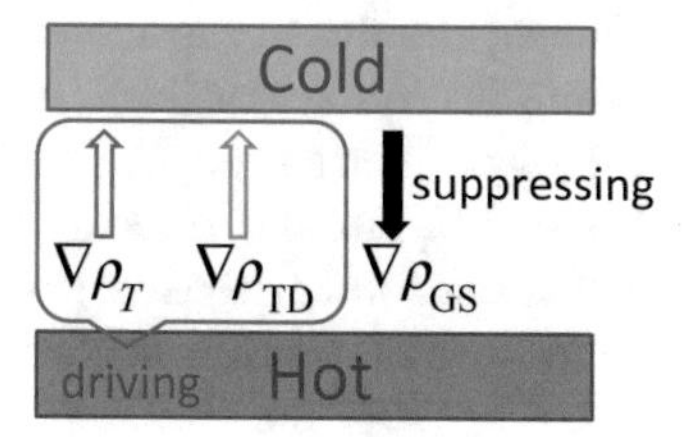

Fig. 5.1 Schematic diagram of the density gradients $\nabla\rho_T$, $\nabla\rho_{\mathrm{TD}}$ and $\nabla\rho_{\mathrm{GS}}$ caused by thermal expansion, thermal diffusion and gravitational sedimentation, respectively, arising in a ferromagnetic nanofluid heated from below [136].

concentrational density gradient $\nabla\rho_{TD}$ is directed downwards and has a stabilising influence, while thermal expansion of fluid leads to the occurrence of the destabilising upward density gradient $\nabla\rho_T$. The onset of convection is then dictated by the relative strength of these two opposing effects. In contrast to binary fluids, the thermodiffusion coefficient in magnetic colloids is positive[1], and both density gradients due to thermal expansion and due to thermodiffusion have a destabilising effect; see Figure 5.1. However, in magnetic fluids these two upward density gradients are opposed by a stabilising negative density gradient $\nabla\rho_{\mathrm{GS}}$ arising due to gravitational sedimentation of solid particles and their aggregates [84, 100]. Moreover, ferrofluids based on organic carrier fluids (kerosene, transformer oil) contain molecules of different masses and sizes as well as molecules of a surfactant (up to 10% by volume). Such different molecules are also subject to thermodiffusive redistribution and can lead to liquid phase stratification. In addition, organic carrier liquids forming the base of ferrofluids can contain contaminants and chemical species that also form insoluble sediment that leads to the fluid density stratification. Fluid mixing caused by fluid motion on the other hand tends to destroy such density gradients resulting in irregular intermittent convection in magnetic nanofluids [38, 102, 136, 200].

A large number of various de-homogenising mechanisms present in realistic non-isothermal ferro-nanofluids and complexity of their analytical description makes it virtually impossible to model them accurately and consistently at

[1] The intensity of thermodiffusion is proportional to the Soret coefficient $S_t = D_T/D$, where D_T and D are the coefficients of thermodiffusion and Brownian diffusion, respectively. It is positive for magnetic fluids in the absence of magnetic field and has the value of approximately $0.1\,\mathrm{K}^{-1}$ [23, 75, 162], which is several orders of magnitude larger than for binary mixtures.

present. For this reason the many theoretical studies of convection in ferrofluids opted to neglect the non-homogeneity of the fluid treating it as a hypothetical single-component medium [12, 88, 89, 209, 222, e.g.]. A standard justification given to such an approach is based on estimating the characteristic time $t_D = h^2/D$ over which substantial changes of fluid density due to diffusion of species occur. For typical experimental and industrial setups, this time is of order of a week. However, this time corresponds to the variation of concentration across a fluid layer of thickness h by a factor of e, which is much larger than that required to initiate macroscopic convection. Indeed, the density gradients due to the thermal expansion and the variation of species concentration C are $\beta\nabla T$ and $\beta_c\nabla C$, respectively, where the typical values of thermal expansion coefficient $\beta \sim 10^{-3}\,\mathrm{K}^{-1}$, temperature gradient $\nabla T \sim 1\,\mathrm{K/cm}$ and solutal expansion coefficient $\beta_c = \frac{1}{\rho}\frac{\partial\rho}{\partial C} \sim 1$ [23, 100, 212]. Thus, the solutal density gradient comparable with thermal gradient required to induce convection occurs when the concentration gradient is of the order of $10^{-4} - 10^{-3}\,\mathrm{cm}^{-1}$. Therefore, the occurrence of solutal convection in magnetic colloids cannot always be ruled out. Yet its numerical modelling remains virtually impossible because it requires the knowledge of various transport properties such as the thermodiffusion coefficient and the rotational viscosity. Their values for practical ferrofluids with moderate and large concentrations of a solid phase are not always known because they depend on the fluid storage and use conditions and can vary even for the same sample of fluid. Thus, a direct evaluation of the governing nondimensional parameters corresponding to experimental conditions remains impossible, which limits the opportunity of a meaningful comparison of experimental and theoretical results. Since accounting of all acting transport mechanisms by a single theoretical model is not always possible, experimental investigation of flow phenomena in non-isothermal magnetic colloids remains an important investigation avenue of their prospective application that will be discussed in detail next.

5.2 Horizontal Layer

5.2.1 Temporal Behaviour of Convection Flows

The main distinction of behaviour of colloidal fluids from their single-component counterparts contained in a layer bounded by horizontal plates is that colloid remains an “active medium” even if it is isothermal and remains at rest at a macroscopic scale. Microscopically, the continuous action of gravity on the contained solid particles and their aggregates leads to their barometric redistribution, which breaks fluid’s homogeneity. If the fluid is kept in non-isothermal conditions, the effects of gravitational sedimentation can be enhanced by thermodiffusion. The overall degree of inhomogeneity depends on the time fluid remains at macroscopic rest. Therefore, the behaviour of a colloid becomes dependent on a pre-history of experiment. We discuss this feature of ferrofluid behaviour in detail below.

Experiments have been conducted in flat cylindrical cavities of diameter $d = 75$ mm and thicknesses $d = 2.0$, 3.5 and 5.0 mm; see Figures 4.4 and 4.10. The solid-phase concentration of experimental fluids varied between $\phi = 4$ and 12%. The properties of ferrofluids[2] are listed in Table 5.1.

In a thin layer ($d = 2$ mm), stationary convection sets as a result of supercritical bifurcation regardless of concentration once the critical temperature difference ΔT_c between the bottom and top walls of the experimental chamber is reached. Figure 5.2 shows the values of Nusselt number Nu as the function of the relative temperature difference $\Delta T/\Delta T_c$ for fluids with various solid-

Table 5.1 Properties of ferrofluids and the corresponding convection threshold temperatures.

ϕ (%)	ρ (10^3 kg/m^3)	M_s (kA/m)	ΔT_c (K) $d = 2.0$ mm	ΔT_c (K) $d = 3.5$ mm	ΔT_c (K) $d = 5.0$ mm
12	1.55	55	25.0	5.1	2.5
8	1.26	37	7.5	—	—
4	0.98	20	4.5	1.5	0.63

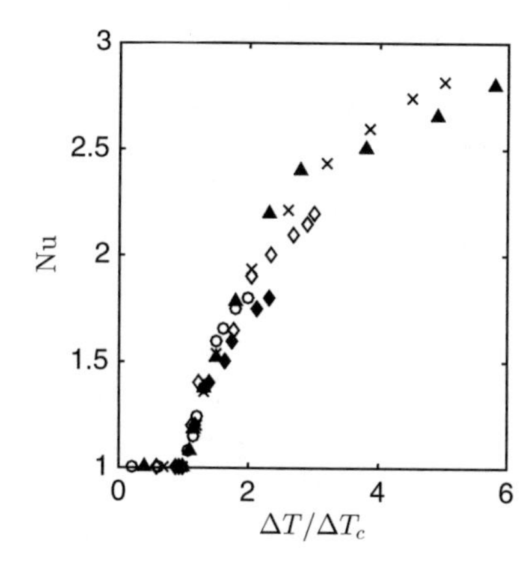

Fig. 5.2 Nondimensional heat transfer in a thin ($d = 2.0$ mm) horizontal ferrofluid layer with solid phase concentration $\phi = 4\%$ (triangles), $\phi = 8\%$ (crosses) and $\phi = 12\%$ (circles) as a function of relative temperature difference. The data for single-component fluids (water (filled diamonds) and helium (empty diamonds)) [79] are also given for comparison.

phase concentrations. In experiments, the maximum temperature difference that can be created across the layer is limited by the heater-cooler specifications and cannot be changed, while the critical temperature for the onset of convection increases with solid-phase concentration ϕ due to the increase of fluid viscosity. Therefore, the experimental values of the maximum degree of supercriticality $\Delta T/\Delta T_c > 1$ achievable in experiments decrease with ϕ from approximately 6 for $\phi = 4\%$ to 5 for $\phi = 8\%$ to 2 for $\phi = 12\%$. The experimental data in Figure 5.2 shows that for $\Delta T/\Delta T_c < 3$ the nondimensional heat flux does not depend on the concentration of solid phase and is similar to that measured in single-component fluids such as water and helium [79].

[2] Synthesised in Ferrofluid Manufacturing Laboratory "Polus", Ivanovo, Russian Federation.

In thicker layers ($d = 3.5$ and $5.0\,\mathrm{mm}$), the onset of convection in ferrofluids has a qualitatively different character to that in single-component fluids. It occurs subcritically and is characterised by a hysteresis; see Figures 5.3 and 5.4. Convection in the initially resting fluid sets and acquires a finite amplitude in an abrupt way as the applied temperature difference ΔT is gradually increased. The critical value of the temperature difference at such an onset depends on pre-history of experiment. In contrast, when the applied temperature difference is gradually decreased, transition to a quiescent state occurs smoothly at a well-defined value of $\Delta T = \Delta T_c$ that does not depend on the history of experiment. The depth of hysteresis, i.e. the distance between

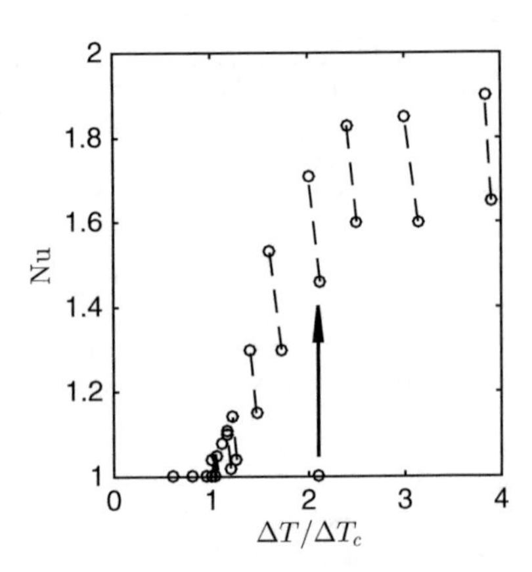

Fig. 5.3 Nondimensional heat transfer in an intermediate ($d = 3.5\,\mathrm{mm}$) horizontal ferrofluid layer with solid phase concentration $\phi = 12\%$ as a function of relative temperature difference. The dashed lines connect the minimum and maximum values of the Nusselt number detected in oscillatory convection regimes. The vertical arrows show hysteresis transitions when the temperature difference applied to the initially quiescent fluid is gradually increased.

the temperature difference values corresponding to the decay of convection in a well-mixed fluid and the convection onset in an initially unmixed fluid, was detected to reach $3\Delta T_c$.

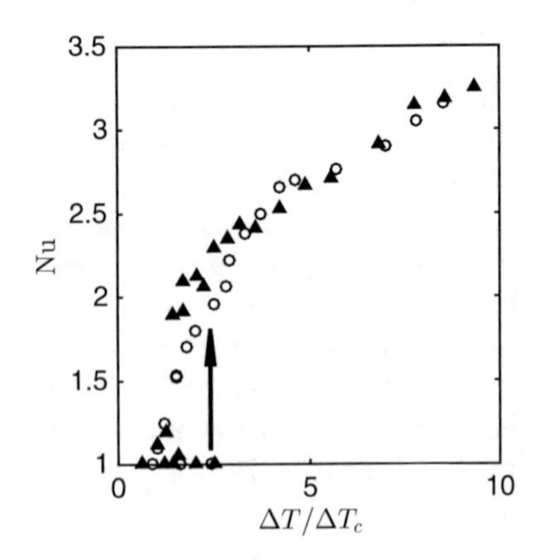

Fig. 5.4 Nondimensional heat transfer in a thick ($d = 5.0\,\mathrm{mm}$) horizontal layer of diluted ($\phi = 4$ % (triangles)) and concentrated ($\phi = 12\%$ (circles)) ferrofluids as a function of relative temperature difference. The vertical arrow shows hysteresis transition when the temperature difference applied to the initially quiescent fluid is gradually increased.

The other important feature of convective flows arising in ferrofluids is that they have oscillatory rather than stationary nature. Registering such flows experimentally faces additional challenges, and both the type, sensitivity and location of sensors must be chosen very carefully. For example, when the local sensor was placed in the centre of the 3.5 mm thick cavity depicted in Figure 4.4, a strong temporal variation of heat flux of up to 25% of the average values was recorded for the complete supercritical experimental range of ΔT as seen from Figure 5.3.

In contrast, the readings of an integral sensor in a 5.0 mm cavity shown in Figure 4.10 appear as noisy data with a well-defined average trend; see Figure 5.4. The depth of hysteresis in this configuration could reach the value of up to $2\Delta T_c$.

The above experimental observations prompt the following conclusion on the physical processes taking place in a horizontal layer of ferrofluid heated from below. The temperature difference applied across a thin layer of fluid results in a large temperature gradient and, thus, stronger thermophoresis of solid particles that lead to the formation of solid-phase concentration gradient [23, 75, 252]. This leads to the establishment of a destabilising density gradient codirected with the density gradient due to the thermal expansion; see Figure 5.1. The combined action of thermophoresis and thermal expansion in thin fluid layers [115, 212] overcomes the effect of gravitational sedimentation. As a result convection sets in thin ferrofluid layers in a qualitatively the same way as in single-component fluids: supercritical transition leads to stationary convection.

In thicker layers the onset of convection occurs at smaller values of the applied temperature difference (as follows from the definition of the critical Rayleigh number $\Delta T_c \sim d^{-3}$), see Table 5.1, corresponding to smaller cross-layer temperature gradients. Thus the role of thermophoresis decreases, while the relative contribution of a stabilising sedimentation effect (see Figure 5.1) increases. The presence of transport mechanisms of comparable strength but of opposite action in thicker layers is the main physical reason for the qualitative change in the character of convection.

5.2.2 Spatial Patterns

The experimental chamber shown in Figure 4.4 was used for visualisation of flows arising in pure transformer oil and kerosene-based ferrofluids with concentration of solid phase $\phi = 4$ and 12%; see Table 5.1. At these concentration values, ferrofluid is characterised by Prandtl number that is approximately equal to that of a pure transformer oil.

Convection patterns observed in a layer of transformer oil consisted of stationary concentric ring rolls shown in Figure 5.5(a). They remained unchanged for tens of hours. The patterns observed in transformer oil are fully consistent with those found in silicon oil in [168], which emphasises their generic nature.

Despite a similar value of Prandtl number, spatio-temporal convection patterns observed in a horizontal layer of ferrocolloid heated from below differ drastically from those observed in ordinary fluids. Their typical evolution is shown in Figure 5.6. As time progresses, the convection rolls seen in the south-east sector of image (a) break into cells due to cross-roll instability and then recombine into bent rolls seen in image (b). The temperature time series with a sampling interval of 5 s recorded for the experiment illustrated in Figure 5.6 is shown in Figure 5.7(a). The corresponding wavelet spectrum [130, 249] shown in Figure 5.7(b) reveals that the dominant period τ of temperature

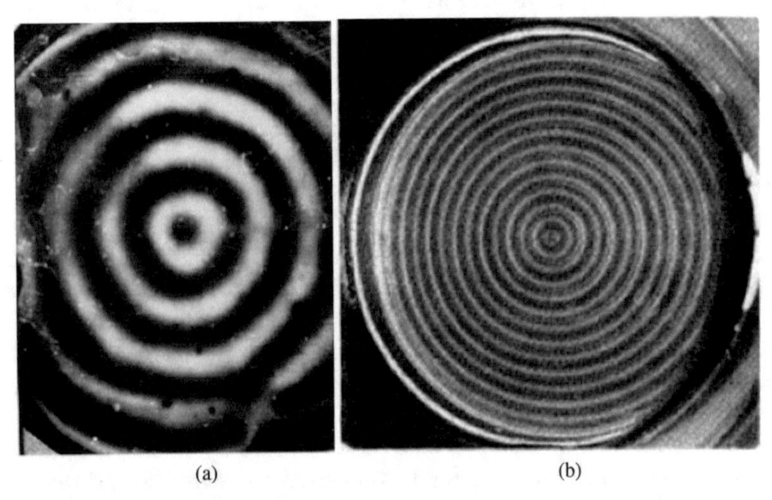

(a) (b)

Fig. 5.5 Stationary convection rolls in (a) transformer oil at $\Delta T/\Delta T_c = 1.7$ and (b) silicon oil [168].

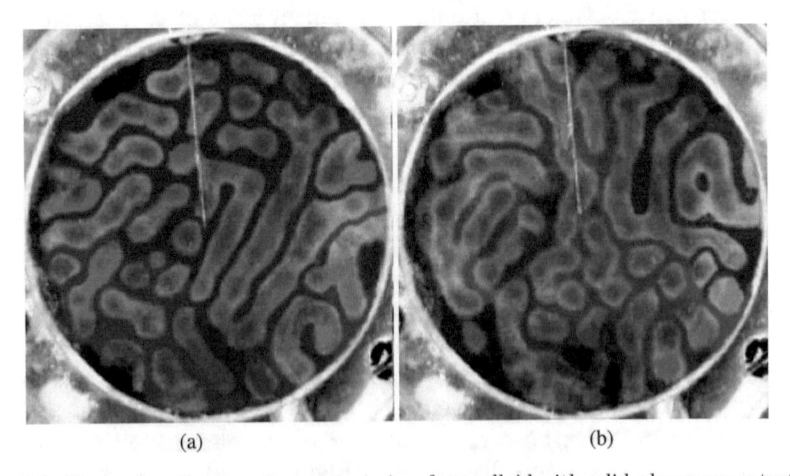

(a) (b)

Fig. 5.6 Unsteady roll convection patterns in a ferrocolloid with solid-phase concentration of $\phi = 12\%$ at $\Delta T = 2.1\Delta T_c$. The time interval between the snapshots (a) and (b) is 30 min.

oscillations (dark regions in the figure) did not remain constant and during different stages of the experiment, various long- (several hours) and short- (up to 40 min) period oscillations were observed. The Fourier spectrum A in Figure 5.7(c) also indicates that several oscillatory modes with periods $\tau = 1/f$ from about 6 h to several minutes persisted throughout the experiment. Fourier spectrum also indicates the existence of bound modes with frequencies differing by an integer factor appearing due to the nonlinearity of the system.

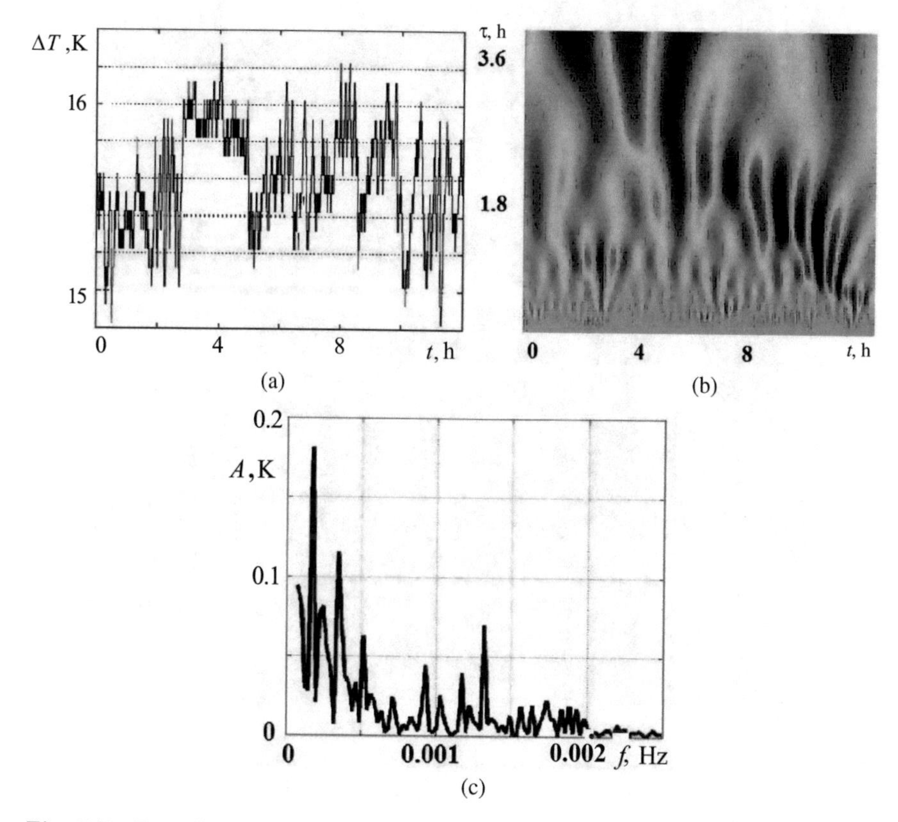

Fig. 5.7 Typical temperature oscillations in ferrofluid convection in a horizontal layer heated from below at $\Delta T = 2.1\Delta T_c$ ($\phi = 12\%$): (a) time series, (b) wavelet and (c) Fourier transforms of the recorded signal [31].

One of the prominent specific features of ferrofluid convection is the spontaneous and irregular appearance of local flow structures in the form of spiral and concentric rolls ("targets") seen in Figure 5.8. Such regimes were observed for the same values of the control parameters as those in Figure 5.6 where straight and bent rolls were detected. In the lower half of Figure 5.8, (a) spiral rolls are seen, while a target pattern is seen in the upper half of the image. As time progresses the spiral disintegrates, and the target pattern is replaced

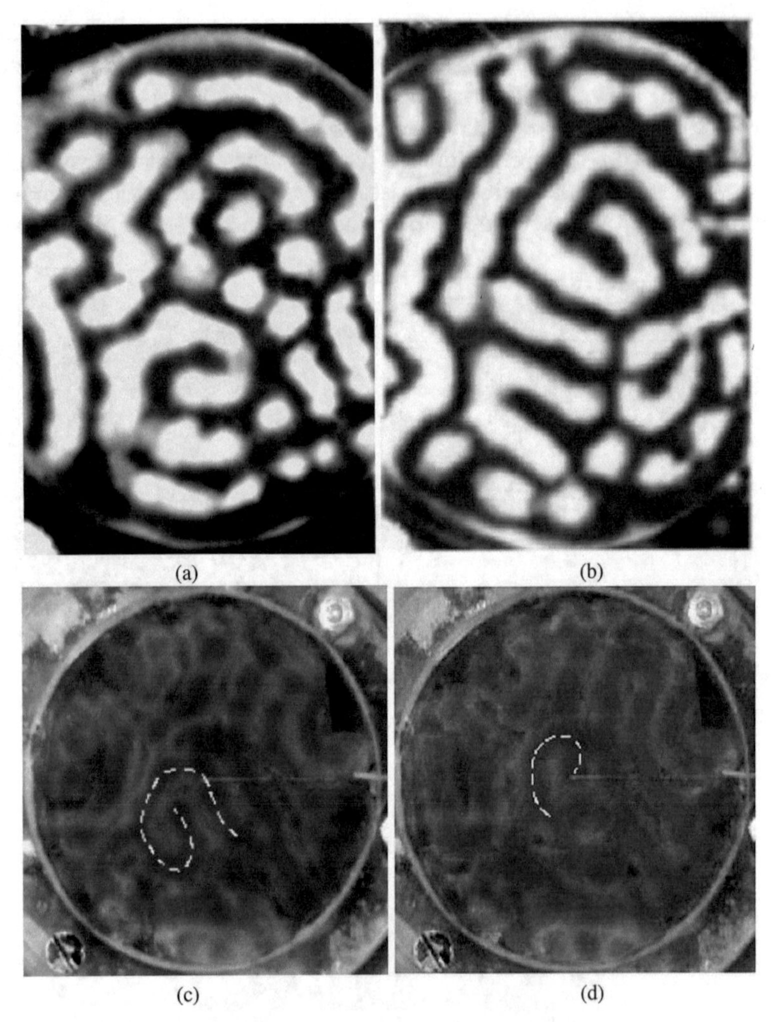

Fig. 5.8 Spiral defect chaos in a horizontal layer of ferrofluid heated from below: (a) and (b) $\phi = 12\%$, $\Delta T = 1.5\Delta T_c$; (c) and (d) $\phi = 4\%$, $\Delta T = 2.7\Delta T_c$. The time interval between the snapshots is 30 min.

by a spiral of the opposite orientation; see Figure 5.8(b). A spontaneous appearance of spiral defects known as spiral defect chaos in low-concentration ferrofluid is demonstrated in Figure 5.8(c) and (d): the spiral of one orientation in image (c) gives way to a shorter spiral of the opposite orientation in image (d). It is of interest to note that previously such spiral defects have only been detected in convection of low-Prandtl-number fluids [99, 189].

Large-scale one- and two-arm spirals occupying the complete experimental chamber occur when the applied temperature difference changes suddenly from a sub- to supercritical value. Such large-scale patterns are found to be structurally unstable: they exist for about an hour and then break into shorter rolls and cells. The evolution of single-arm spiral that appeared after a stepwise increase of the applied temperature difference is illustrated in Figure 5.9. A large spiral with an anticlockwise orientation seen in image (a) breaks down into cells starting from the centre in image (b). Subsequently, these cells self-organise into another spiral with a clockwise orientation seen in image (c). Similar unstable spirals have been observed in gas layers heated in a stepwise manner [189] or when the edges of the layer were subject to additional heating [163].

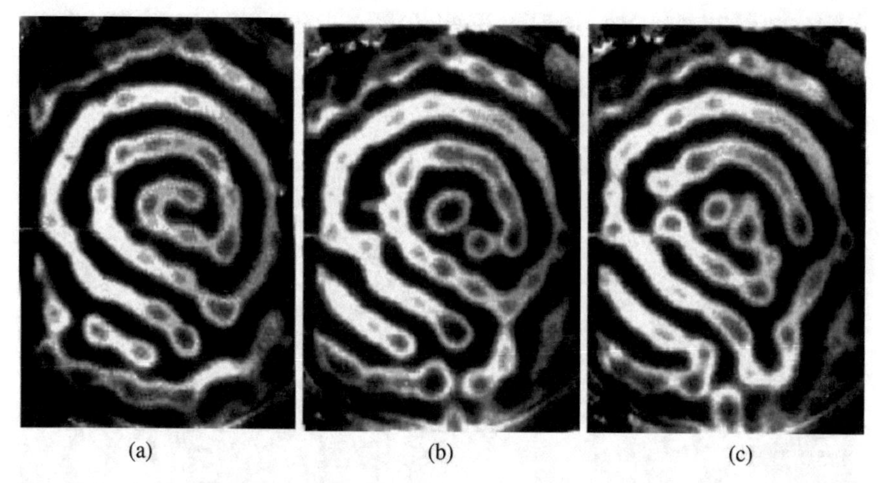

Fig. 5.9 Evolution of a large-scale spiral structure appearing in a horizontal layer of ferrofluid ($\phi = 12\%$) heated from below after a sudden increase of the applied temperature difference from $\Delta T < \Delta T_c$ to $\Delta T = 4\Delta T_c$. Snapshots (a)–(c) are taken 3, 18 and 23 min after the stepwise increase of heating, respectively.

When the supercritical temperature difference applied to concentrated ferrofluid increases gradually, convection patterns assume a different shape of rolls shown in Figure 5.10. Such rolls are observed for all values of the supercritical temperature difference and convection remains unsteady. The characteristic time of the reorganisation of convection structures is of the order of tens of minutes. The typical scenario is illustrated in Figure 5.10 for $\Delta T = 4\Delta T_c$. The straight and horseshoe-shaped convection rolls seen in image (a) start breaking into small cells in the central part of the experimental chamber: see image (b). Subsequently, a new roll system emerges with axes rotated by approximately 60° with respect to the original rolls (see image (c)) and the process repeats.

In a diluted ferrofluid, convection has yet another character. Initial small-scale patterns illustrated in Figure 5.11(a) undergo coarsening with individual small convection cells coalescing into rolls (see Figure 5.11(b)) eventually resulting in large stationary rolls shown in Figure 5.11(c).

The experimental results discussed in this chapter demonstrate that convection occurring in ferrocolloids differs drastically from that in single-component homogeneous fluids. The observed flow patterns depend on the concentration of solid phase and generally are unsteady with chaotic spatio-temporal evolution. The transition between various spatial patterns is driven by the cross-roll instability of convection rolls that leads to their breakup into smaller cells. Such cells subsequently recombine to give rise to a different coarser patterns. As time progresses, these patterns become unstable again, and the process repeats in a temporally irregular and spatially chaotic way. Such a behaviour is known as the zipper state [21, 129]. In concentrated ferrofluids it was observed in all investigated experimental regimes up to $\Delta T = 4\Delta T_c$. In contrast, the behaviour of diluted ferrofluids approached that of single-component fluid at large $\Delta T > 3.7\Delta T_c$: the appearing large

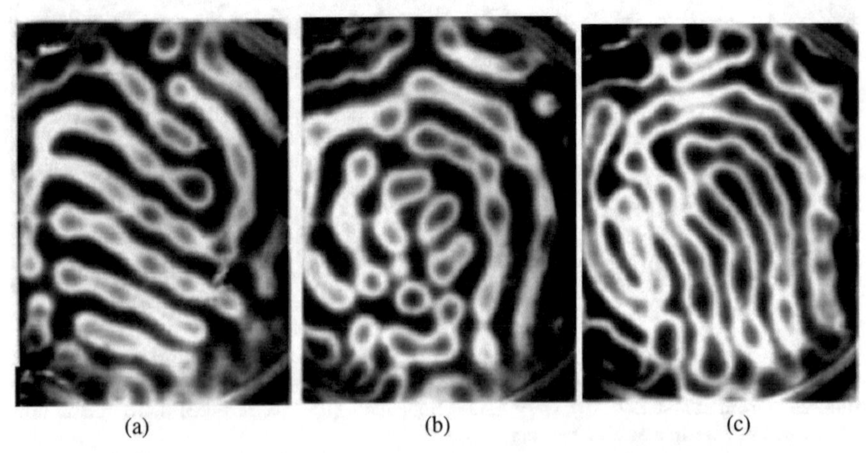

Fig. 5.10 Instability of convection rolls at $\Delta T = 4\Delta T_c$ in concentrated ferrofluid (ϕ = 12%). The time interval between snapshots is 20 min.

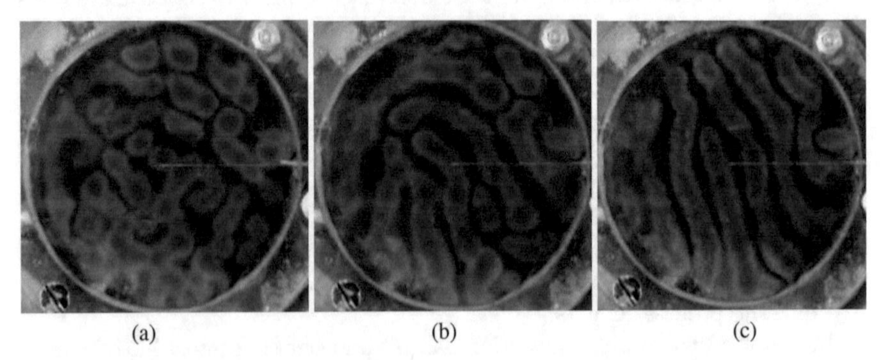

Fig. 5.11 Transition to steady convection in a diluted ferrofluid (ϕ = 4%) at $\Delta T = 3.7\Delta T_c$. The time interval between snapshots is 1 h.

convection rolls stabilised with time, and convection became steady. Such a behavioural difference is due to a stronger mixing in diluted ferrofluids that are less viscous and are characterised by a more intensive convection flows in similar thermal conditions.

5.3 Vertical Layer: The Influence of Sedimentation

The main focus of the experimental investigation reported in this book is on the influence of magnetic field on the arising convection patterns. In order to maintain such a focus in experiments conducted in the normal gravity conditions, a special care has been taken to avoid the onset of thermogravitational instability that can obscure thermomagnetic effects. For this reason the used experimental setup described in Chapter 4 has been designed to correspond to sufficiently small gravitational Grashof numbers (see Section 2.2) below the critical values for the onset of thermogravitational waves. A parallel flow with fluid rising along the heated wall and descending along the cold one is expected to exist in the absence of a magnetic field if the fluid is homogeneous, which is the case when ferrofluid is well mixed. However, if it is left at rest for a considerable time, the gravitational sedimentation of a solid phase will necessarily result in the fluid density stratification that can have a strong effect on the flow at the initial stages of experiment.

This sedimentation process can be modelled by a standard nondimensional advection-diffusion equation

$$\frac{\partial C}{\partial t} = V\frac{\partial C}{\partial z} + \frac{\partial^2 C}{\partial z^2}, \tag{5.1}$$

where the vertical coordinate z and time t are scaled using the vertical extent of the layer d and diffusion time d^2/D, respectively. Here $V = vd/D$ is the nondimensional separation parameter, and v is Stokes speed of sedimentation. Assuming that initially the fluid is homogeneous and taking into account that solid particles cannot penetrate through the enclosure walls, Equation (5.1) is solved subject to the following nondimensional initial and boundary conditions

$$C(z,0) = 1\,,\; 0 \le z \le 1\,, \quad CV + \frac{\partial C}{\partial z} = 0 \text{ at } z = 0, 1\,. \tag{5.2}$$

A standard method of separation of variables results in

$$\begin{aligned} C &= \frac{Ve^{-Vz}}{1-e^{-V}} + 16\pi^2 V^2 \\ &\times \sum_{n=1}^{\infty} n \frac{1-(-1)^n e^{\frac{V}{2}}}{(V^2+4\pi^2 n^2)^2} e^{-\frac{V^2+4\pi^2 n^2}{4}t - \frac{V}{2}z} \left(\sin(n\pi z) - \frac{2\pi n}{V}\cos(n\pi z)\right). \end{aligned} \tag{5.3}$$

The first term in Equation (5.3) is a well-known steady equilibrium barometric distribution of solid particles. For small values of the separation parameter V, this distribution approaches linear $C/C_0 = 1+(1/2-z)V$. The first term in the sum represents the slowest decaying mode that gives the estimate of the characteristic time required to reach this equilibrium in the

$$t^* = \frac{4}{V^2+4\pi^2}\frac{d^2}{D} \approx \frac{d^2}{\pi^2 D}. \tag{5.4}$$

Regimes of full and partial stratification are distinguished. The corresponding typical solid-phase concentration profiles are illustrated in Figure 5.12.

5.3.1 Fully Stratified Fluid

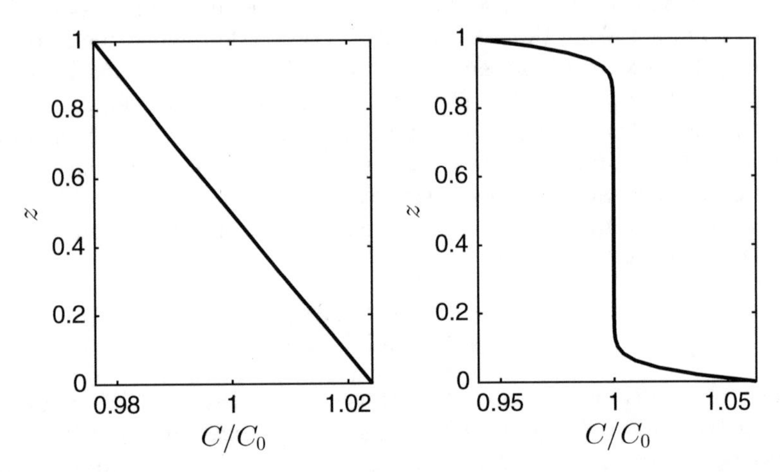

Fig. 5.12 Distribution of the relative concentration C/C_0 of solid particles in regimes of full (left) and partial (right) stratification.

The characteristic time t^* given by (5.4) for establishing the equilibrium stratification depends quadratically on the vertical extent d of a fluid layer. Thus, the fastest way of achieving it is by leaving the fluid layer at rest laying horizontally on its wide face. For the experimental cavity of thickness $d =$ 6 mm the values of V and t^* then are estimated as 0.048 and 54 h, respectively, which are computed using the independently measured diffusion coefficient $D \approx 1.9\times10^{-11}$ m^2/s and Stokes speed of sedimentation $v \approx 1.52\times10^{-10}$ m/s [101]. This time is sufficient for the concentration and thus density profiles across the 6 mm thick layer to approach a linear distribution shown in the left panel in Figure 5.12 that corresponds to the maximum separation of 4.8% in concentration between the fluid layers located near the top and bottom walls.

At the start of the main experiment just before the heating of one of the cavity walls started, the layer was rotated to its upright position so that the layer of the fluid with a larger density quickly slid towards the bottom edge of the cavity and displaced the less concentrated fluid near the opposite wall upwards creating the vertical density stratification. The flow patterns in such a vertically stratified layer of ferrofluid differ drastically from their counterparts seen in homogeneous fluids. Namely, the flow domain breaks into several cells with sharp horizontal boundaries as seen in Figure 5.13. The flows within each cell are qualitatively similar to those existing in a cavity filled with a uniform fluid. Namely, in a vertical layer of a homogeneous fluid heated from one side, a local thermal expansion leads to the appearance of the upward buoyancy force that drives heated fluid up displacing cool fluid down along the opposite wall. As a result a parallel up-down flow establishes

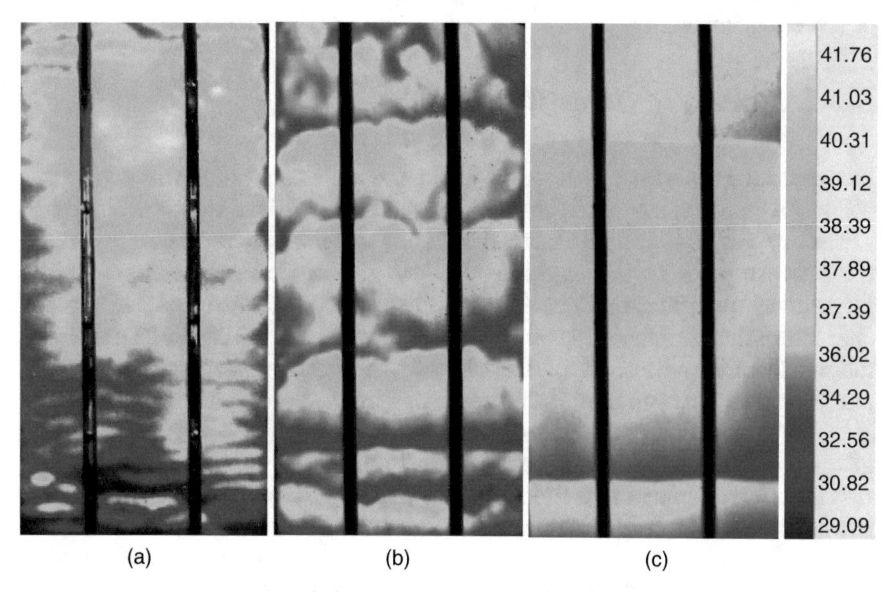

Fig. 5.13 Infrared images of convection structures in a fully stratified rectangular vertical ferrofluid ($M_s = 43\,\mathrm{kA/m}$) layer with $d = 6\,\mathrm{mm}$, $l = 375\,\mathrm{mm}$ at $\Delta T = 28\,\mathrm{K}$ in the absence of a magnetic field. Fluid layer laid horizontally on its wide face at $\Delta T = 0\,\mathrm{K}$ for a month prior to the start of experiment. The time interval between images (a) and (b) is 7 min and between images (b) and (c) is 27 min [41]. The colours correspond to the temperature measured in degrees Celsius. The vertical black lines are the shadows of the externally attached ribs used to prevent the observation wall from bulging under magnetic pressure (see Figure 3.4 and discussion in Section 3.5.5).

sufficiently far away from the edges of the layer and remains stable until the critical value of the governing Grashof number is reached. In contrast, in a vertically stratified fluid, the thermal buoyancy force competes with the gravitational pull preventing denser fluid near the lower edge of the layer from rising all the way to the top edge. This leads to an instability of the expected parallel shear flow that exhibits itself in the form of a large number of horizontal rolls with the wavelength of the order of the layer thickness (6–12 mm). This is seen in Figure 5.13(a) taken 10 min after the layer was rotated from its initial horizontal position. As time progressed, the number of the observed horizontal rolls decreased indicating the flow regularisation due to fluid mixing. This is seen in Figure 5.13(b): 17 min after the start of the experiment, only six horizontal rolls were still observed even though their boundaries remained irregular. After 27 min the formation of larger cells with sharp horizontal boundaries was observed; see Figure 5.13(c). Such cellular structures are very similar to those observed earlier in vertically stratified layers of saline solutions heated from the side [59, 62, 146, 183]. However, convective fluid mixing erodes the individual cells and a unicellular flow typical for homogeneous fluids eventually established about 6.7 h after the start of the experiment.

5.3.2 Partially Stratified Fluid

In a separate experiment, the same enclosure was used, but it was rotated by 90° in its own vertical plane so that the effective width of the layer increased at the expense of its height. Instead of laying horizontally, it was kept vertical for 4 weeks prior to the experiment. For this configuration the separation parameter appearing in Equation (5.1) became $V = vw/d \approx 1.44$, and the characteristic equilibrium time increased to $t^* \approx 4.6 \times 10^4$ h. Therefore, as follows from Equation (5.3), in this case the vertical distribution of solid particles was far from equilibrium and was only non-uniform in two regions near the top and bottom edges of the enclosure each occupying about 12% of the total height; see the right panel in Figure 5.12.

The infrared image of a fluid layer heated from the back and cooled from the facing side is shown in Figure 5.14. Since the fluid layer remained vertical,

Fig. 5.14 Infrared image of the primary thermogravitational flow in a partially stratified rectangular vertical ferrofluid layer. The temperature difference between the heated and cooled walls of the layer is 23°C. The fluid layer is 6 mm thick and 180 mm high.

no initial gravity flow caused by a sudden rotation of the density-stratified fluid was generated. As a result the multiple roll instability described in Section 5.3.1 is not visible in Figure 5.14. However, the two density-stratified regions near the top and bottom edges of the layer are clearly identifiable in the photograph.

5.4 Inclined Layer

Convection in an inclined layer of a single-component fluid bounded by two parallel plates maintained at constant different temperatures is a classical problem well-studied both theoretically and experimentally; see [52, 70, 98, 109, 140, 220] and references therein to name a few. When the plates are not horizontal, the motionless basic state is replaced with a parallel flow where

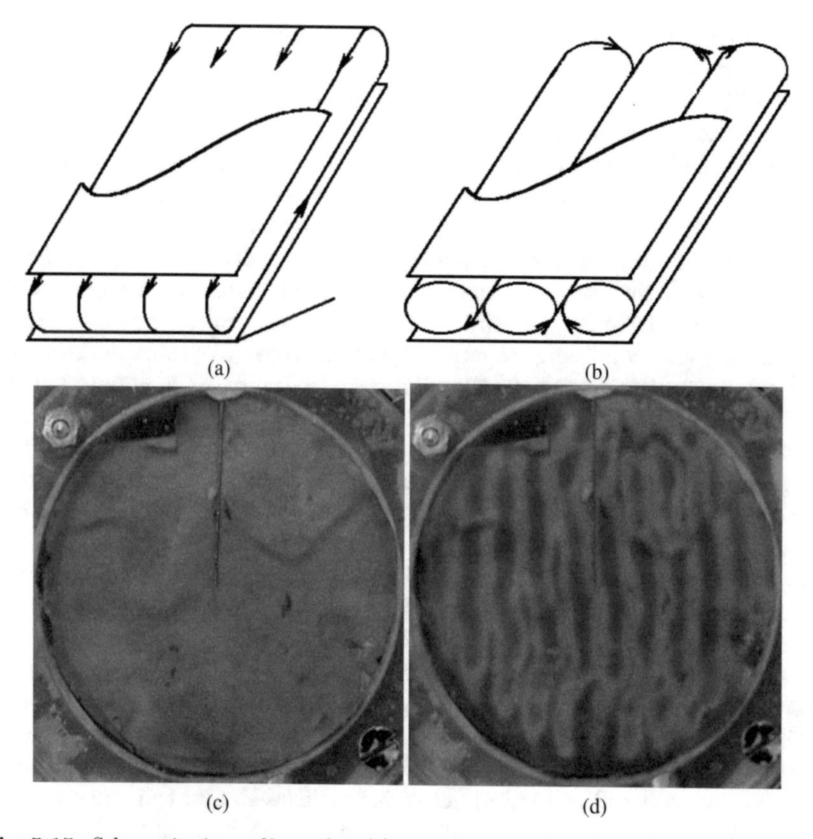

Fig. 5.15 Schematic views of basic flow (a) and convection rolls (b) and the corresponding visualisations (c and d) using thermosensitive liquid crystal film. The photographs show temperature distributions at the cooled upper plate bounding the 3.5 mm thick layer of fluid inclined at 10° with respect to the horizontal. Images (c) and (d) correspond to the applied temperature differences ΔT between the plates of 8 and 16 K, respectively.

fluid rises along a heated wall and descends along the cooled one as shown schematically in Figure 5.15(a). If the lower plate has a higher temperature, a buoyancy-driven instability occurs once the temperature difference between the plates reaches the critical value that depends on the layer inclination angle [98, e.g.]. This instability leads to the formation of convection rolls aligned with the basic flow as shown in Figure 5.15(b) (longitudinal rolls). The photographs of the corresponding flow visualisations using thermosensitive film are shown in Figure 5.15(c) and (d).

The visualisation of such flows has been performed in the disk-shaped experimental chamber shown in Figure 4.4, which may appear somewhat unnatural. The reason for choosing such a geometry is that the same experimental chamber had to be used in experiments in magnetic field. While a rectangular chamber might be better suited for natural convection studies, the presence of corners would lead to a strong and unpredictable distortion of the applied magnetic field in magnetoconvection experiments. Thus, to avoid the ambiguity of results obtained in a magnetic field and to enable a straightforward comparison between natural and magnetoconvection observation, a round chamber was used.

Stability diagrams for various convection regimes observed in an inclined layer of a concentrated ferrofluid ($\phi = 12\%$; see Table 5.1) heated from below are presented in Figures 5.16(a), 5.17(a) and 5.20(a). They represent the same experimental parametric space but are given separately for different convection patterns to avoid overcrowding the plot. Separation of the diagrams by the depicted regimes also enables us to clearly identify the observed distinct regimes even if they occur at the same parametric values (e.g. when the observed regime depends not only on the parametric values but also on the pre-history and conditions of experiment). In these diagrams ΔT_c denotes the reference critical temperature difference for the onset of convection in a horizontal layer heated from below (0° angle). The filled circles in Figure 5.16 show the convection thresholds at the selected layer inclination angles. It increases monotonically as previously shown in [97]. The undisturbed basic parallel flow exists below them; see Figure 5.15(a) and (c). Above them, convection rolls and cells superpose onto the basic flow; see Figure 5.15(b) and (d). In contrast to convection in single-component fluids [98], convection in a concentrated ferrofluid was found to remain irregular in both space and time for all experimental conditions. The vertical bars in Figure 5.16 represent the amplitude of the temperature difference oscillations recorded at the onset of convection. In addition and in contrast to the predictions of linear stability theory for single-component fluids [98], at small layer inclination angles (empty circles in Figure 5.16(a)), convection rolls were found to be influenced by cross-roll and spiral defect instabilities similar to those observed in a horizontal ferrofluid layer; compare Figures 5.16(b, c) and 5.6. The stars in Figure 5.16(a) denote zipper state regimes, where convection rolls aligned with the basic flow break up into cells and then reappear again as shown in Figure 5.16(b) and (c).

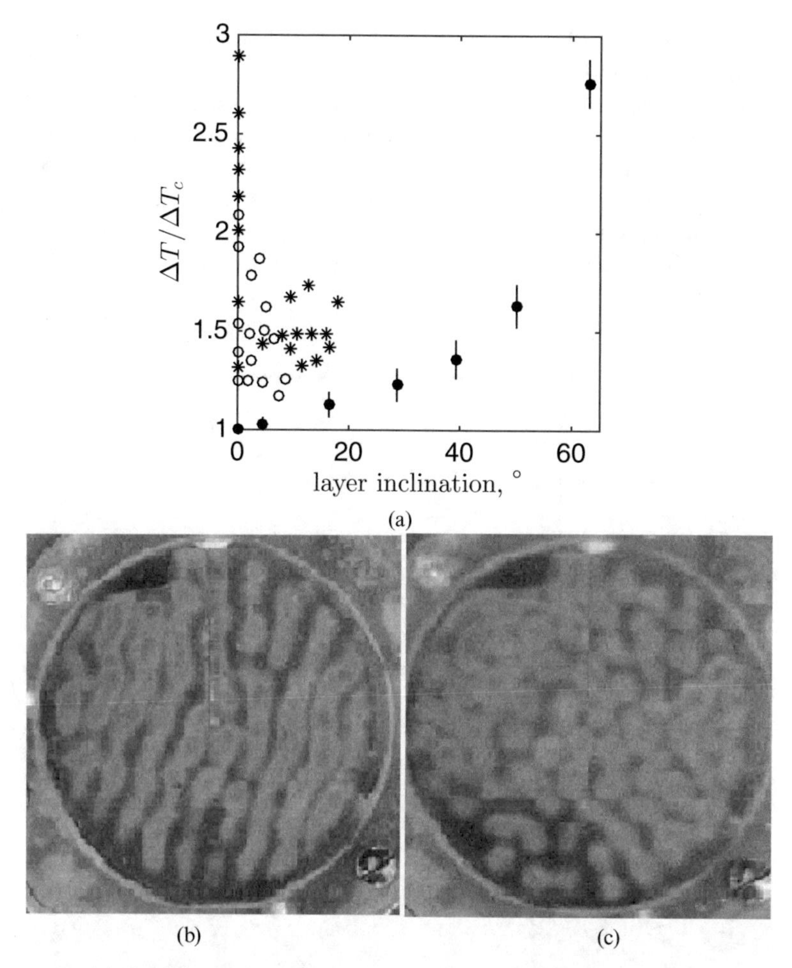

Fig. 5.16 (a) Experimental stability diagram for convection in an inclined layer of ferrofluid heated from below: filled black circles, the average temperature difference at which oscillatory convection occurs (the vertical bars show its amplitude); empty circles, spiral and target-like patterns; stars, elongated convection rolls, (b) and (c) images of typical spatial patterns observed in a layer inclined at 5° with respect to the horizontal at $\Delta T = 1.4\Delta T_c$. The time interval between snapshots (b) and (c) is 6 min.

With the increase of the layer inclination angle and the applied temperature difference, the strength of the basic flow increases and so does its aligning influence. As a consequence, more longitudinal rolls appear, and they become the dominant convection pattern. However, in contrast to convection in single-component fluids, such rolls undergo an amplitude modulation that could be sufficiently strong to temporarily suppress them locally or in a complete layer. Such regimes were observed at parametric points marked by diamonds and squares, respectively, in Figure 5.17(a). Typical experimental images of wandering spots with locally suppressed convection (also referred to as localised states in literature [4, 21, 129, 164, e.g.]) are shown in Figure 5.17(b) and (c). In image (b) convection is fully suppressed in the south of the layer. Six minutes later convection is restored there but gets suppressed in the north-west part of image (c).

At the layer inclination angles greater than $\sim 15°$, spontaneous appearance and decay of convection affect the complete longitudinal rolls. An example of such a regime is illustrated in Figure 5.18. In image (a) convection rolls occupy the complete layer. A minute later rolls almost completely disappear in the left and central parts of the layer; see image (b). After another minute convection rolls are restored in the left part of the layer and decay in its right part; see image (c).

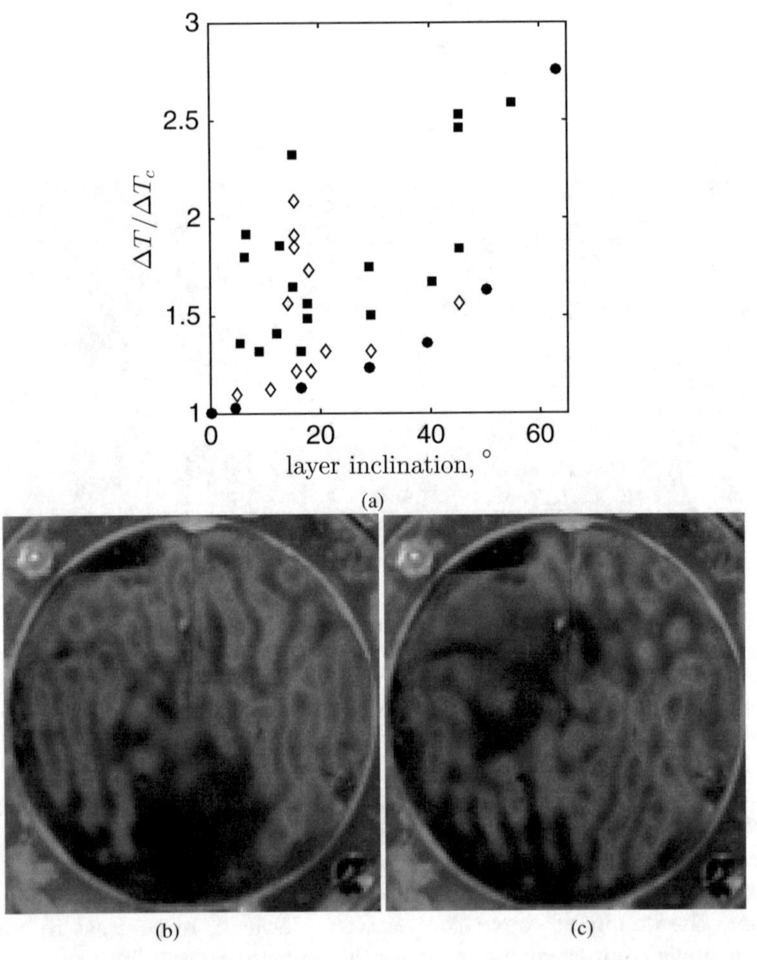

Fig. 5.17 (a) Regimes of amplitude modulation of convection rolls in an inclined layer of ferrofluid heated from below: diamonds, spontaneous appearance and decay of convection in a complete layer; squares, wandering convection spots, circles, the onset of convection, (b) and (c) images of a wandering spot with suppressed convection observed in a layer inclined at 10° with respect to the horizontal at $\Delta T = 2.3\Delta T_c$. Time interval between snapshots (b) and (c) is 6 min.

The regime where convection was intermittently suppressed over the complete layer is illustrated in Figure 5.19 and is marked by diamonds in Figure 5.17(a). Initially, convection rolls are seen in the top part of the layer (image (a)). Subsequently, convection decays there but arises in the lower part of the layer (image (b)) and finally decays completely (image (c)). The variation of the local Nusselt number at the centre of the layer is shown in Figure 5.19(d). It shows that convection there is completely suppressed

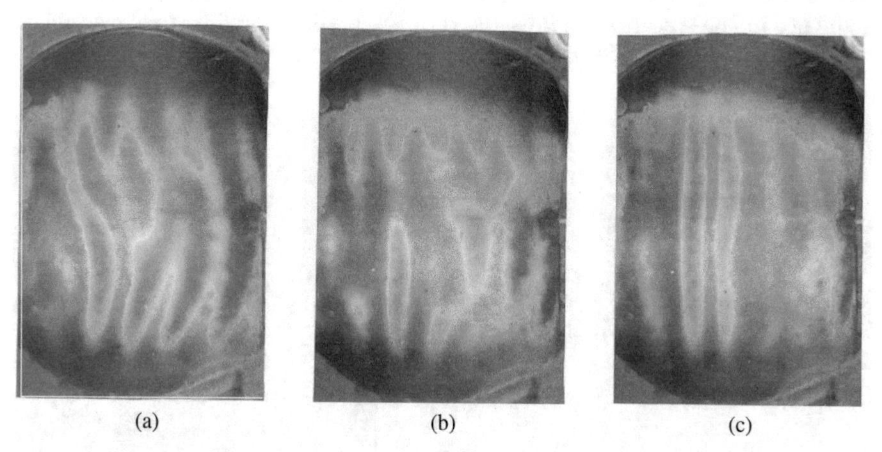

Fig. 5.18 Localised states in a ferrofluid layer inclined at the angle 45° with respect to the horizontal at $\Delta T = 2.4\Delta T_c$. Time interval between the snapshots is 1 min.

(Nu $\approx$ 1) every 30–40 min. As time progresses localised convection spots can appear randomly at any part of the layer before being fully suppressed. Such convection states can be classified as regimes with propagating convection fronts [99] or associated with the motion of the boundary of a convection domain [98, 153]. In an inclined layer of ferrocolloid, the motion of a convection front from one side of a chamber to the other occurs spontaneously. It does not decay throughout the duration of experiment and leads to several dozens of cycles of decaying/re-appearing convection. Similar irregular transitions between conduction and convection states have also been observed in water-ethanol mixtures [4], which indicate that its nature is related to double diffusion arising at certain values of the separation ratio.

In the parametric region marked by the triangles in Figure 5.20(a), modulated travelling waves were detected that are referred to as blinking states in literature [21, 129]. The prominent characteristic of such regimes is the irregular variation of both the amplitude and spatial period of convection structures. This is illustrated in Figure 5.20(b–d). Three vertical blue stripes can be seen in the right half of image (b) that correspond to six convection rolls. As time progresses, an additional pair of rolls arises near the top of the layer. It corresponds to the thin blue stripe in the centre of image (c). This pair of additional rolls is a result of a climbing dislocation [99, 191] moving along the main rolls. The new pair of rolls displaces the original rolls to the right by a half of their wavelength as seen in image (d). Such a scenario evolves over the temporally modulated (blinking) thermal field in the left part of the layer.

At large values of the applied temperature difference convection patterns are never suppressed and are predominantly defined by the motion of convection rolls and their defects. Such regimes correspond to parametric points denoted by empty squares in Figure 5.20(a). The typical pattern evolution sequence in these regimes is illustrated in Figure 5.21. Similar to the scenario illustrated in Figure 5.20(b–d), a pair of additional rolls appears in the top

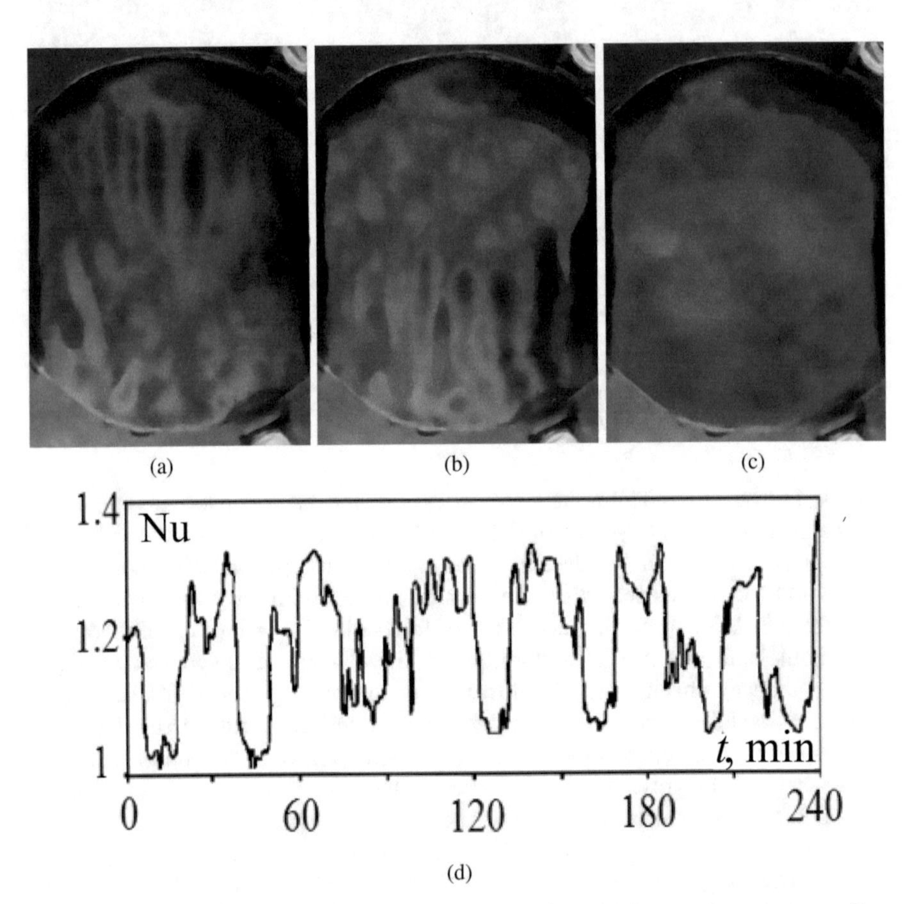

Fig. 5.19 (a–c) Irregularly decaying spatio-temporal convection patterns corresponding to parametric points shown by diamonds in Figure 5.17(a) (time interval between snapshots is 15 min), (d) temporal evolution of the local Nusselt number at $\Delta T = 1.8\Delta T_c$ and the layer inclination angle of 15°.

part of the layer and starts moving down the layer. Initially such a motion is seen as a climbing dislocation; see image (a). However, once this dislocation approaches the centre of the layer, it coalesces with one of the original

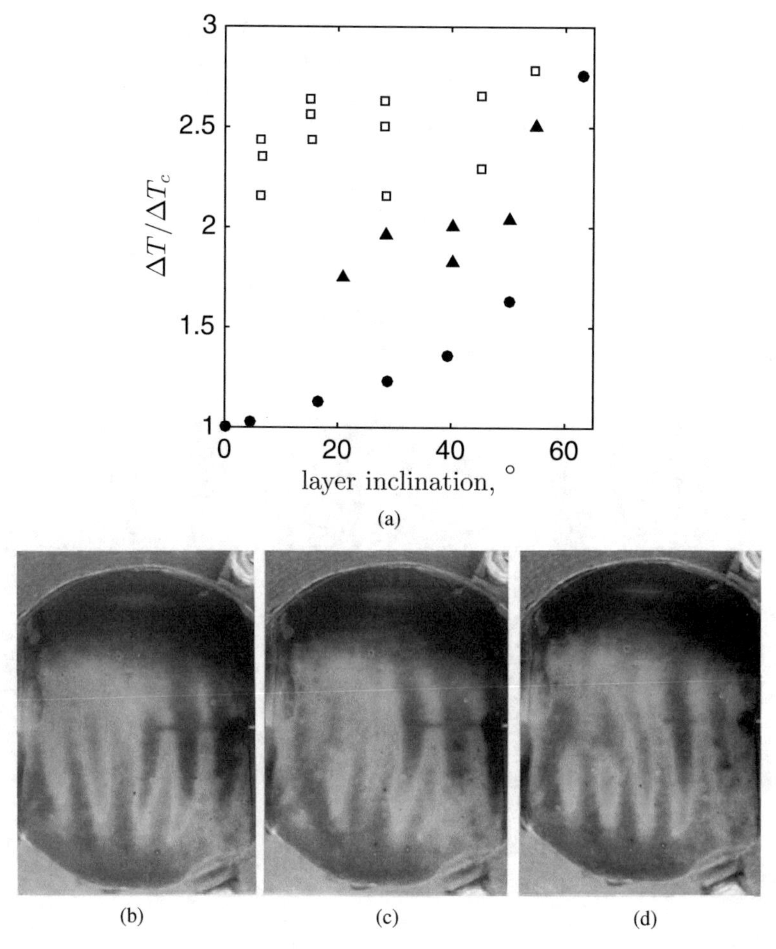

Fig. 5.20 (a) Regimes of travelling convection rolls in an inclined layer of ferrofluid heated from below: triangles, blinking states; squares, rolls with dislocations; circles, the onset of convection, (b), (c) and (d) images of a blinking state observed in a layer inclined at 55° with respect to the horizontal at $\Delta T = 2.5\Delta T_c$. Time interval between snapshots (b) and (c) is 2 min, between snapshots (c) and (d) 12 min.

longitudinal convection rolls (see image (b)) creating the so-called pinning effect [99, 265]. Subsequently, the defect slides in the direction perpendicular to the rolls (gliding dislocation); see image (c). As a result the total number of convection rolls in the layer changes with time.

Experimental images shown in Figures 5.18, 5.19, 5.20 and 5.21 also indicate the presence of temperature stratification in an inclined layer of finite size that is created by the basic fluid flow [140, 221]. The accumulation of warm and cool fluid in the upper and lower parts of the layer, respectively, results in the longitudinal temperature gradient. In the images corresponding to small layer inclination angles smaller than 15°, the existence of such a stratification is indicated by the presence of narrow blue and brown zones near the top and bottom edges of the chamber, respectively. At larger inclination angles, the extent of temperature stratification zones increases and can reach up to a half of the diameter in a vertical chamber. Such stratification breaks the longitudinal symmetry of the observed convection patterns.

To conclude, unlike in single-component fluids, convection in inclined layers of ferrocolloids always has an irregular oscillatory character. The existence of states where convection can be partially or fully suppressed indicates the presence of an essential density non-homogeneity that appears to be stronger than in a horizontal configuration. This is due to the two main factors. First,

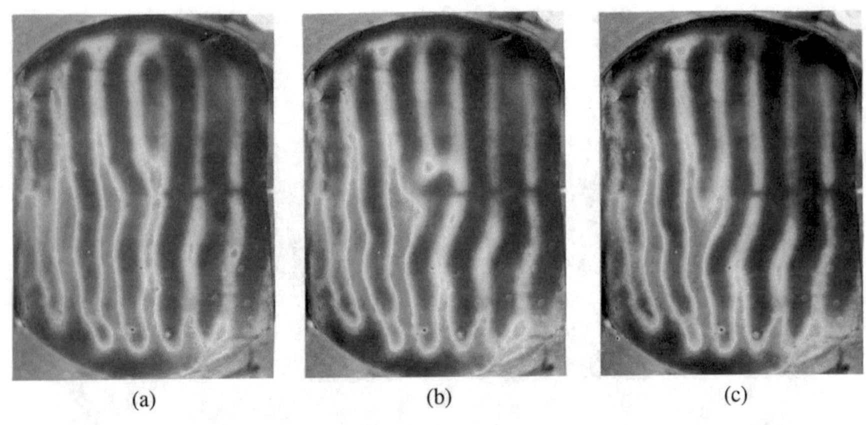

Fig. 5.21 Modulated convection rolls observed at the 15° layer inclination at $\Delta T = 2.4\Delta T_c$. Time interval between the snapshots is 1 min.

the layer inclination leads to the increase of its characteristic barometric height, which in turn enhances the gravitational sedimentation effect. Second, the species first separated in a cross-layer direction by thermodiffusion are subsequently separated in a longitudinal direction by the fluid flowing in the opposite directions near the hot and cold walls thus intensifying a longitudinal stratification—the process similar to that occurring in thermodiffusion columns [23].

5.5 Sphere

In this section we describe experiments conducted with a spherical cavity carved in a Plexiglas block as shown in Figure 4.12 and heated from below and cooled from the top. When the applied temperature difference is sufficiently small, the temperature in a sphere varies linearly in the vertical direction but is horizontally uniform [97, 180], and the fluid remains at rest. However, when the temperature difference exceeds the critical value, this mechanical equilibrium is destabilised, and fluid's conduction state bifurcates to convection. The first mode of convection in such a geometry in homogeneous fluids is a single convection vortex (Figure 4.13) with a stationary axis oriented arbitrarily in the equatorial plane of the cavity [97, 176, 260, 267]. The visualisation of such a flow in water was performed in experiments reported in [180]. In ferrocolloid it is found to exist for a sufficiently wide range of the governing parameters before the next instability mode sets in. Yet in ferrofluids the axis of the convection vortex rotates in the equatorial plane [136]. Thus the specific behavior of the induced convection flow observed near the convection onset in a sphere is not associated with the competition of various instability modes or with the interaction of multiple convection cells as is the case in flat layers [21, 28, 79]. Since nonlinear convection mode interaction is excluded as a potential explanation for such a peculiar oscillatory flow behaviour, the hypothesis has been put forwards that the complex fluid composition could be held responsible for the precession of a convection vortex.

According to estimations for an experimental spherical cavity of the diameter $d = 16\,\mathrm{mm}$ filled with a kerosene-based ferrofluid, the critical temperature difference ΔT_c between the poles at which convection sets is $\sim 0.1\,\mathrm{K}$. Such a small temperature difference is very close to the practical resolution limit of experimental measurements. Thus more viscous ferrofluids based on transformer oil (FF-TO) and polyethylsiloxane (FF-PES) (their viscosities are 7.7 and 41.8 times larger than that of the kerosene-based ferrofluid, respectively; see Table 5.1) were used for detailed studies of convection near the onset. The critical temperature differences for these fluids are 1.9 K and 12.8 K, respectively. However, since the typical flow regimes observed in FF-TO and FF-PES are qualitatively the same, in this section we will mostly focus on experiments with FF-TO. In addition, to establish the physical reasons for long-term oscillations in ferrofluids [38, 136, 238], comparative experiments involving fluids with progressively less complex structure (industrial and purified transformer oil (TO and CTO, respectively)) will be discussed.

Note that the flow phenomena observed in experiments occur on very different time scales: momentum diffusion time $t_\nu = \rho d^2/(\pi^2\eta)$, thermal diffusion time $t_t = d^2/(\pi^2\kappa)$ and particle diffusion time $t_D = d^2/(\pi^2 D)$. For FF-TO these are estimated as $t_\nu \sim 10^0\,\mathrm{s}$, $t_t \sim 10^2\,\mathrm{s}$ and $t_D \sim 10^7\,\mathrm{s}$. The shortest of these scales controls the measurement sampling interval, and the longest determines the total duration of experiments. The large difference between these physical time scales makes the experimental observations very time-demanding.

Several characteristic types of oscillatory flow regimes are observed near the onset of convection in a sphere. The parametric ranges of their existence in FF-TO are summarised in Figures 5.22 and 5.25. Figure 5.22 presents the dependence of nondimensional heat flux (Nusselt number Nu) on the relative temperature difference between the poles of a sphere. The black squares along the horizontal axis in this figure correspond to regimes where the abrupt transition was detected. The empty circles correspond to self-induced oscillations that have been found to be associated with the precession of the axis of the convection vortex in the equatorial plane. The filled circles depict the regimes of stationary single-vortex convection when the orientation of the flow axis did not change in time. The values of the Nusselt number are shown for oscillatory and stationary regimes that lasted from several days to several months at fixed thermal conditions.

As shown in, for example, [97, 180] convective motion in a spherical cavity filled with a one-component Newtonian fluid heated from below arises in the form of a single stationary vortex as a result of a supercritical bifurcation when the applied temperature difference between the poles exceeds the critical value ΔT_c. In contrast, in a ferrocolloid that remained at rest for at least several hours prior to the start of experiment convection sets as a result of subcritical transition, and a hysteresis occurs when the transition between convection and conduction states is observed at continuously increasing (black squares in Figure 5.22) or decreasing (empty circles in Figure 5.22) temperature difference between the poles. To eliminate hysteresis col-

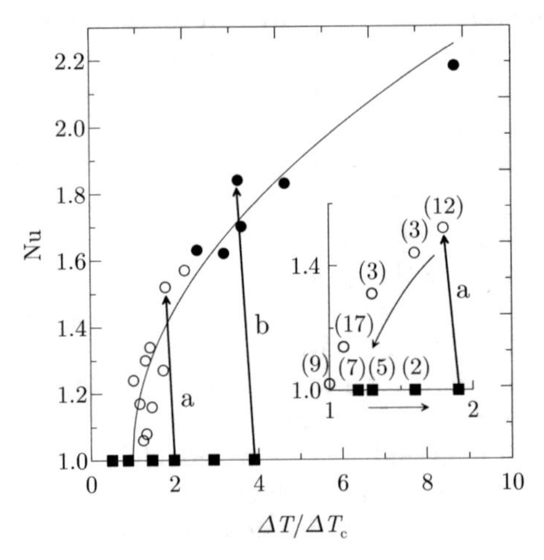

Fig. 5.22 The dependence of the Nusselt number on the relative temperature difference $\Delta T/\Delta T_c$ between the poles of the sphere. The solid curve shows the least squares fit to experimental data given by $\text{Nu} = 1 + 0.44\sqrt{\Delta T/\Delta T_c - 1}$. The number of days the temperature difference was maintained before the heat flux measurements were performed is given in the inset parenthetically. The symbols are defined in the text.

loid had to be thoroughly mixed before an experimental run. This was done by turning the experimental setup sideways prior to the main observation and positioning the heat exchanger plates vertically. They were then maintained at the maximum possible temperature difference of 40 K for an hour.

This induced a strong convective motion inside the sphere that ensured good mixing of the fluid. In experiments with such a homogeneous ferrocolloid, the convection threshold value of $\Delta T_c = 1.8 \pm 0.1\,\mathrm{K}$ was reproduced in several independent runs, and it has been used to construct Figure 5.22.

If preliminary mixing is not performed, the convection was found to establish abruptly. Hysteresis was observed when a gradual increase of the applied temperature difference had been reversed. The arrows labelled "a" and "b" in Figure 5.22 show the examples of such abrupt transitions from a mechanical equilibrium to convection that were detected when the fluid remained static for about a month prior to the start of experiments. The close-up shows the typical hysteresis when a sequence of the stepwise increments of the temperature difference between the poles of the sphere was followed by that of stepwise decreases. The figures in parentheses give the observation time (in days) in each thermal regime. The finite amplitude convection flow was abruptly induced at $\Delta T = 1.9\Delta T_c$; see arrow "a". The time series record shown in Figure 5.23 provides details of this transition. The fluid remained isothermal and at rest for 25 days before the temperature difference $\Delta T = 1.2\Delta T_c$ was applied between the poles of the sphere. This regime was maintained for 7 days with no convection detected (see the first horizontal line segment in Figure 5.23). Subsequently, the temperature difference was increased to $\Delta T = 1.3\Delta T_c$, which was observed for 5 days, and then to $\Delta T = 1.6\Delta T_c$ for 2 days. No convection was detected in these regimes either (see the second and third horizontal line segments in Figure 5.23). However, as soon as the applied temperature difference was set to $\Delta T = 1.9\Delta T_c$ (day 15 in Figure 5.23), the finite amplitude oscillatory convection started. For the first several days, its amplitude was approximately 1.5 times smaller than that in the subsequent established oscillations regime. This means that a medium intensity convection requires a significant time to mix the fluid and eliminate concentrational non-homogeneity caused by sedimentation.

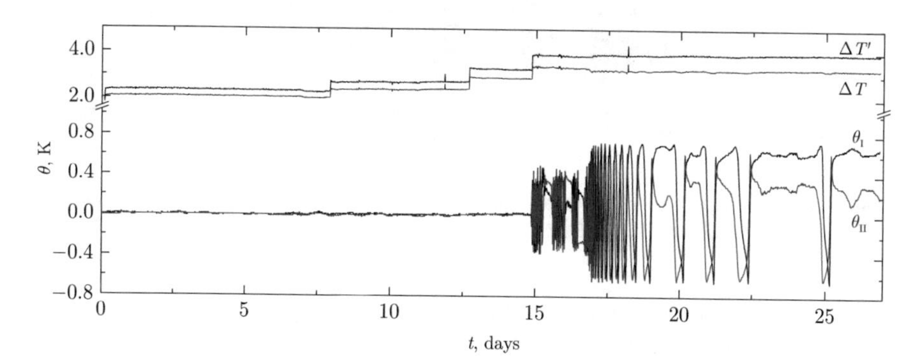

Fig. 5.23 Temperature time series for $\theta_{\mathrm{I,II}}$ defined in Section 4.6 for a sequence of stepwise increments of the temperature difference between the poles of a sphere. Convection was first detected once the relative temperature difference of $\Delta T = 1.9\Delta T_c$ was applied.

The thermogram corresponding to the transition shown by arrow "b" in Figure 5.22 is given in Figure 5.24. In this case, the fluid remained at rest in isothermal conditions for 34 days prior to the start of experiment. Then the temperature difference between heat exchangers was applied in a stepwise manner with four equal temperature increments of 2 K each. After each temperature difference increase, the system was left to adjust for 24 h. The convection starts the form of quasi-harmonic oscillations 10 h after the temperature difference of 8 K between the heat exchangers was applied (the left edge of Figure 5.24), which corresponded to $\Delta T = 3.9\Delta T_c$. Their period increased from 6 min during the first few hours after the onset to around 3 h a day later. Such oscillations were observed for 27 h before they gave way to a steady convection in the form of a single vortex with a fixed orientation of its axis in the equatorial plane. As seen from the thermal records presented in Figure 5.24, the variation of the amplitude of oscillations detected by the equatorial thermocouples leads to the change of the temperature difference between the poles of the cavity, which in turn results in the variation (of the order of several percent) of the heat flux through the fluid. In contrast, in the reverse transition, when the temperature difference between the poles is being reduced, the transition to a motionless state is characterized by a gradual decrease of the flow amplitude to zero at $\Delta T = \Delta T_c$; see the curve shown by the solid line in Figure 5.22. The experiments with a non-premixed ferromagnetic nanofluid indicate that the particle concentration gradient arising due to the gravitational sedimentation enhances the stability of a mechanical

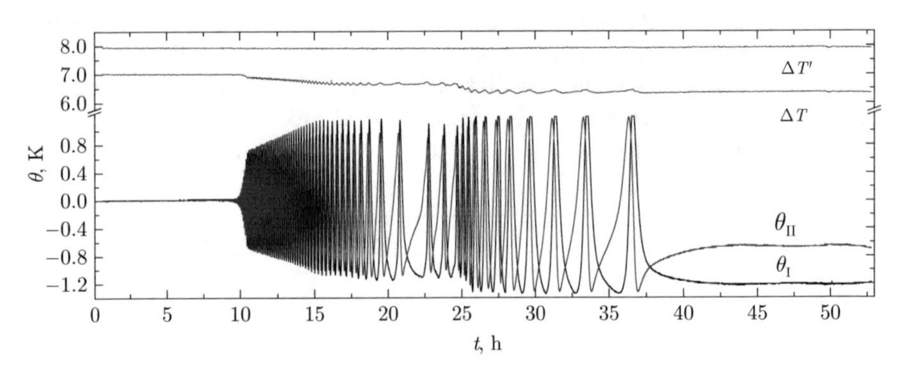

Fig. 5.24 Temperature time series $\theta_{\text{I,II}}$ defined in Section 4.6 illustrating the onset of oscillatory convection at $\Delta T = 3.9\Delta T_c$.

equilibrium in such a fluid. Even if a large destabilising temperature gradient is applied to non-premixed fluid (see arrow "b" in Figure 5.22), it takes a significant amount of time for convection to start. Such a delay is not observed in single-component fluids or well-mixed nanofluids. This indicates that the gravitational sedimentation of particles is counteracted by a thermodiffusion that gradually reduces the stabilizing density gradient and eventually leads to the onset of convection in the initially gravity-stratified ferromagnetic nanofluids. It is convenient to illustrate the details of the convection onset using the nondimensional amplitude

$$A = \frac{\theta}{\Delta T}, \tag{5.5}$$

of a thermal signal, where θ was previously defined in Equation (4.6). It remains zero for pure conduction states. When convection sets in, three regimes are clearly distinguished in a premixed fluid; see Figure 5.25. For sufficiently large values of the applied temperature difference, the convection begins in the form of a stationary vortex with a fixed amplitude and orientation of the axis. At the moderately supercritical values of the temperature difference, the axis of convection vortex precesses in the equatorial plane and the vortex amplitude vary within the range shown by vertical bars in Figure 5.25. Perhaps, the most peculiar behaviour is observed near the threshold of convection. In these regimes, convection has an intermittent character. It arises spontaneously and leads to a sharp increase of the amplitude from $A = 0$ (conduction regime) to $A \approx 0.1$ (convection regime); see the upward arrow in Figure 5.25. Subsequently, such convection decays, as shown in the figure by the downward arrow. However, approximately 6 h later, convection arises again, and the cycle repeats as seen from Figure 5.26(a) (nine cycles are shown). Therefore, in the vicinity of the convection threshold, intermittent regimes, in which oscillatory convection spontaneously arises and completely decays, are detected. The details of one such cycle are demonstrated in Figure 5.26(b). When both thermal readings θ_{I} and θ_{II} are zero (e.g. time moment **A** in Figure 5.26(b)), no convection exists in the cavity. When $\theta_{\mathrm{I}} > 0$ (moments **B** and **F**), the fluid rises at the location of thermocouple 1 (see Figure 4.13) and sinks near thermocouple 3. The flow direction is reversed if $\theta_{\mathrm{I}} < 0$ (moment **D**). Similarly, the direction of the flow induced by the second component vortex is determined by monitoring the sign of θ_{II}: if $\theta_{\mathrm{II}} > 0$, the fluid rises near thermocouple 2 and sinks near thermocouple 4 (moment **E**) and vice versa (moment **C**). The orientation of the vector of the angular velocity ω of the convection vortex estimated from the thermocouple data for time moments **B**–**F** in Figure 5.26(b) is shown in Figure 5.27. This diagram demonstrates that the axis of the convection vortex rotates in the equatorial plane of the sphere.

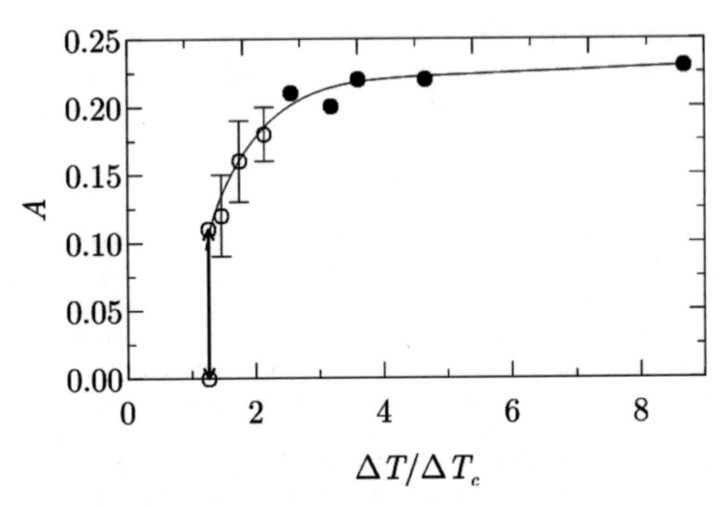

Fig. 5.25 Nondimensional amplitude of the signal recorded by the equatorial thermocouples as a function of the relative temperature difference. Solid and empty circles correspond to stationary and oscillatory regimes of convection in a premixed fluid, respectively. Vertical bars show the amplitude ranges detected for oscillatory convection.

Fourier and Morlet wavelet-based analyses applied to the signals registered by the equatorial thermocouples enable determining their detailed spectral characteristics. For example, a Fourier spectral density distribution computed for a thermocouple signal θ_{I} recorded for $\Delta T = 1.2\Delta T_c$ is shown in Figure 5.28(a). It indicates the presence of the four main frequencies $\nu_{1,2,3,4} \approx (0.6, 0.8, 1.2, 1.4) \times 10^{-5}\,\mathrm{Hz}$ corresponding to periods of 1.9, 1.4, 1.0 and 0.8 days, respectively.

The coefficients of the wavelet transform of a discrete time series θ_k sampled at times $t_k = k\Delta t_s$, $k = 1, 2, \ldots, n$, where n is the total number of readings, have been computed as

$$W_{u,s} = \frac{1}{\sqrt{s}} \sum_{k=1}^{n} \theta_k \psi\left(\frac{k-u}{s}\right). \tag{5.6}$$

Here $\psi\left(\frac{k-u}{s}\right)$ is the base wavelet of scale $s\Delta t_s$ evaluated at time $t_k = k\Delta t_s$ and shifted by $u\Delta t_s$ [130]. Each such wavelet represents a contribution to the overall signal that is observed at $t = (k-u)\Delta t_s$ and has a characteristic time span of $\tau = s\Delta t_s$. The wavelet transform of a thermocouple signal θ_{I} recorded for $\Delta T = 1.2\Delta T_c$ is shown in Figure 5.28(b). The dark regions indicate the presence of temporal structures with the characteristic periods ranging from 1 to 2 days during days 2–5 and 9–11 and from 3.5 to 4.5 days during days 7–12 of the experiment. Comparison of the Fourier and wavelet spectra of the signal shown in Figure 5.26 indicates that wavelet analysis contains more accurate information about the structure of the time

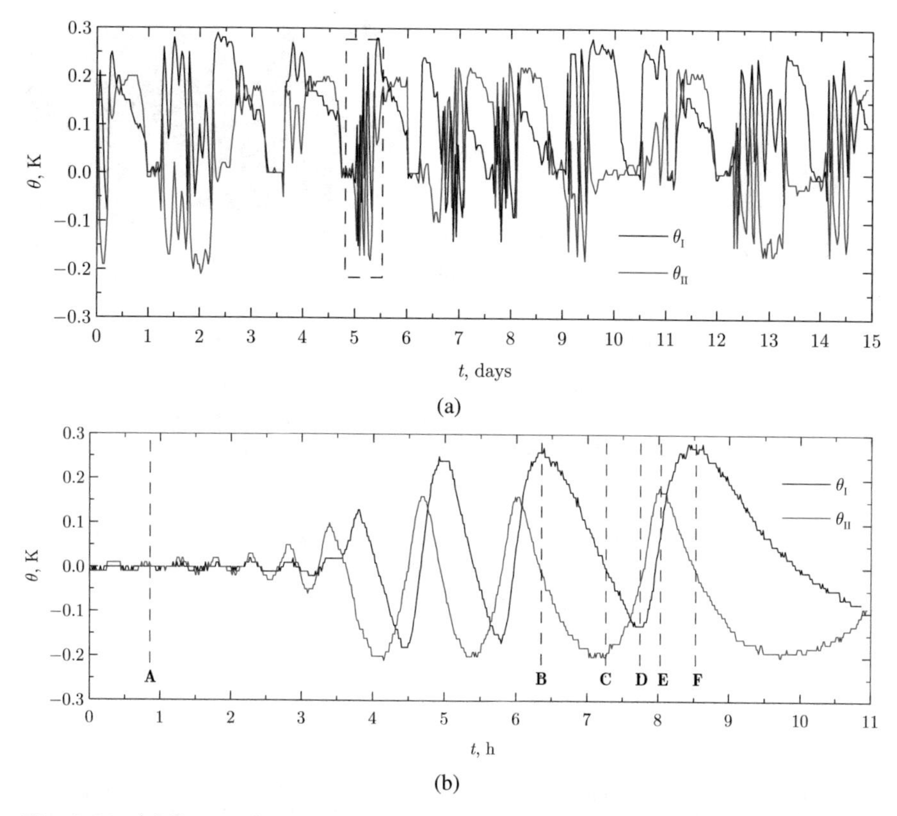

Fig. 5.26 (a) Temporal evolution of the temperature response $\theta_{\mathrm{I,II}}$ defined in Section 4.6 at $\Delta T = 1.2\Delta T_c$ and (b) close-up of a region enclosed by a dashed frame in panel (a).

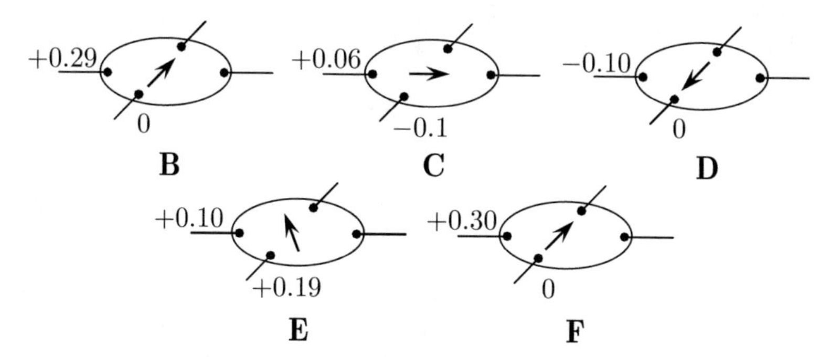

Fig. 5.27 The orientation of the angular velocity vector ω of the convection vortex for time moments **B**–**F** in Figure 5.26(b). The shown numerical values correspond to the thermocouple readings $\theta_{\mathrm{I,II}}$ defined in Section 4.6.

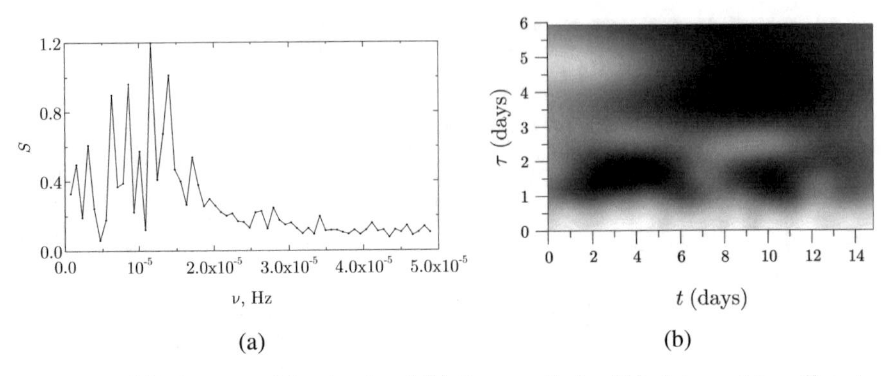

Fig. 5.28 (a) The spectral density S and (b) the magnitude of Morlet wavelet coefficients for the thermocouple signal θ_{I} defined in Section 4.6 and recorded for $\Delta T = 1.2\Delta T_c$. The darker shade corresponds to the larger value of the coefficient amplitude.

series. In particular, Fourier spectrum does not resolve the slowest oscillations with a period of approximately 4 days, the presence of which is informed by the wavelet analysis. Therefore, in the subsequent analysis of experimental results, only wavelet decomposition will be used.

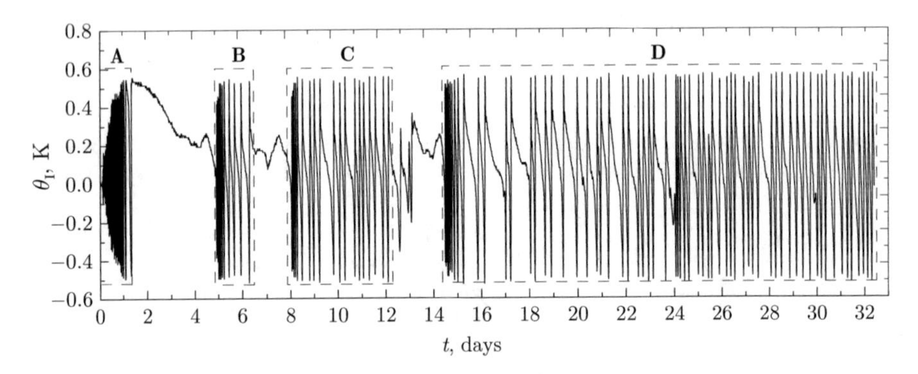

Fig. 5.29 The temporal evolution of the temperature response θ_{I} at $\Delta T = 1.8\Delta T_c$.

Further away from the critical point (for $\Delta T > 1.2\Delta T_c$), the qualitative behavior of the system changes. A typical record of a thermocouple signal is shown in Figure 5.29 for the applied temperature difference $\Delta T = 1.8\Delta T_c$. The regimes of slow variation over the time intervals lasting for 1–2 days when the axis of the convection vortex rotates by a small angle in the equatorial plane alternate with regimes of relatively fast precession of the vortex axis with the period ranging from tens of minutes to several hours. Such regimes are shown by empty circles in Figures 5.22 and 5.25. The Fourier spectrum of the full record discussed in [136] is rather noninformative and only indicates the existence of the main frequency $\nu \approx 0.7 \times 10^{-6}$ Hz (the period of 16.5

days). On the other hand, the Morlet wavelet transform shows that the overall signal initially contained dominant oscillations with a period of approximately 5 days that have been replaced with the slower oscillations with periods in the range between 12 and 17 days in later observations. Thus, it is beneficial to analyse the signal partitioned into segments corresponding to sections labelled in Figure 5.29.

There are four oscillatory bursts labelled **A** (day 1), **B** (days 5–7), **C** (days 8–12) and **D** (days 15–35) separated by the intervals of a slow rotation of the axis of a convection vortex observed. Their details are shown in Figure 5.30.

During the first day after the start of the experiment, the quasi-harmonic oscillations establish. Such a behaviour is typical for systems with weak dissipation [201]. The wavelet coefficients of the time series corresponding to interval **A** in Figure 5.30(a) are shown in Figure 5.31(a). The figure indicates that the dominant period of a signal during this time is approximately 50 min. The amplitude of the oscillatory component of a convection flow initially increases approximately linearly with time until it saturates after about 13 h. The plot of Morlet wavelet transform coefficients (Figure 5.31(a)) shows that from this point the period of oscillations starts slowly increasing, which is indeed seen in Figure 5.30(a).

For the second nonharmonic oscillatory burst **B** shown in detail in Figure 5.30(b), the wavelet analysis summarised in Figure 5.31(b) reveals that two oscillatory modes are present. Their periods monotonically increase from approximately 1 to 4 and from 5 to 8 h over the intervals of 2–37 and 10–35 h, respectively. The wavelet transform presented in Figure 5.31(c) for the third oscillatory burst in Figure 5.30(c) shows a qualitative difference in the signal behavior during interval **C**: in contrast to the regimes observed at earlier times, the period of oscillations does not increase monotonically but rather increases and decreases periodically around the average value (oscillation period varies between 4 and 7 h). The main feature of the oscillatory convection observed during interval **D** (see Figure 5.30(d)) is that the flow becomes less regular. As evidenced by Figure 5.31(d), the oscillation periods tend to decrease from about 1 day in the beginning of the burst to less than half a day by the end of the recorded experimental run.

Thus, when a moderately supercritical temperature difference is applied, convection starts in the form of oscillations that are nearly harmonic in the beginning of the experiment but become progressively irregular as time progresses. The regimes of well-defined oscillations initially alternate with the nearly stationary convection stages. The duration of the latter decreases with time. Eventually, irregular unsteady convection sets in after a number of discrete oscillatory bursts.

A summary diagram of the dependence of the nondimensional heat transfer coefficient (Nusselt number) on the relative temperature difference between the poles of a sphere is shown in Figure 5.32. The data is collected for four types of fluids: ferrofluids based on polyethylsiloksan (FF-PES) and transformer oil (FF-TO), industrial transformer oil (TO) and centrifugally

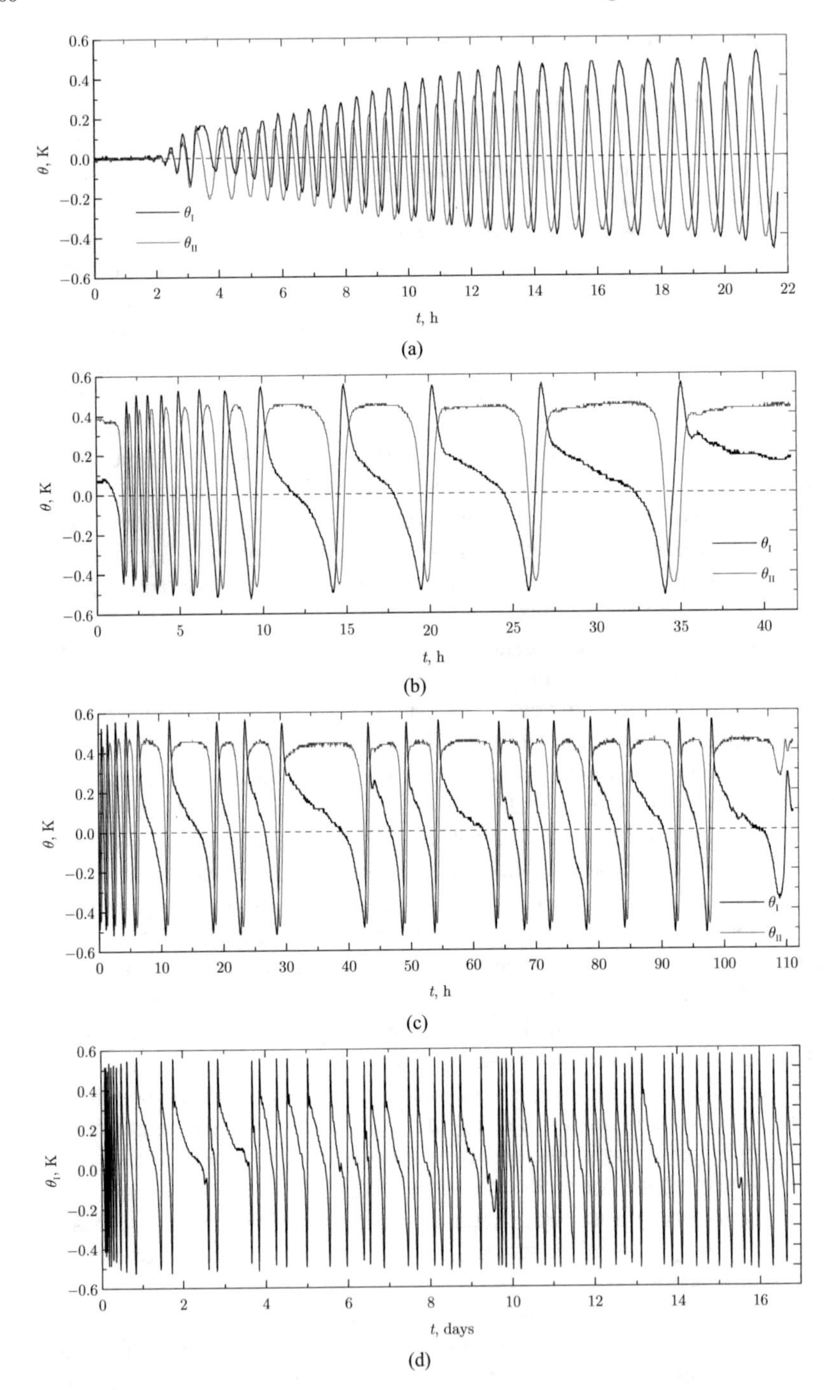

Fig. 5.30 Details of the temporal evolution of the temperature response $\theta_{\rm I}$ and $\theta_{\rm II}$ at $\Delta T = 1.8\Delta T_c$ for intervals (a) **A** (day 1), (b) **B** (days 5–7), (c) **C** (days 8–12) and (d) **D** (days 15–35).

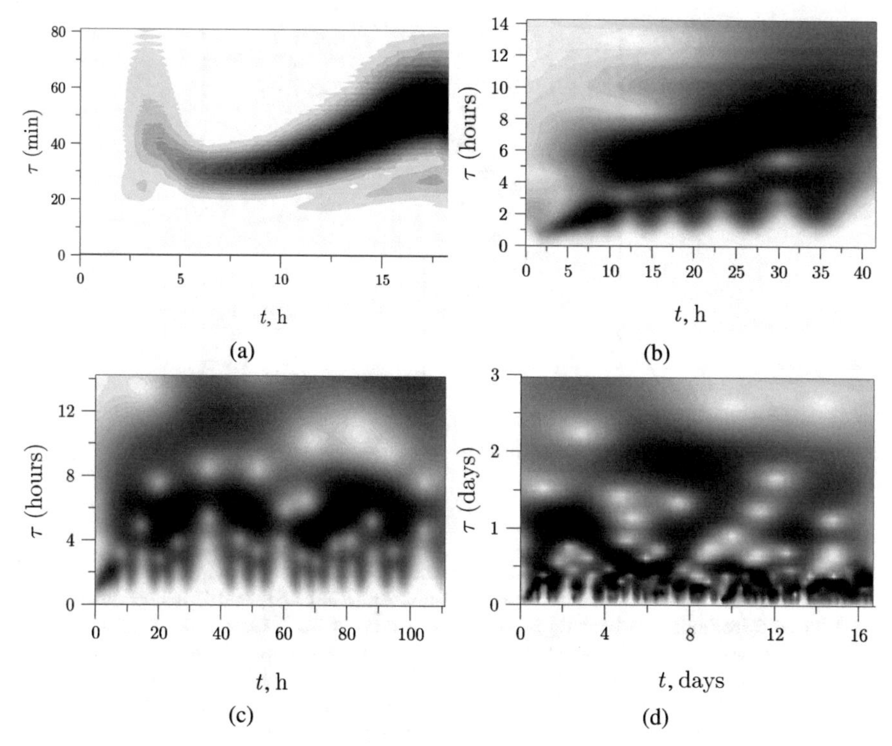

Fig. 5.31 The magnitude of Morlet wavelet coefficients for the thermocouple signal θ_I defined in Section 4.6 and recorded for $\Delta T = 1.8\Delta T_c$ for intervals **A–D** in Figure 5.29. The darker shade corresponds to the larger value of the coefficient amplitude.

purified transformer oil (CTO). Note that TO not only serves as the base fluid in manufacturing ferrofluids but also is widely used as a heat carrier fluid. Industrial TO contains up to 95% of saturated hydrocarbons (paraffins, naphthenes, aromatic hydrocarbons) as well as asphalt-resinous substances and naphthenic acid that can lead to the formation of insoluble residues. To eliminate their influence on the observed flows, a sample of transformer oil was purified by centrifuging it at 1500 rpm for 15 min, which removed solid residues from it[3].

If the fluids are premixed, then the values of Nu are independent of the type of fluid used in experiments: the experimental points follow the same trend, which is typical for ordinary fluids. However, if the fluids are left at rest before the start of the experiment for a sufficiently long time (from 1 day

[3] The efficiency of centrifuging is characterised by separation factor F that is the ratio of the centripetal acceleration to the gravity acceleration g: $F(z) = F_0(1 + (L - z)/r_0 \cos\gamma$, where $L = 115$ mm is the length of the sample container tube, $F_0 = (\omega^2 r_0 \cos\gamma)/g = 38$, $r_0 = 17$ mm is the distance between the top of the tube and the axis of the centrifuge and $\gamma = 30°$ is the angle inclination angle with respect to the horizontal.

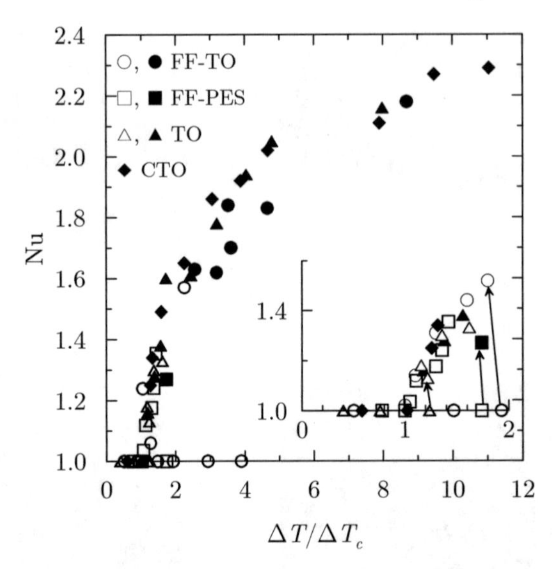

Fig. 5.32 The dependence of the Nusselt number on the relative temperature difference. Critical temperature differences for polyethylsiloksan (FF-PES)- and transformer oil (FF-TO)-based ferrofluids and transformer oil (TO) are $\Delta_c = 12.8$, 1.8 and 0.5 K, respectively. The inset shows a hysteresis near the convection onset: an abrupt excitation in an initially stratified fluid and a smooth decay after the fluid are convectively mixed. For FF-PES and FF-TO, ferrofluids and TO, the depth of the hysteresis was 89%, 72% and 24% when they remained isothermal and at rest for 25, 3 and 7 days, respectively. The empty symbols correspond to conduction regime (Nu = 1) and oscillatory convection (Nu > 1). The filled symbols represent stationary convection.

to a month), then onset of convection in FFs and TO is delayed compared to that in ordinary fluids; see the empty symbols for Nu = 1 in the inset in Figure 5.32. Such a delay is not detected in CTO, which indicates that a stable density stratification establishes in resting multicomponent fluids over time due to the gravitational sedimentation of a heavy solid fraction. The arrows in the inset in Figure 5.32 show that the delayed convection starts abruptly acquiring a finite amplitude immediately at its onset. However, the backward transition to a conduction regime with Nu = 1 occurs smoothly: convection amplitude gradually decreasing to zero when the applied relative temperature difference is reduced towards the critical value corresponding to ordinary fluids. Thus, convection sets abruptly, and hysteresis is observed in the reversed transition not only in FFs but also in non-purified TO.

Similar to ferrofluids, convection in non-purified TO observed at $\Delta T < 2\Delta T_c$ had an oscillatory character; see the empty symbols in Figure 5.32. However, the oscillations in ferrofluids were observed regardless of the prehistory of experiment, while in TO they occurred only if it was left at rest for at least a day (for 3 days in Figure 5.32).

Figure 5.33 shows the typical time series of the thermocouple outputs $\theta_{\rm I}$ and $\theta_{\rm II}$ registering thermal perturbations in the equatorial plane of a sphere filled with FF-TO and its carrier fluid, TO. Such perturbations are caused by the horizontal precession of the axis of the convection vortex that is not observed in single component fluids. The top panel in Figure 5.33 depicts the oscillatory thermogram (see the close-up for the details of nonharmonic oscillations) recorded in ferrofluid over the first 7 days of a 2-week experiment conducted at $\Delta T = 1.4\Delta T_c$. During the observations the period of oscillations was changing spontaneously between 1.8 and 2.7 h. The bottom panel in Figure 5.33 shows fragments of time series recorded for CTO and TO after the

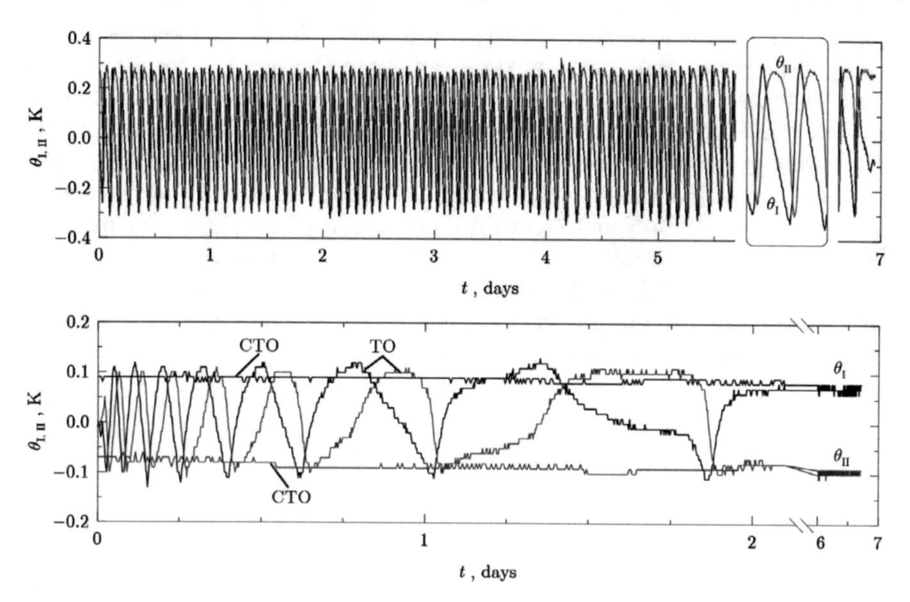

Fig. 5.33 The temporal evolution of the temperature response in FF-TO at $\Delta T = 1.4\Delta T_c$ (top) and in TO, CTO at $\Delta T = 1.6\Delta T_c$ (bottom). The inset shows the details of the response for a ferrofluid.

fixed temperature difference of $\Delta T = 1.6\Delta T_c$ was applied. Only a stationary convection was observed in CTO in this case. In contrast, the equatorial precession of a convection vortex was observed in TO for the first 2 days of the experiment. The first full rotation took about 1 hour, and the last one occurred over a period of almost 20 h. After that a stationary convection was observed for the rest of the experimental run.

The analysis of several similar experimental runs shows that the irregular oscillations alternating with intermittent quasi-stationary intervals were observed in all types of ferrofluids over the complete duration of experiments lasting from 1 to several weeks. The period of such oscillations varied from tens of minutes to several hours. In contrast, oscillations in TO had a transient character: they lasted from 10 to 48 h from the initiation of convection with the period of oscillations gradually increasing from several minutes to several days. The likely reason for such a peculiar behaviour of flows observed in both FF and TO is the dynamic competition between the thermogravitational convection, positive thermodiffusion and gravitational sedimentation caused by a complex composition of such fluids; see Figure 5.1.

5.6 Conclusions

For well-mixed ferrofluids with different bases, the values of Nusselt number are independent of the fluid type and follow the same trend as single-component liquids and gases. In thin ($d \approx 2\,\mathrm{mm}$) layers of ferrocolloids convection arises at large temperature gradients when destabilising thermodiffusion, and thermal expansion effects always play a stronger role than the stabilising gravitational sedimentation. As a result the onset of stationary convection is observed. In thicker horizontal or inclined layers and in a spherical geometry, gravitational sedimentation of solid phase plays a comparable role to that of thermodiffusion and thermal expansion, which leads to oscillatory convection. Alternating dominance of these opposite effects in stratified fluids creates conditions in which convection is spontaneously excited or forced to decay and hysteresis is observed.

Chapter 6
Experimental Investigation of Thermomagnetic Convection in Ferrofluids

Abstract The thermomagnetic mechanism of convection is considered at various orientations of the applied uniform magnetic field with respect to the temperature gradient and of the fluid layer with respect to the gravity. Its interaction with the thermogravitational mechanism is discussed. Experimental maps of stability convection-free states are obtained. Heat transfer in a magnetic field is studied as a function of fluid properties and experimental control parameters. The dependence of spatio-temporal evolution of convection flows on the strength of the applied field, the magnitude of the temperature gradient and fluid layer inclination with respect to the gravity is investigated. The conditions leading to stabilisation and destabilisation of convection-free states in non-isothermal ferrofluids placed in a magnetic field are determined. Factors influencing the onset of convection and convection flows in finite enclosures filled with realistic ferrofluids are listed, and their overall effect that is currently beyond the reach of theoretical treatment is demonstrated experimentally.

6.1 Magnetic Control of Magneto-Polarisable Media

Flow and heat transfer control using magnetic forcing in electrically non-conducting magneto-polarisable media has been the subject of numerous studies over several decades [2, 26, 44, 88, 144, 208]. The appearance of a ponderomotive force is associated with the action of a magnetic field on molecular dipoles (in natural dia- and paramagnetic fluids) or single-domain magnetic nanoparticles (in synthesised ferrofluids). In the pioneering work [2]

See Appendix B for the list of previously published materials reused in this chapter with permission.

A. A. Bozhko, S. A. Suslov, *Convection in Ferro-Nanofluids: Experiments and Theory*, Advances in Mechanics and Mathematics 40,
https://doi.org/10.1007/978-3-319-94427-2_6

the intensity of magnetoconvection arising in a volume of air placed in a magnetic field was used to determine the concentration of oxygen in the air. Later the influence of the applied magnetic field on the buoyancy-driven convection in a cubic enclosure filled with oxygen and heated from a side was studied in [54]. It was demonstrated that the arising ponderomotive forces are strong enough to reverse the direction of the primary circulation. The onset of magnetically driven convective motion in a stably stratified volume of air heated from above was observed in [123]. The principle of magnetic promotion of combustion by a magnetically induced flow of air was demonstrated in [255] offering a means of combustion control in microgravity conditions [124].

Table 6.1 Magnetic susceptibility of fluids in normal conditions.

Medium	Magnetic susceptibility, χ
Oxygen	1.8988×10^{-6}
Air	3.7×10^{-7}
Protein solutions in water	$(-7 \text{ to } -9) \times 10^{-6}$
Gadolinium nitrate solution	1.63×10^{-4}
Magnetic fluids	~ 1

The use of magnetic ponderomotive forces has been suggested for compensating gravitational buoyancy and suppressing thermal convection, which, if not controlled carefully, can adversely affect the quality of the final product in crystal growth from dia- and paramagnetic melts [81] and protein solutions [74, 185, 206, 214, 261]. Since the values of the magnetic susceptibility χ for dia- and paramagnetic fluids are very small (see Table 6.1), ponderomotive forces of the required magnitude can only be created in strong (up to 20 T) fields of superconducting magnets. Their experimental studies have been recently conducted in laboratories in France, Japan and Poland [15, 44, 90, 125]. The main focus of these studies has been on the use of strongly non-uniform external fields [114] when the variation of an internal magnetic field due to the dependence of fluid magnetisation on the temperature could be neglected.

The magnetic susceptibility of ferrofluid is many orders of magnitude larger than that of dia- and paramagnetic fluids (see Table 6.1). Therefore, ponderomotive forces comparable in magnitude with gravitational buoyancy can be created in magnetic fields of ordinary permanent or electromagnets (0.01–1 T). Because of that ferrofluids can be effectively used to enhance heat transfer in various devices such as energy converters and transformers that generate magnetic fields as a by-product of their operation [26, 80, 173, 209]. Thermomagnetic convection also offers an efficient mechanism of heat transfer enhancement in situations where natural convection cannot exist due to either the extreme confinement, e.g. in microelectronics [165], or the lack of gravity as on orbital stations [40, 169]. Thermosensitive ferrofluids with large pyromagnetic coefficients (see Section 2.1) are used in such applications [91, 127, 156, 262].

Ponderomotive forces arising in ferrofluids can also be used for physical modelling of centrosymmetric radial gravitational forces. For example, it was suggested in [177] to model oceanic currents by ferrofluid flows created in spherically symmetric magnetic fields in microgravity conditions. Such a field can be experimentally created by arranging a system of several permanent magnets in way described in [210]. Further application of magnetoconvection for registering microaccelerations in microgravity conditions is discussed in [205].

A wide range of developed and potential applications warrants detailed experimental studies of magnetoconvection that will be discussed in subsequent sections. In a vast majority of theoretical studies, a distortion of magnetic field caused by the non-uniformity of fluid magnetisation brought about by spatial variations of the temperature or concentration of magnetic phase is assumed small and is neglected. However, theoretical results obtained for a flat layer [13, 202, 203, 236] and a sphere [34] filled with ferrofluid and placed in a uniform external magnetic field showed that such a distortion can lead to the change of the character and structure of a convective motion. The use of a uniform magnetic field is preferred from a practical point of view since it offers a means of convection control that is not obscured by the uncertainty of the local direction of a non-uniform magnetic field and its magnitude. Therefore, the major focus of experiments discussed next is on the processes taking place in a ferrofluid placed in uniform fields created along the axes of solenoidal electromagnets or within Helmholtz coils. Despite this apparent simplification, the observed convection flows have a very complex spatial structure and temporal dynamics. In particular, these experiments enable us to make a definitive conclusion that weak fluid density variations caused by gravitational sedimentation and thermo- and magnetophoresis of solid nanoparticles and their aggregates can fundamentally change the flow behaviour near the onset of convection [32, 37, 38, 43, 136, 200].

6.2 Horizontal Layer

6.2.1 Historical Overview and the Current State of Knowledge

Theoretical studies of thermomagnetic convection in a horizontal layer placed in a non-uniform external magnetic field were first undertaken in [143] and [69]. Results obtained for the field created by a magnetic dipole were later found to be in a good agreement with experiments reported in [1]. Theoretical and numerical investigation of thermomagnetic convection in a horizontal layer with non-magnetic boundaries placed in a uniform magnetic field [89] and in a layer confined by perfect permanent magnets [105] indicated that in these cases, thermomagnetic effects become dominant at normal conditions when the layer thickness is reduced to about a few millimetres. In such thin layers thermomagnetic convection was predicted to set in a layer heated from below at lower values of Rayleigh number than the critical value for Rayleigh-Bénard convection. However, previous attempts to verify such theoretical predictions experimentally led to ambiguous results. In early experiments [30] as well as in some later studies [37, 86], the onset of convection in strong magnetic fields was delayed compared to that observed when the field was switched off. In contrast, experiments reported in [37, 38, 217–219] confirmed theoretical results of [89] demonstrating a destabilising effect of a uniform magnetic field applied normally to the layer. It was suggested that the stabilising effect of magnetic field detected in earlier experiments could be attributed to the increase of fluid viscosity with the field intensity [30, 86]. However, the measurements reported in [30] showed that the value of the critical temperature difference for the onset of convection increased much faster than fluid viscosity measured in the same magnetic fields. For example, while the critical temperature difference increased by a factor of 2.5 in the applied field, the viscosity grew only by about 60%. In contrast, as reported in [86], the application of a magnetic field increased the convection threshold only by about 10% despite the manyfold increase of viscosity.

These observations indicate that magnetoviscous effect alone cannot explain the behaviour of non-isothermal ferrofluid placed in an external magnetic field, and the non-uniformity of fluid magnetisation is the key factor in defining it. The volume distribution of magnetic phase has a great influence on the magnetic properties of the fluid. However, due to the present lack of adequate models enabling its accurate description, the majority of theoretical studies (see Chapter 2) assume its uniformity and thus take into account only thermomagnetic and thermogravitational mechanisms of convection as is done in Chapter 3. They, however, cannot explain the observed variation [30, 37, 38, 86, 217–219] of the convection threshold either. Thus, other processes such as thermophoresis, thermodiffusion and gravitational sedimentation play an important role in defining the overall behaviour of ferrofluid. Some of these processes nontrivially affected by the application of magnetic field. For example, depending on the composition of ferrofluid and the concentration of magnetic phase and magnetoviscosity can increase nonlinearly with magnetic field leading to quenching convection in strong magnetic fields. The Soret coefficient of ferrofluids is found to change its sign from positive to negative as the magnitude of the magnetic field increases [25, 252, 253]. Thus thermophoresis in a horizontal ferrofluid layer heated from below and placed in a vertical magnetic field leads to the fluid density stratification that is destabilising in weak magnetic fields but is stabilising in strong fields. Another factor influencing ferrofluid behaviour is magnetophoresis [26, 223]. Unfortunately, the coefficients of diffusion, thermodiffusion, viscosity, etc. depend strongly on the presence of particle aggregates, their sizes and number and other factors and are not known quantitatively. For this reason the values of the nondimensional governing parameters corresponding to experimental conditions can only be estimated approximately. Because of this difficulty the results are presented in this chapter in terms of the experimentally measured values of the applied temperature difference ΔT across the layer and the external magnetic field H or the corresponding fluid magnetisation M. The experimental runs were conducted following two scenarios: (a) increasing the cross-layer temperature difference while keeping the external magnetic field and the average fluid temperature T_* fixed and (b) strengthening the field while keeping the temperature difference fixed. The first scenario corresponds to increasing both gravitational and magnetic Rayleigh (Grashof) numbers. However, since $\mathrm{Ra} \sim \mathrm{Gr} \sim \Delta T$ and $\mathrm{Ra}_m \sim \mathrm{Gr}_m \sim \Delta T^2$, the intensification of magnetic effects is much stronger than that of gravitational buoyancy. Such runs correspond approximately to the square-root trajectories in the $\mathrm{Gr}_m - \mathrm{Gr}$ plane sketched in Figure 6.1(a).

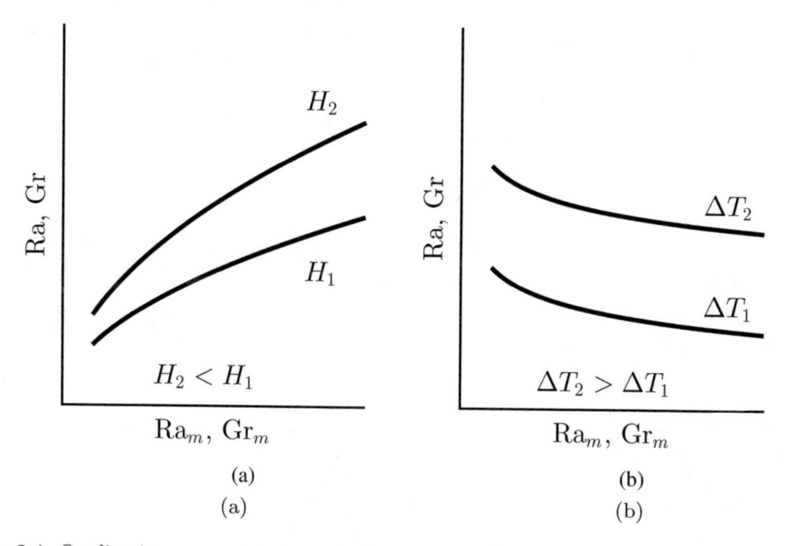

Fig. 6.1 Qualitative parametric trajectories corresponding to two experimental scenarios: (a) increasing temperature difference in constant magnetic field; (b) increasing magnetic field at fixed temperature difference across the fluid layer.

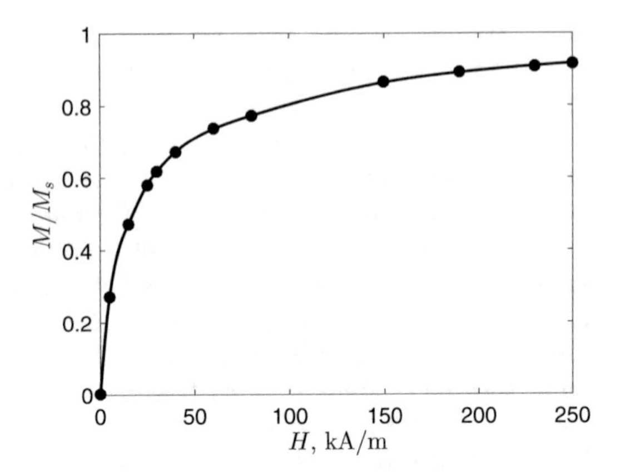

Fig. 6.2 Magnetisation curve for a concentrated ferrofluid with saturation magnetisation $M_s = 55\,\mathrm{kA/m}$.

When the temperature difference is kept constant and magnetic field is increased, the magnetoviscous effect makes fluid more viscous so that the effective Rayleigh (or Grashof) number becomes smaller. The influence of the applied magnetic field on the effective value of Ra_m (or Gr_m) is more complicated. According to [13, 23] when the fluid is far from magnetic saturation, see the left end of the magnetisation curve in Figure 6.2, as was the case in experiments described here, its pyromagnetic coefficient entering the definition of Ra_m, see (2.26), or Gr_m, see (2.29), is given by

$$K = M\left(T_*^{-1} + 2\beta_{m*} + \beta_*\right) , \tag{6.1}$$

where $\beta_{m*} \approx 0.5 \times 10^{-3}\ \mathrm{K}^{-1}$ is the relative pyromagnetic coefficient of magnetite. The three terms in parentheses in (6.1) correspond to the three mechanisms reducing fluid magnetisation at higher temperatures: Brownian motion of magnetic particles that disorients their individual magnetic moments in the

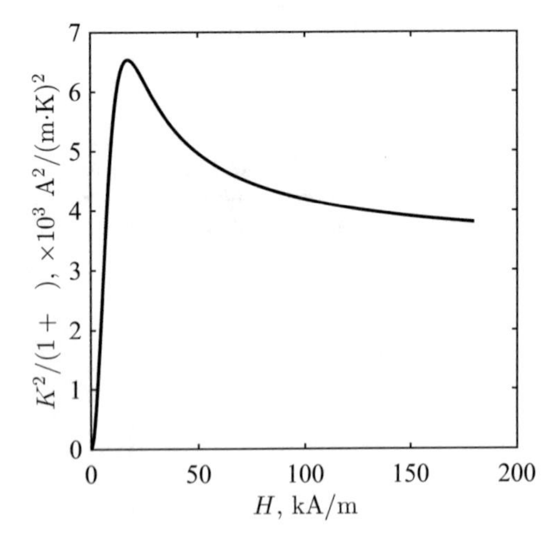

Fig. 6.3 Typical variation of magnetic sensitivity parameter of ferrofluid with the strength of the applied magnetic field.

bulk of a fluid, the thermal reduction of magnetic moments of individual particles and the thermal expansion of a carrier fluid that leads to the reduction of magnetic phase concentration. All of these effects are proportional to fluid magnetisation M which is a linear function of the applied field, $M = \chi H$. Therefore at a fixed temperature the fluid's pyromagnetic coefficient increases with the field. In our experimental conditions this increase was found to be stronger than that of the fluid's viscosity so that the ratio K^2/η^2 and thus Ra_m (Gr_m) were increasing function of H. The parametric trajectories corresponding to this experimental scenario are sketched in Figure 6.1(b).

Both experimental scenarios described above correspond to strengthening of thermomagnetic effects provided that ferrofluid magnetisation is far from saturation. However, in strong magnetic fields the magnetic sensitivity $K^2/(1+\chi)$ of ferrofluids starts decreasing as shown in Figure 6.3 so that a further increase of magnetic Rayleigh or Grashof number governing the onset of magnetoconvection (see Definitions (2.26) and (2.29)) becomes impossible.

In the following sections we discuss results of experiments aiming at quantifying the conditions under which the transition between the destabilising and stabilising influences of a uniform transverse magnetic field on the onset of convection occurs in a classical Rayleigh-Bénard configuration. Ferrofluids with different concentrations of magnetic phase (see Table 5.1) were used. This enabled the variation of the effective Rayleigh number range (see Definition (2.34)) while keeping the thickness of a fluid layer, which is dictated by the design of the experimental chamber, fixed at $d = 2$, 3.5 or 5 mm.

6.2.2 Convection and Heat Transfer

We start with the discussion of typical experimental results for a 2 mm thick layer of concentrated magnetic fluid ($M_s = 55\,\mathrm{kA/m}$). The critical cross-layer temperature difference ΔT_c corresponds to the onset of convection when ferrofluid is heated from below in the absence of magnetic field ($\mathrm{Ra}_c = 1.7 \times 10^3$ [57]). As seen from Figure 6.4, convection arising in a transverse uniform magnetic field sets as a result of a supercritical bifurcation when the applied temperature difference increases, and no hysteresis is detected. Because of a

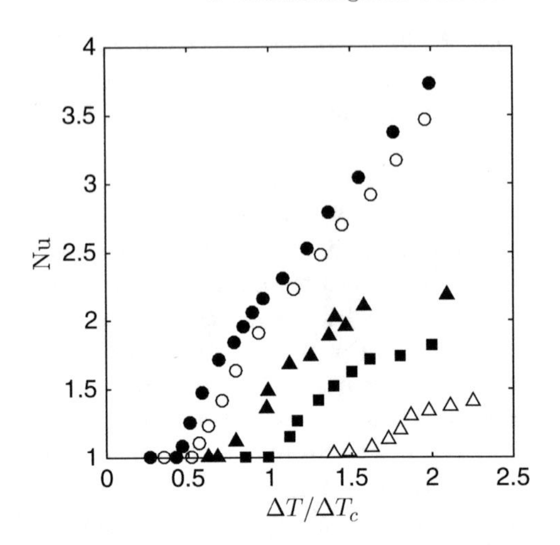

Fig. 6.4 Nondimensional heat transfer across a horizontal ferrofluid layer placed in a uniform constant vertical magnetic field H and heated from below (filled squares, triangles and circles: $H = 0$, 10 and 70 kA/m, respectively) or above (empty triangles and circles: $H = 10$ and 70 kA/m, respectively). $M_s = 55\,\text{kA/m}$, $\Delta T_c = 25\,\text{K}$, $d = 2.0\,\text{mm}$.

strong magnetisation of this concentrated ferrofluid, large values of the magnetic Rayleigh number up to $\text{Ra}_m \sim 3 \times 10^3$ were achieved even in moderate magnetic fields $H \sim 10\,\text{kA/m}$. In these conditions thermomagnetic effects are destabilising and convection sets for $\Delta T < \Delta T_c$; see the filled triangles and circles in Figure 6.4. These are in agreement with the results reported in [89, 217–219]. Note that thermomagnetic effects are strong enough to overcome the stabilising effect of a vertical density stratification and initiate convection even when the layer is heated from above; see the empty triangles and circles in Figure 6.4.

It is seen from Figure 6.4 that the dependence of Nusselt number on the applied cross-layer temperature difference in small fields is rather weak. For this reason in experiments with small magnetic fields, it is more convenient to determine the convection threshold from the Nusselt number measurements performed for a fixed cross-layer temperature difference by gradually varying the field intensity. The results obtained following this procedure are presented in Figure 6.5.

The data presented in Figures 6.4 (empty triangles) and 6.5 (empty squares) show that when the fluid is heated from above Nusselt number experiences a weak initial growth not exceeding a few percent. Such a weak variation of heat flux is a consequence of a localised convective motion near the edges of the experimental chamber; see Figures 4.6 and 4.7. It is a con-

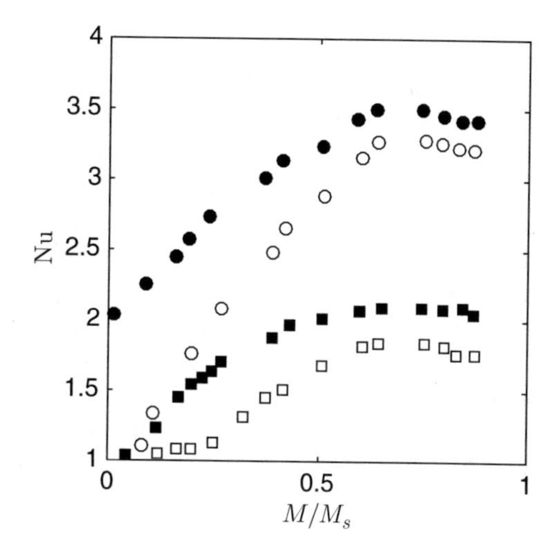

Fig. 6.5 Nondimensional heat transfer across a horizontal ferrofluid layer placed in a uniform vertical magnetic field and heated from below (filled symbols) and above (empty symbols) for $\Delta T = 19\,\mathrm{K}$ (squares) and $\Delta T = 38\,\mathrm{K}$ (circles). $M_s = 55\,\mathrm{kA/m}$, $\Delta T_c = 25\,\mathrm{K}$, $d = 2.0\,\mathrm{mm}$.

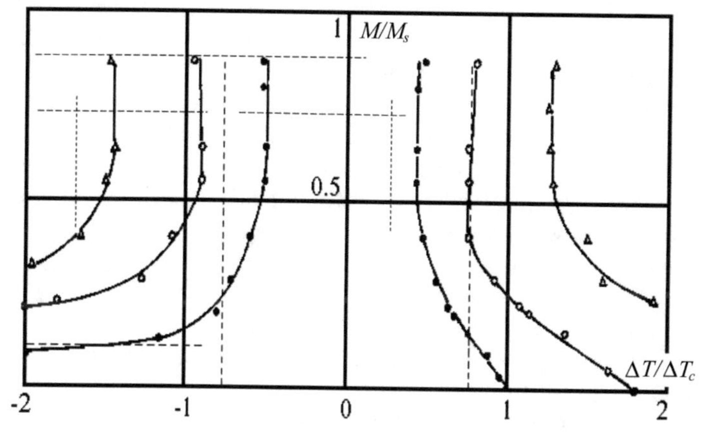

Fig. 6.6 Isolines of Nusselt number (the solid lines) for Nu = 1 (filled circles), Nu = 2 (empty circles) and Nu = 3 (triangles) [38]. The dashed lines correspond to experimental runs the data from which is presented in Figures 6.4 (horizontal lines) and 6.5 (vertical lines). $M_s = 55\,\mathrm{kA/m}$, $\Delta T_c = 25\,\mathrm{K}$, $d = 2.0\,\mathrm{mm}$.

sequence of unavoidable magnetic field non-uniformity associated with the refraction of magnetic field lines at the boundaries separating media with different magnetic properties. It arises as soon as magnetic field is switched on. A further increase of the temperature difference in Figure 6.4 or the magnetic field in Figure 6.5 leads to a rapid increase of Nusselt number due to the onset of magnetoconvection in the bulk of the ferrofluid layer.

The cumulative Nusselt number diagram is presented in Figure 6.6. It is more convenient to use the relative magnetisation M/M_s of a fluid than the magnitude of the applied magnetic field H [37] along one of the parametric axes because as seen from Figure 6.2, this stretches the weak-to-moderate field region where the most rapid changes occur and compresses the large field region where the Nusselt number variation is quenched by magnetic saturation of the fluid. The positive and negative values of $\Delta T/\Delta T_c$ correspond to heating from below and above, respectively.

Figure 6.6 shows that in strong magnetic fields when $M/M_s > 0.5$, the value of Nusselt number reaches maximum, and it cannot be increased further by strengthening the applied field. Such a saturation is likely to be related to the fact that the factor $K^2/(1+\chi)$ entering the definition of magnetic Rayleigh number Ra_m (2.26), on which the intensity of the convective heat transfer depends, reaches its physical maximum for a given fluid; see Figure 6.3.

The region of mechanical equilibrium where the bulk of a fluid remains at rest and heat is transferred by conduction is limited by the curves corresponding to $\mathrm{Nu} = 1$ in the left and right parts of Figure 6.6. Thermomagnetic convection occurs in the region located above the $\mathrm{Nu} = 1$ curve for $\Delta T/\Delta T_c < 0$. The combination of thermomagnetic and thermogravitational convection is observed above the $\mathrm{Nu} = 1$ curve for $\Delta T/\Delta T_c > 0$. Typical convection patterns arising in these regimes are illustrated in Figures 6.7 and 6.8, where the photos are presented of an experimental chamber shown in Figure 4.4. Similar to thermogravitational convection in the absence of a magnetic field convection patterns arising in a weak field contained spontaneously appearing spiral rolls that subsequently disintegrated resulting in convection cells as seen in Figure 6.7. In stronger magnetic fields convection cells formed as a result of disintegration of straight rolls due to the cross-roll instability ("zipper state" as classified in [21, 129]) illustrated in Figure 6.8. The wavenumber of the observed patterns was found to increase with the magnitude of the applied magnetic field, which is consistent with previous observations reported in [217].

In ferrocolloid with a smaller concentration of magnetic phase ($M_s = 20\,\mathrm{kA/m}$), experimental values of the magnetic Rayleigh number Ra_m were two orders of magnitude smaller than those achieved for a concentrated ferrofluid discussed above. Similar smaller values of Ra_m were also achieved in previous experiments reported in [30], where the increase of the critical temperature difference for the onset of convection was observed in a magnetic field. Our experiments confirmed previous observations: indeed the onset of convection in a magnetic field occurred at larger values of ΔT than in a zero field; see filled circles and squares in Figure 6.9, respectively. Moreover, a hysteresis was observed near the onset of convection in a strong magnetic field when the fluid was heated from below. The transition from a motionless state to convection shown by the vertical arrow in Figure 6.9 occurred abruptly when the applied temperature difference was gradually increased, while the reverse transition from convection to a quiescent state was observed to be smooth and occurred at smaller values of $\Delta T_c(H)$ taken as critical temperature differences for the convection threshold at various magnitudes H of the applied magnetic field. The depth of a hysteresis increased with H. At the

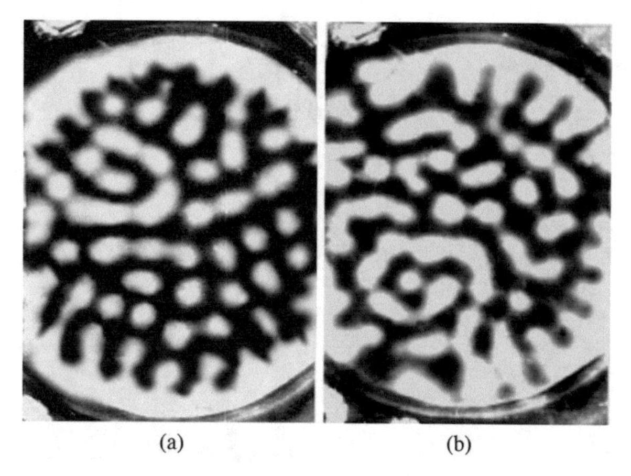

Fig. 6.7 Convection patterns in a horizontal ferrofluid layer heated from below and placed in a uniform vertical magnetic field $H = 10\,\mathrm{kA/m}$ at $\Delta T/\Delta T_c = 1.5$. The time interval between snapshots (a) and (b) is 10 min. $M_s = 55\,\mathrm{kA/m}$, $\Delta T_c = 5.1\,\mathrm{K}$, $d = 3.5\,\mathrm{mm}$.

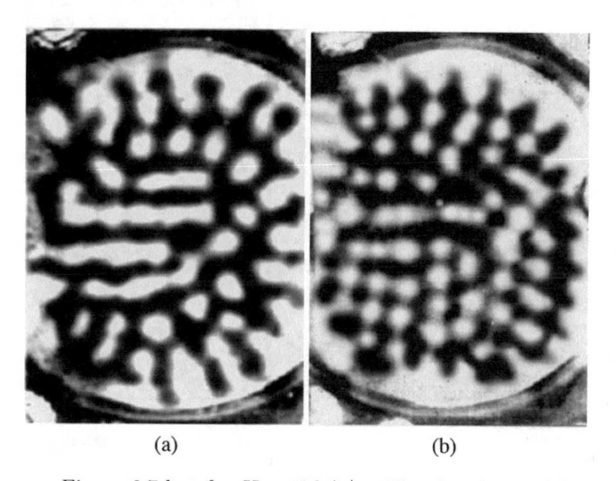

Fig. 6.8 Same as Figure 6.7 but for $H = 18\,\mathrm{kA/m}$. The time interval between snapshots (a) and (b) is 5 min.

same time, no hysteresis was observed in the onset of thermomagnetic convection in a ferrofluid layer heated from above; see empty circles in Figure 6.9.

The experimentally determined dependence of Nusselt number on the applied magnetic field expressed in terms of the relative magnetisation M/M_s of ferrofluid shown in Figure 6.10 reveals a noteworthy feature: the value of Nusselt number initially increases with field intensity up to $M/M_s \approx 0.6$ but then starts decreasing sharply. In a layer heated from below, the value of Nusselt number recorded at $M/M_s \approx 0.98$ (corresponding to $H = 200\,\mathrm{kA/m}$) is approximately equal to that determined in the absence of the field. In other words, a very strong magnetic field has no influence on the intensity of con-

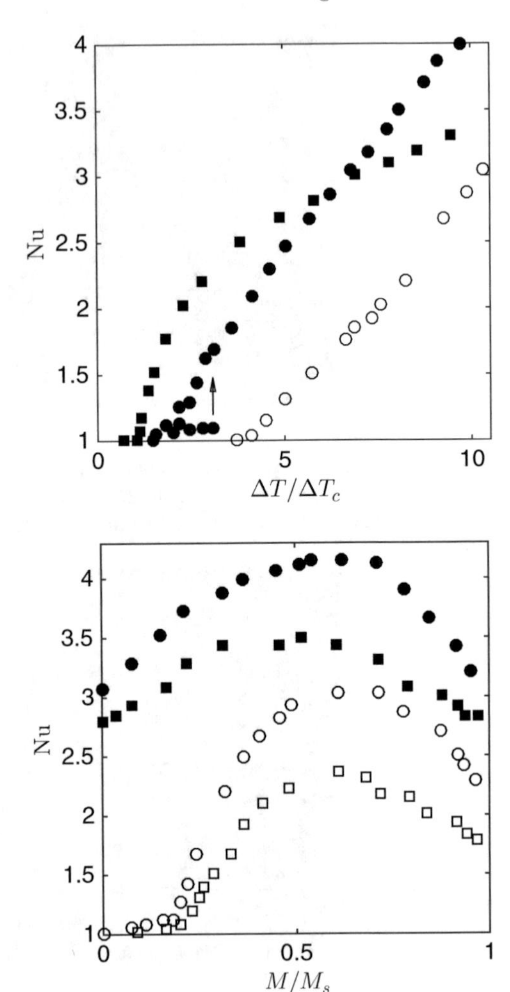

Fig. 6.9 Nondimensional heat transfer across a horizontal ferrofluid layer placed in a uniform constant vertical magnetic field and heated from below (filled symbols) and above (empty symbols) for $H = 0$ (squares) and $H = 70\,\mathrm{kA/m}$ (circles). $M_s = 20\,\mathrm{kA/m}$, $\Delta T_c = 4.5\,\mathrm{K}$, $d = 2.0\,\mathrm{mm}$.

Fig. 6.10 Nondimensional heat transfer across a horizontal ferrofluid layer placed in a uniform vertical magnetic field and heated from below (filled symbols) and above (empty symbols) for constant $\Delta T = 28\,\mathrm{K}$ (squares) and $\Delta T = 37\,\mathrm{K}$ (circles). $M_s = 20\,\mathrm{kA/m}$, $\Delta T_c = 4.5\,\mathrm{K}$, $d = 2.0\,\mathrm{mm}$.

vection. Such a surprising effect detected in weakly concentrated ferrofluids approaching magnetic saturation needs to be accounted for in practical design of heat exchangers using ferrofluids as heat carriers.

The cumulative Nusselt number map for convection in a horizontal layer of ferrofluid with small concentration of magnetic phase ($M_s = 20\,\mathrm{kA/m}$) is shown in Figure 6.11. Because of the hysteresis the Nu $= 1$ curve for a layer heated from below ($\Delta T/\Delta T_c > 0$) is obtained from experimental data corresponding to measurements at the gradually decreasing temperature differences. The shape of the Nusselt number isolines demonstrates that when the applied magnetic field exceeds a certain value, further intensification of heat transfer can only be achieved by simultaneous increase of the magnitude of the applied field and of the applied temperature difference. In the strongest magnetic fields applied in experiments, the temperature difference required for the onset of convection exceeded that in the absence of the field by a factor of two.

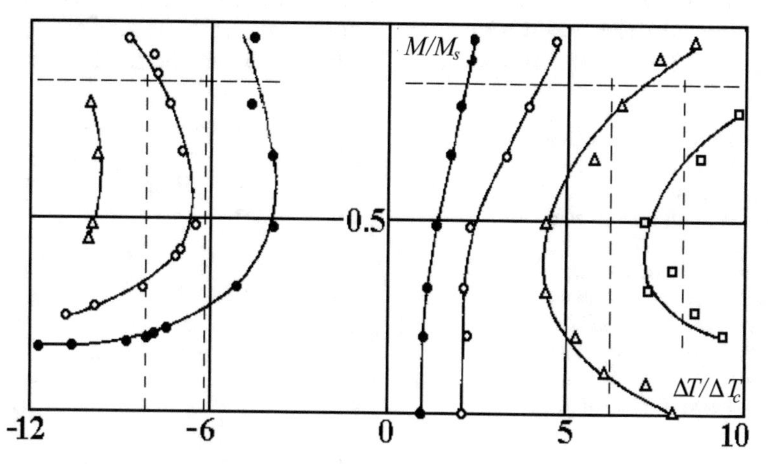

Fig. 6.11 Isolines of Nusselt number (the solid lines) for Nu = 1 (filled circles), Nu = 2 (empty circles), Nu = 3 (triangles) and Nu = 4 (squares). The dashed lines correspond to experimental runs the data from which is presented in Figure 6.9 (horizontal lines) and 6.10 (vertical lines). $M_s = 20\,\mathrm{kA/m}$, $\Delta T_c = 4.5\,\mathrm{K}$, $d = 2.0\,\mathrm{mm}$.

Fig. 6.12 Nondimensional heat transfer across a horizontal ferrofluid layer placed in a uniform vertical magnetic field and heated from below (filled symbols) and above (empty symbols) for $H = 0$ (squares), $H = 35\,\mathrm{kA/m}$ (circles) and $H = 70\,\mathrm{kA/m}$ (triangles). $M_s = 37\,\mathrm{kA/m}$, $\Delta T_c = 7.5\,\mathrm{K}$, $d = 2.0\,\mathrm{mm}$.

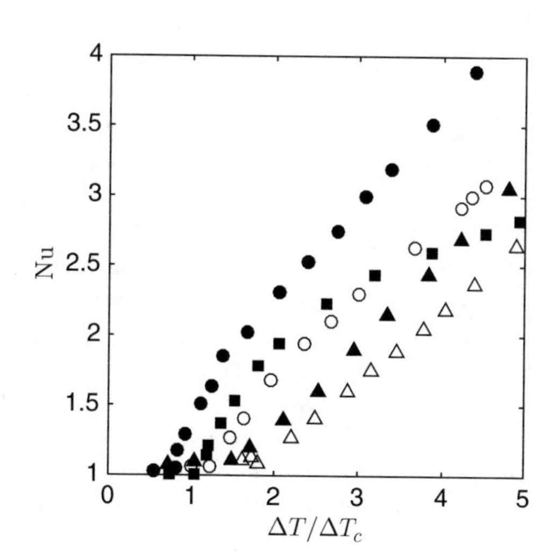

The experimental results discussed above show that the thermo-transport behaviour depends strongly on the concentration of magnetic phase in a particular ferrofluid and how close it is to the magnetic saturation. In strong magnetic field two fluids representing the limiting cases of low and high concentration behave in qualitatively opposite ways: the application of a magnetic field enhances heat transfer in concentrated fluids and suppresses it in diluted ones. To trace the transition between these two limiting behaviours, experiments with a ferrofluid with an intermediate concentration of magnetic particles ($M_s = 37\,\mathrm{kA/m}$) were performed with results presented in Figures 6.12 and 6.13. As seen from Figure 6.12 when the fluid layer heated

Fig. 6.13 Nondimensional heat transfer across a horizontal ferrofluid layer placed in a uniform vertical magnetic field and heated from below (filled symbols) and above (empty symbols) for constant $\Delta T = 22\,\mathrm{K}$ (squares) and $\Delta T = 36\,\mathrm{K}$ (circles). $M_s = 37\,\mathrm{kA/m}$, $\Delta T_c = 7.5\,\mathrm{K}$, $d = 2.0\,\mathrm{mm}$.

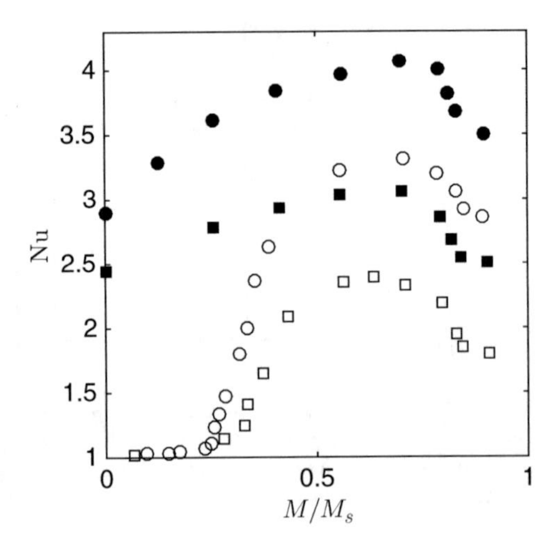

from below is placed in the vertical magnetic field $H = 35\,\mathrm{kA/m}$ (filled circles), convection sets at the cross-layer temperature difference that is twice as small as that required for the onset in zero field (filled squares). Such a behaviour is in agreement with theoretical prediction of [89]. However, in a stronger field $H = 70\,\mathrm{kA/m}$ (filled triangles), the opposite is observed: convection sets at the temperature difference that is twice as large as that required in the zero field. Thus the destabilising effect of weak magnetic fields is replaced with a stabilising influence in strong fields in a fluid with an intermediate concentration of magnetic phase ($M_s = 37\,\mathrm{kA/m}$). Such a behaviour can be explained with a reference to a particular behaviour of the parameter $K^2/(1+\chi)$ that enters the definition of the magnetic Rayleigh number. As discussed above (see Figure 6.3), it grows approximately linearly in weak fields but then reaches maximum and starts decreasing in stronger fields. In addition, the suppression of convection in strong magnetic fields can be caused by the increase of magnetoviscosity, sedimentation of particle aggregates (their number and sizes increase with the growth of a magnetic field [53, 187, 198]) and the change of sign of Soret coefficient and the associated reversal of thermodiffusive fluxes [253] in strong magnetic fields. Figure 6.13, where the dependence of the measured cross-layer heat flux on the magnetic field is shown, confirms the observed non-monotonic effect of an increasing magnetic field on convection in a ferrofluid layer. The qualitative behaviour of heat flux depicted there is similar to that observed in a diluted ferrofluid; compare Figures 6.13 and 6.10.

The Nusselt number isolines for a fluid with an intermediate concentration of magnetic phase are shown in Figure 6.14. It is evident from this figure that in all thermal regimes the heat flux achieves its maximum value at $M/M_s = 0.5 - 0.6$ and starts decreasing beyond this.

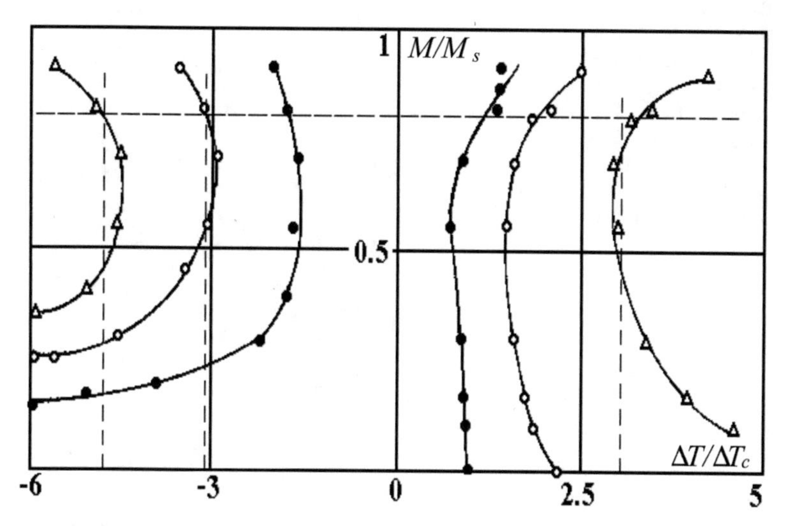

Fig. 6.14 Isolines of Nusselt number (the solid lines) for Nu = 1 (filled circles), Nu = 2 (empty circles) and Nu = 3 (triangles). The dashed lines correspond to experimental runs, the data from which is presented in Figure 6.12 (horizontal lines) and 6.13 (vertical lines). $M_s = 37\,\mathrm{kA/m}$, $\Delta T_c = 7.5\,\mathrm{K}$, $d = 2.0\,\mathrm{mm}$.

Thus, depending on the value of the magnetic Rayleigh number (2.26), which in a layer of a constant thickness maintained at a fixed cross-layer temperature difference is determined by the magnetic parameter $K^2/(1+\chi)$ (see Figure 6.3), the influence of the applied magnetic field can be either stabilising or destabilising. While destabilising effects are accurately described by theory [89], the stabilising effects such as magnetoviscous effect, sedimentation of solid phase as well as (negative) magnetophoretic and thermodiffusive fluxes [25, 252, 253] are hard to quantify and have no accurate theoretical description to date.

Since as follows from (2.26)

$$\frac{\mathrm{Ra}_m}{\mathrm{Ra}} = \frac{\mu_0 K^2 \Delta T}{(1+\chi) g \beta d},$$

in thicker layers the thermomagnetic forces play a smaller relative role, which is further reduced by the barometric effects due to the sedimentation of solid phase that is proportional to the thickness d of the layer. Besides, the edge effects leading to the distortion of a magnetic field near the layer boundaries intensify as the layer aspect ratio d/L, where L is its characteristic width, is decreased. The appearance of essentially non-uniform magnetic field leads to the onset of convection in thicker layers immediately after the magnetic field is switched on. This is seen in Figure 6.15. As soon as magnetic field

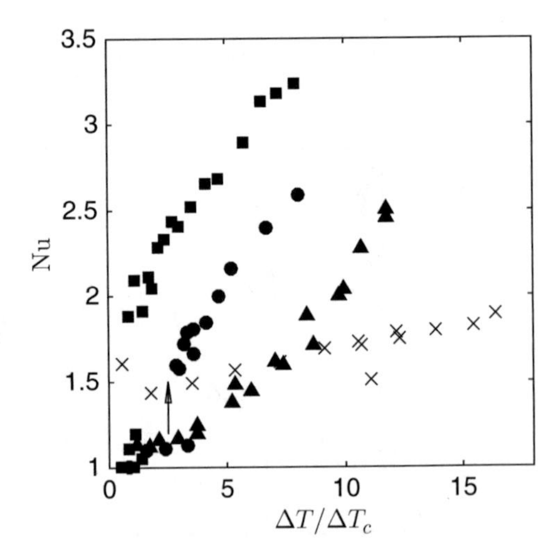

Fig. 6.15 Nondimensional heat flux in a horizontal layer of ferrofluid heated from below in constant vertical magnetic fields $H =$ 0 (squares), 16 (circles), 36 (triangles) and 144 (crosses) kA/m. $M_s =$ 20 kA/m, $\Delta T_c = 0.75$ K, $d = 5.0$ mm.

of magnitude $H = 16$ kA/m is applied, the near-edge convection arises in a thicker layer ($d = 5$ mm) that increases the value of the measured Nusselt number from 1 to approximately 1.15 (circles in Figure 6.15). The initial value of Nusselt number at $H = 144$ kA/m increases to 1.6 (crosses in Figure 6.15). The intensification of heat transfer in layers of finite aspect ratio due to the edge effects complicates the comparison of experimental results with theory [89] developed using an infinite-layer assumption. Thus both from practical and fundamental points of view, it is important to establish when the edge effects become dominant. In addition to shedding light onto this aspect of the problem, Figure 6.15 demonstrates a noteworthy fact that the application of a magnetic field to weakly concentrated ferrofluids plays a dual role. On one hand it intensifies heat transfer due to the edge effects. On the other it delays the onset of the bulk convection. Namely, the bulk convection sets at $\Delta T_c \approx$ 0.75 K in the absence of magnetic field (squares in Figure 6.15), but at $H =$ 16 kA/m the onset shifts to $\Delta T_c(H) \approx 2.5$ K (circles in Figure 6.15). Note also that the value of the gravitational Rayleigh number at the convection onset in this thicker layer is two orders of magnitude larger than that of the magnetic Rayleigh number so that the bulk convection remains primarily of buoyancy type with magnetic field just shifting the bifurcation point. A similar increase of the critical temperature difference in a magnetic field was observed in earlier experiments in a fluid with $M_s = 26.8$ kA/m [30]. It is also noted that in weak fields the bulk convection sets abruptly (see the arrow between circles in Figure 6.15), which was also observed in [30]. In stronger magnetic fields, the heat flux through a thicker layer is completely dominated by the edge effects, and the measured Nusselt numbers tend to fall onto a continuous curve; see the data for $H = 36$ (triangles) and $144\,\mathrm{kA/m}$ (crosses) in Figure 6.15.

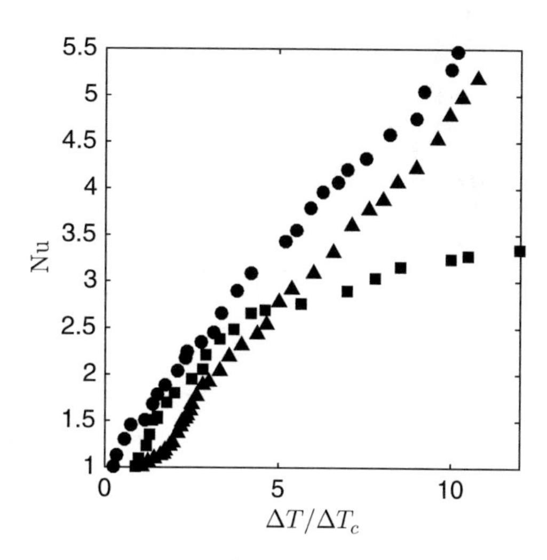

Fig. 6.16 Nondimensional heat flux in a horizontal layer of ferrofluid heated from below in vertical magnetic fields $H = 0$ (squares), 52 (triangles) and 216 (circles) kA/m. $M_s = 55\,\mathrm{kA/m}$, $\Delta T_c = 2.5\,\mathrm{K}$, $d = 5.0\,\mathrm{mm}$.

The values of magnetic Rayleigh number Ra_m that are achieved in a concentrated ferrocolloid ($M_s = 55\,\mathrm{kA/m}$) at $\Delta T < 12\,\mathrm{K}$ and $H < 55\,\mathrm{kA/m}$ are of the order $10^2 - 10^3$ and similar to a low-concentration fluid convection there sets at larger values of the cross-layer temperature difference (triangles in Figure 6.16) than in the zero field ($\Delta\, T_c|_{H=0} = 2.5\,\mathrm{K}$, squares in Figure 6.16) due to gravitational buoyancy. However, as the applied temperature difference ΔT, and thus the magnetic Rayleigh number $\mathrm{Ra}_m \sim \Delta T^2$, see (2.26), increases, the thermomagnetic contribution to heat transfer grows and at $\Delta T \approx 25\,\mathrm{K}$ (the rightmost triangle in Figure 6.16) leads to the 30% increase over the total heat flux compared to that measured in the zero field. In large magnetic fields, the role of the associated physical factors destabilising the motionless state becomes dominant, and at $H > 55\,\mathrm{kA/m}$, the onset of convection shifts to lower values of ΔT (circles in Figure 6.16) than $\Delta\, T_c|_{H=0}$. Such a shift can be associated with both the development of convection in the bulk of the fluid and the intensification of edge effects associated with finite thickness and lateral dimensions of the enclosure.

Figure 6.17 demonstrates yet another ambiguity in interpreting the influences of magnetic field on convection in a horizontal layer. In this figure the heat transfer results are presented for weak (circles) and strong (triangles) magnetic fields and compared with those for the zero field (squares). It is seen from the figure that near the onset of convection, the strength of the applied magnetic field has virtually no effect on the intensity of heat transfer. Moreover, the onset of convection in a magnetic field is not sharply defined due to the edge effects. In thick layers the influence of the edge effects propagates deep inside the fluid domain so that they affect the flow structure in a complete layer. Only in strongly supercritical regimes where the bulk convection

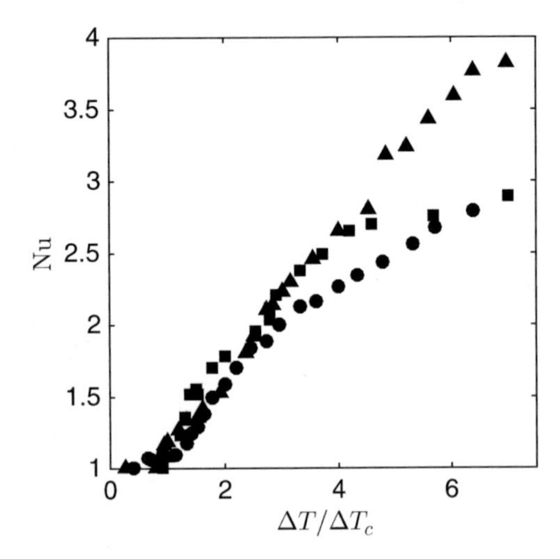

Fig. 6.17 Nondimensional heat flux in a horizontal layer of ferrofluid heated from below in vertical magnetic fields $H = 0$ (squares), 8 (circles) and 104 (triangles) kA/m. $M_s = 55\,$kA/m, $\Delta T_c = 2.5\,$K, $d = 5.0\,$mm.

caused by buoyancy and ponderomotive forces acquires strength comparable to that of the near-edge rolls arising due to the non-uniformity of a magnetic field there can the effects of various fields be distinguished. Even then, it remains non-monotonic and depends on the magnitudes of both the applied magnetic field and the cross-layer temperature difference. The application of weak fields suppresses the overall heat flux across a thick layer of ferrofluid compared to that measured in the zero field while strong fields promote it.

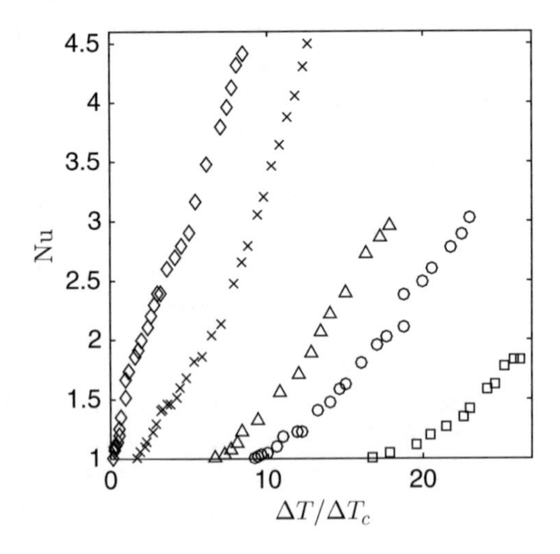

Fig. 6.18 Nondimensional heat flux in a horizontal layer of ferrofluid heated from above in vertical magnetic fields $H = 13$ (squares), 18 (circles), 27 (triangles), 52 (crosses) and 204 (diamonds) kA/m. $M_s = 55\,$kA/m, $\Delta T_c = 2.5\,$K (when heated from below), $d = 5.0\,$mm.

When a horizontal layer of ferrofluid is heated from above, the gravitational component of convection is precluded and the enhancement of the cross-layer heat transfer is achieved solely due to magnetic forcing. Figure 6.18 demonstrates that in this case strengthening of the applied magnetic field leads to the monotonic increase of heat flux and the decrease of the threshold temperature difference at the convection onset: at $H = 13\,\mathrm{kA/m}$ convection arises only at $\Delta T \approx 42\,\mathrm{K}$ (squares in Figure 6.18), while at $H = 216\,\mathrm{kA/m}$, it sets at $\Delta T \approx 0.34\,\mathrm{K}$. However, these values cannot be treated as critical for the onset of pure thermomagnetic convection because they are still influenced by the edge effects that depend on the shape and the aspect ratio of the experimental chamber. Indeed the comparison with Figure 6.4, where similar results for thin layer are shown, demonstrates a qualitative change in the behaviour of the Nusselt number curves: they change their shape from the square-root- to the parabolic-type indicating a strong influence of the edge effects on flow structure in thick layers.

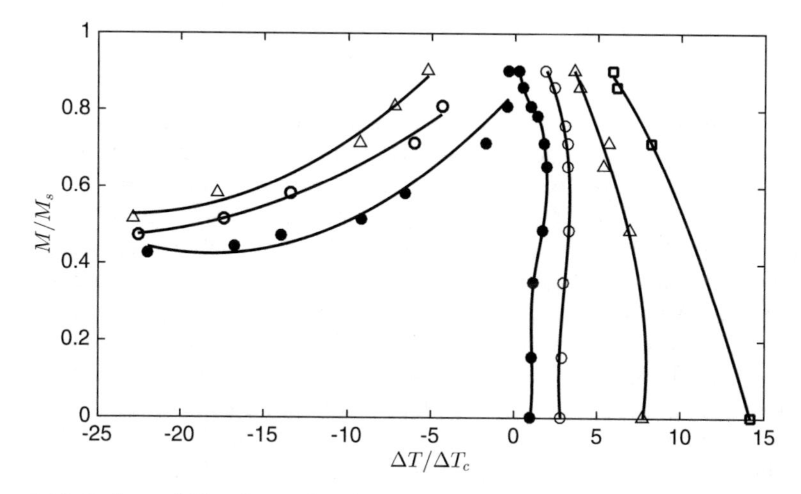

Fig. 6.19 Isolines of Nusselt number for a thick layer ($d = 5\,\mathrm{mm}$) of concentrated ferrofluid: Nu $= 1$ (filled circles), Nu $= 2$ (empty circles), Nu $= 3$ (triangles) and Nu $= 4$ (squares). $M_s = 55\,\mathrm{kA/m}$, $\Delta T_c = 2.5\,\mathrm{K}$.

The comparison of the overall Nusselt number diagrams for thin and thick layers of concentrated ferrofluid shown in Figures 6.6 and 6.19 also demonstrates significant qualitative differences. In strong fields ($H > 100\,\mathrm{kA/m}$) the deviation of the measured Nusselt number from unity occurs in thick layers at temperature differences of only several tenths of a degree both when the layer is heated from below and from above. This heat flux enhancement not detected in thin layers is attributed to the near-edge toroidal flows illustrated in Figures 4.6 (left) and 4.7. In smaller fields ($H < 100\,\mathrm{kA/m}$) the onset of convection in the bulk of the layer is shifted towards larger temperature differences regardless of the direction of heating.

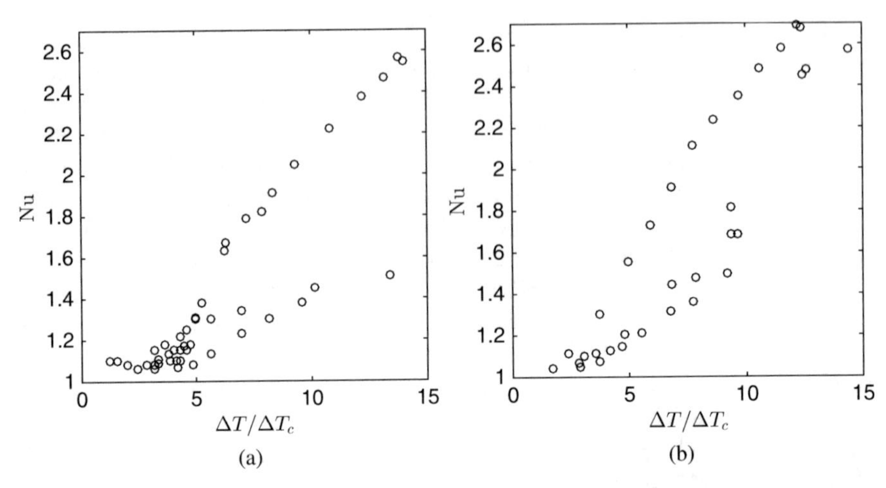

Fig. 6.20 Nondimensional heat flux in a horizontal layer of ferrofluid heated from above in vertical magnetic fields (a) $H = 26\,\mathrm{kA/m}$ and (b) $H = 36\,\mathrm{kA/m}$. $M_s = 20\,\mathrm{kA/m}$, $\Delta T_c = 0.75\,\mathrm{K}$, $d = 5.0\,\mathrm{mm}$.

The ability to excite convection and increase heat transfer rate manyfolds even in weakly concentrated ferrofluid in conditions such as heating from above where natural gravitational convection is impossible is the major motivation for using magnetically controllable media in heat management. Figure 6.20 confirms that indeed this is practically possible. However, it also shows yet another unexpected effect. If the applied cross-layer temperature difference ΔT increases gradually, the observed growth of Nusselt number is relatively slow; see points along the lower branches in diagrams in Figure 6.20(a) and (b). The system was observed in each of the states corresponding to points marked by symbols for at least an hour without noticeable variation of the heat transfer rate. At the same time if magnetic field was switched on after a large temperature difference $\Delta T \sim 30\,\mathrm{K}$ was established, the measured Nusselt number values fell onto the upper branches in the diagrams. The transitions between the two branches were observed when the applied temperature difference was changed abruptly by several degrees. A similar bistate heat transfer was also observed in experiments with nanofluids [79] and binary mixtures [4]. Weak heat fluxes in these experiments were due to the existence of the localised-state regimes in which regions with a developed convection alternated with quiescent conduction regions. Large heat flux regimes were observed when bulk convection existed in a complete fluid layer. In binary mixtures and nanofluids, the existence of localised states was found to be due to the stabilising density gradient induced by the negative thermodiffusion. In weakly concentrated ferrofluids heated from above, the stabilising density gradient is likely to result from the action of gravitational sedimentation and thermophoresis (when $S_T > 0$) of a solid phase.

6.2.3 Convection in a Horizontal Layer Placed in a Magnetic Field Parallel to the Layer

Since the tangential component of a magnetic field is continuous across the boundary separating media with different magnetic properties, the magnitude of a magnetic field inside a ferrofluid layer placed in an external uniform magnetic field parallel to its plane remains constant (with the exception of the near-edge regions; see Section 4.4.1). Therefore, the ponderomotive magnetic force that is proportional to the gradient of a local magnetic field is absent, and such a magnetic field cannot cause any fluid motion on its own. However, if the fluid is set in motion by other physical forces such as gravitational buoyancy that creates thermal non-uniformity in the direction of the field, the ponderomotive force does occur and influences the structure of a fluid flow. As the computational results for convection in an obliquely applied magnetic field presented in Figures 3.32, 3.33 and 3.34 demonstrate, rolls aligned with the direction of the in-layer field component become preferred. A brief theoretical discussion of the situation when the field is parallel to the layer can be found in [13]. In particular, it is noted that in contrast to flows of conducting fluids [51], the application of such a field to electrically non-conducting ferrocolloids placed in a magnetic field parallel to the layer does not change the convection onset but leads to the alignment of rolls with the field as shown schematically in Figure 6.21. The qualitative

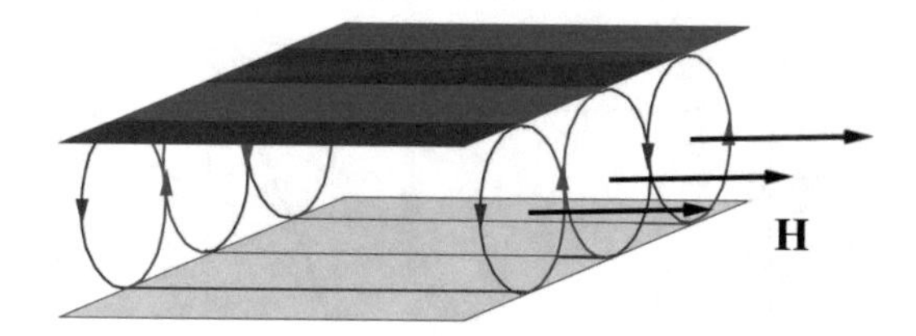

Fig. 6.21 Schematic view of the alignment of convection rolls by a magnetic field parallel to a horizontal ferrofluid layer heated from below.

diagram given in Figure 6.22 explains why rolls not aligned with the applied magnetic field are suppressed by it. A convection roll appearing in a fluid layer heated from below can be partitioned into four quarters labelled 1–4 in Figure 6.22 and characterised by different average temperatures T_1, T_2, T_3 and T_4, respectively. Quarter 2 is the warmest, quarter 3 is the coolest and quarters 1 and 4 have some intermediate temperatures so that $T_1 < T_2$ and $T_4 > T_3$. Therefore, the temperature gradient with a component parallel to the applied field and perpendicular to the roll axis is established. On average, the fluid magnetisation in quarters 1 and 3 (the downflow part of the roll) is larger than that in quarters 2 and 4 (the upflow part of the roll). Similar to the situation with a flat vertical ferrofluid layer heated from a side considered in Chapter 3, the thermally induced variation of the fluid magnetisation in the direction of the applied magnetic field leads to the appearance of the magnetic field gradient ∇H that has the right-to-left component in

Figure 6.22. Subsequently, the right-to-left ponderomotive force arises that tends to drive cooler and stronger magnetised fluid from quarters 1 and 3 to the left (towards quarters 2 and 4). However, the average temperature of the lower half of the roll that is closer to the heated boundary is higher than that

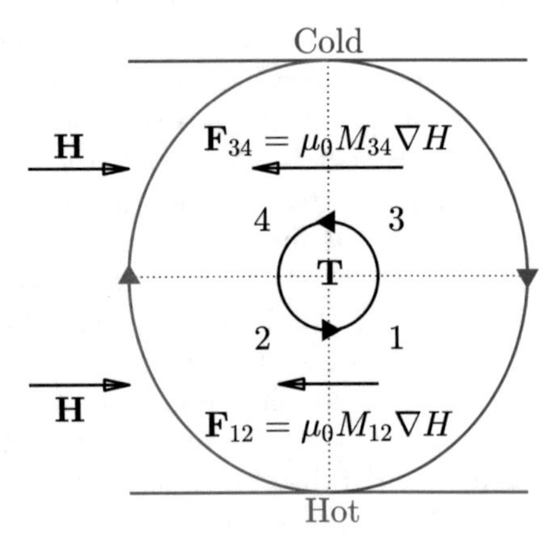

Fig. 6.22 Schematic force diagram explaining the suppressing influence of the applied magnetic field onto convection rolls with axes perpendicular to the field.

of the upper part, and, subsequently, the magnetisation M_{12} of the lower half is weaker than M_{34} of the upper half. The resulting difference between the corresponding ponderomotive forces F_{12} and F_{34} effectively creates a force moment $\mathbf{T}$ that tends to rotate fluid anticlockwise in Figure 6.22 and against

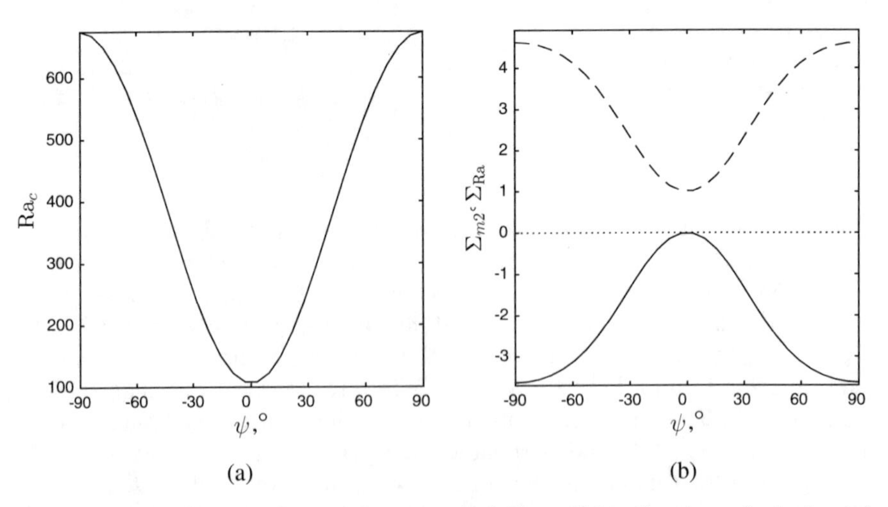

Fig. 6.23 The aligning influence of a magnetic field parallel to the plane of a horizontal ferrofluid layer heated from below: (a) critical Rayleigh number for convection rolls and (b) perturbation energy components as functions of the angle ψ between the axes of convection rolls and the magnetic field vector. The representative results are computed for $\mathrm{Ra}_m = 825$ and $\mathrm{Pr} = 55$.

the primary rotation of the roll. As a result the intensity of rolls not aligned with the field is suppressed. On the other hand, if the applied field is parallel to the rolls, the temperature in its direction remains constant and the ponderomotive force moment does not occur. Such rolls are not affected by the applied magnetic field.

Figure 6.23 contains further computational confirmation of the aligning influence of the in-layer magnetic field on convection rolls. It is seen from Figure 6.23(a) that the critical Rayleigh number computed as described in Section 3.4 for convection rolls forming angle ψ with the vector of the applied magnetic field increases with this angle. In other words, at the same values of the magnetic and gravitational Rayleigh numbers, convection rolls aligned with the field have a larger growth rate. Figure 6.23(b) also demonstrates that convection rolls remain of purely thermogravitational nature: Σ_{Ra} is the only positive contribution into the perturbation energy balance equation (see Sections 3.3.3 and 3.4.3), which in the present case is written as

$$\sigma^R \Sigma_k = \Sigma_{\mathrm{Ra}} + \Sigma_{m2} - 1\,, \tag{6.2}$$

where

$$\begin{aligned}
\Sigma_k &= \int_{-1}^{1} \left(|u|^2 + |v|^2\right)\, \mathrm{d}x > 0\,,\\
\Sigma_{\mathrm{Ra}} &= \mathrm{Ra}\,\mathrm{Pr} \int_{-1}^{1} \Re(\theta\bar{u})\, \mathrm{d}x\,,\\
\Sigma_{m2} &= \mathrm{Ra}_m\, \mathrm{Pr}\, \alpha \sin\psi \int_{-1}^{1} \theta_0 \Re(\alpha\bar{v}\phi - \mathrm{i}\bar{u}D\phi)\, \mathrm{d}x\,.
\end{aligned}$$

As in ordinary non-magnetic fluids, thermal expansion is responsible for the onset of convection, and magnetic field plays a stabilising role ($\Sigma_{m2} < 0$) unless the rolls align with it, in which case it has no effect.

Experimentally, the alignment of convection rolls with the in-layer magnetic field was first reported in [217]. However, several important physical effects that occur in this configuration have not been mentioned there. This section is devoted to their discussion. We also note that the orienting influence of magnetic field parallel to a ferrofluid layer is in many regards similar to that of a buoyancy-driven basic shear flow in an inclined fluid layer heated from below considered in Section 3 and a vertical layer heated from a side discussed in Chapter 3 and Section 6.3. The experiments discussed here have been performed with a setup shown in Figure 4.4.

The observed convection pattern maps are presented in Figure 6.24 (see Figure 6.2 for the $M(H)$ dependence). As seen from this figure, the onset of convection shown by vertical lines (fluid remains at rest to the left of the lines) is independent of the applied horizontal magnetic field, which is consistent with a theoretical discussion given in [13]. The horizontal line segments in panel (a) show the variation of the measured cross-layer temperature differ-

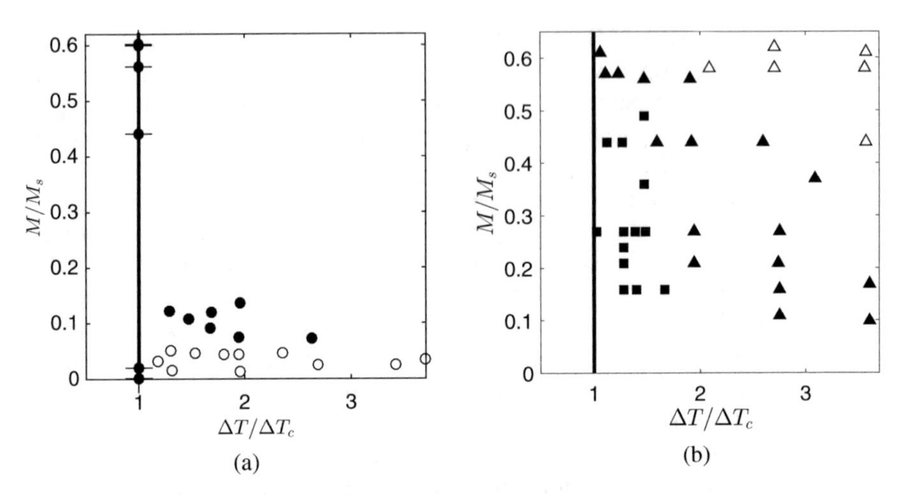

Fig. 6.24 Flow regime diagrams for a horizontal ferrofluid layer placed in a magnetic field parallel to the layer. In panel (a), filled circles with horizontal bars denote the stability boundary of a motionless state (bars show the amplitude of the temperature oscillations detected at the convection onset), empty circles correspond to spiral and target-like convection patterns, and filled circles—elongated rolls of arbitrary orientations. In panel (b), symbols denote squares, confined states; filled triangles, blinking states with rolls moving away from the centre of the layer; and empty triangles, blinking states converging towards the centre of the layer ($M_s = 55\,\mathrm{kA/m}$, $d = 3.5\,\mathrm{mm}$, $\Delta T_c = 5.1\,\mathrm{K}$).

ence that evidences the fact that unlike in ordinary single-component fluids oscillatory convection sets in a horizontal layer of a ferrofluid heated from below.

The oscillations of a cross-layer heat flux in the absence and presence of the magnetic field parallel to the layer are shown in Figure 6.25. The readings were recorded at fixed temperature differences between heat exchangers. As seen from the figure, the increase of heat flux near the convection threshold is faster in magnetic field. The amplitude of heat flux oscillations in magnetic field $H = 25\,\mathrm{kA/m}$ is about 2.5 times larger than in zero field for $\Delta T/\Delta T_c \leq 1.5$. It is likely that the amplification of oscillations is associated with dehomogenisation of the fluid by gravitational sedimentation and thermodiffusion that are intensified in magnetic field due to the magnetically induced formation of particle aggregates and the increase of their sizes [8, 118, 187, 193, 197, 199]: the magnitude of Soret coefficient, which is positive in the absence of magnetic field and when the field is perpendicular to the applied temperature gradient, increases with the field [25, 253]. At $H = 25\,\mathrm{kA/m}$ the Nusselt number reaches its maximum at $\Delta T/\Delta T_c = 1.6$ and then saturates. The saturated value of Nu is about 10% smaller than that in the absence of the field.

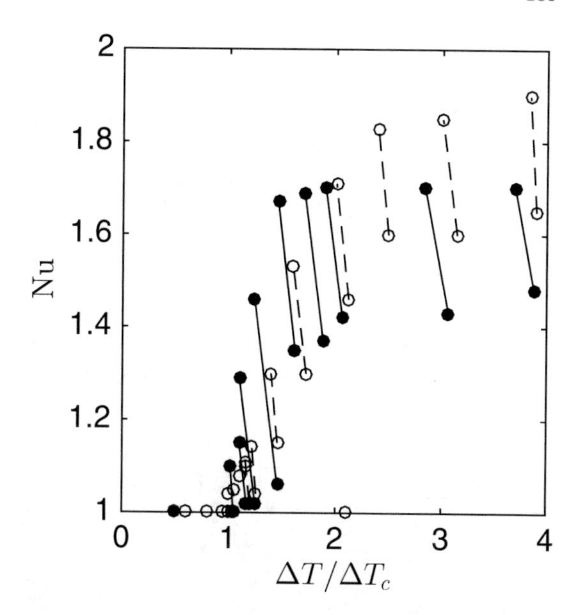

Fig. 6.25 The variation of nondimensional heat flux across a horizontal ferrofluid layer heated from below in the absence of magnetic field (empty circles) and when a uniform magnetic field $H = 25\,\mathrm{kA/m}$ is applied along the layer (filled circles). The lines show the amplitude of heat flux oscillations (vertical extent) and the associated variation of the measured temperature difference (horizontal extent): ΔT decreases as Nu increases.

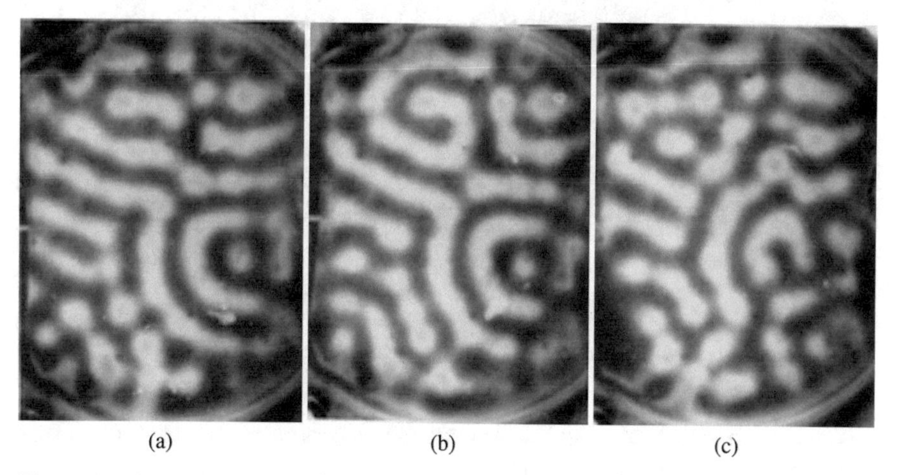

Fig. 6.26 Spiral defect chaos [28, 99] in a weak magnetic field $H = 0.4\,\mathrm{kA/m}$ parallel to a horizontal layer (left to right) of ferrofluid heated from below at $\Delta T/\Delta T_c = 2$. The time interval between snapshots is 10 min.

Of interest is the evolution of oscillatory spatio-temporal structures in an increasing magnetic field. In weak fields (Figure 6.24(a)) convection structures experience cross-roll instability. At moderate values of the applied temperature difference $\Delta T/\Delta T_c = 2$, the aligning influence of the magnetic field is not well pronounced, and convection structures in the shape of spiral and target-like rolls are observed (the empty circles in Figure 6.24(a)). They are illustrated in Figure 6.26. In the lower right quarter of Figure 6.26(a), two rolls form a spiral. Some time later (see image (b)), they transform into a target pattern, and another spiral of the opposite orientation is formed in the upper left quarter of the layer. Subsequently, both spiral and target break into cells as a result of a cross-roll instability; see image (c).

The pattern evolution observed at a larger cross-layer temperature difference $\Delta T/\Delta T_c = 4$ is illustrated in Figure 6.27. In this case the patterns tend to align with the applied magnetic field (from left to right in image (a)) as the interaction between them and magnetic field strengthens as the applied temperature difference increases [13]. The remaining structures that are not aligned with the field break due to a cross-roll instability as seen in image (b) and result in predominantly horizontal rolls illustrated in image (c). They occupy the complete layer apart from the close vicinity of the edges

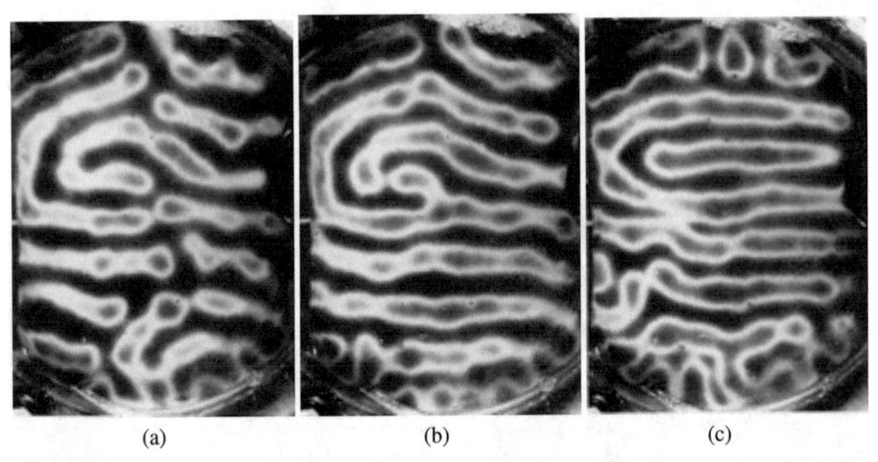

(a) (b) (c)

Fig. 6.27 Evolution of a spiral structure in a weak magnetic field $H = 0.4\,\mathrm{kA/m}$ parallel to a horizontal layer of ferrofluid heated from below at $\Delta T/\Delta T_c = 4$. The time intervals between snapshots (a)–(b) and (b)–(c) are 10 and 25 min, respectively.

of the experimental chamber where multiple defects caused by their curvature dominate the convection pattern.

In the parametric region marked by solid circles in Figure 6.24(a), convection rolls align with the applied magnetic field; see Figure 6.28(a). However, they are spontaneously destroyed in different parts of the layer by a cross-roll instability; see Figure 6.28(b) just to reform some time later.

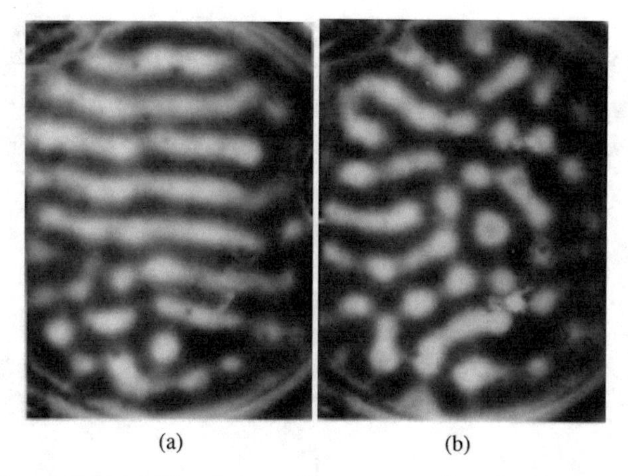

Fig. 6.28 Cross-roll instability of convection patterns aligned with magnetic field $H = 1.0\,\mathrm{kA/m}$ parallel to a horizontal layer of ferrofluid heated from below at $\Delta T/\Delta T_c = 2$. The time interval between snapshots is 15 min.

The parametric region in which convection rolls aligned with the applied field were observed are marked by squares in Figure 6.24(b). Such rolls appeared and disappeared spontaneously in various parts of the layer. A similar convection regime has also been observed in binary mixture convection where it is referred to as confined states [21, 129, 164]. Localised convection states are also observed in an inclined ferrofluid layer in the absence of magnetic field; see Figures 5.18 and 5.19. Such a "wandering" convection is illustrated in a series of photographs in Figure 6.29. Five light stripes corresponding to ten convection rolls are seen in image (a). Since convection patterns occupy a complete layer, the heat flux between its faces is at its maximum. It is seen from image (b) that convection starts decaying in the central part of the layer and the heat flux across the layer decreases. In the series of images (c)–(e), the intensity of convection rolls increases in the lower part of the images and decreases in the upper. Convection rolls start reappearing in the upper and central parts of image (f) and eventually fill the complete layer in image (g), where the white arrow indicates the direction of motion of the central roll. As a result of this motion, the total number of the observed rolls reduces to eight in image (h). Such an evolution of convection patterns led to the variation of the cross-layer temperature difference of up to 2.5 K measured at the centre of the layer over 30–40 min convection decay/reappearance cycle.

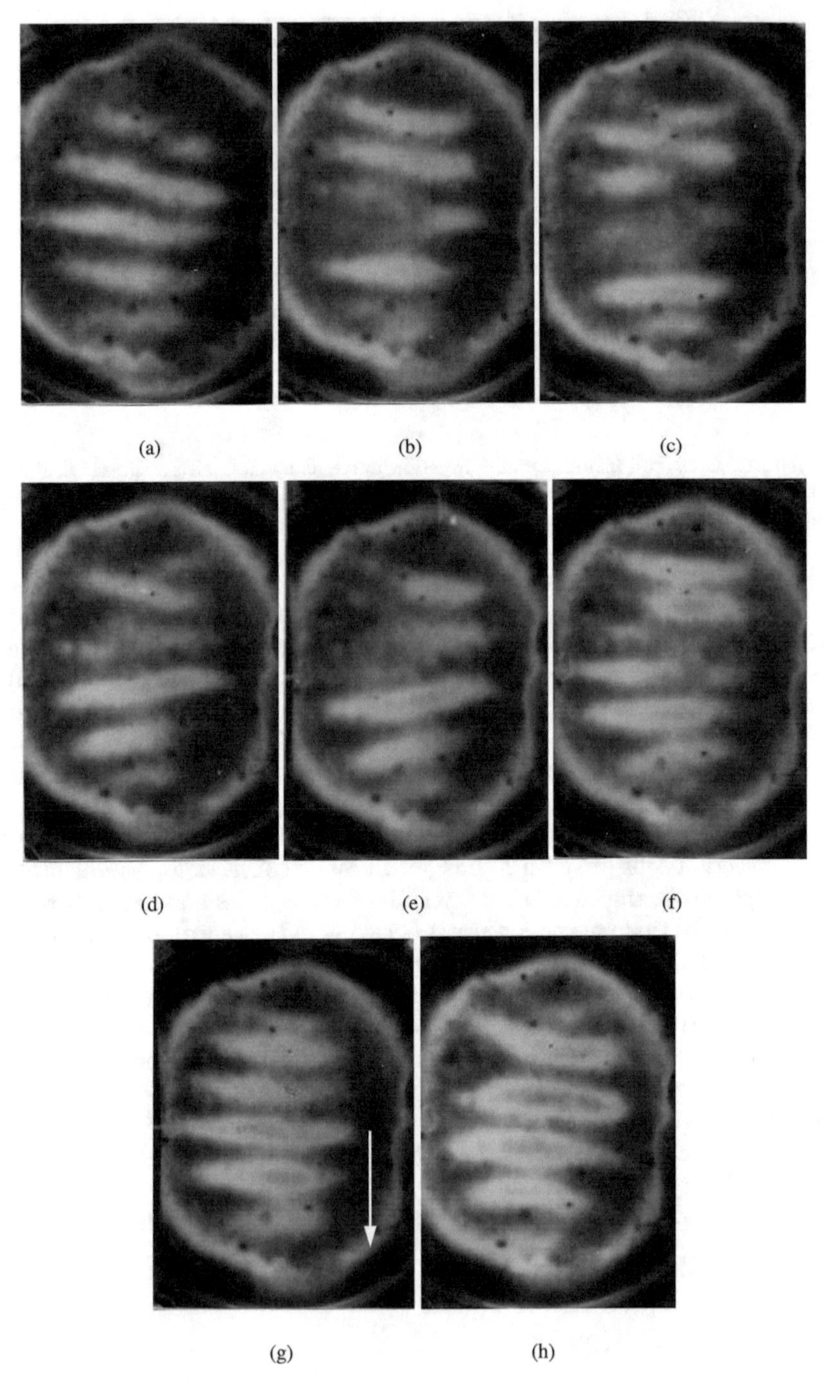

Fig. 6.29 Wandering convection in a horizontal ferrofluid layer heated from below at $\Delta T/\Delta T_c = 1.3$ and placed in magnetic field $H = 17\,\mathrm{kA/m}$ parallel to the layer. The time intervals between consecutive snapshots are 5, 8, 4, 7, 5, 5 and 11 min.

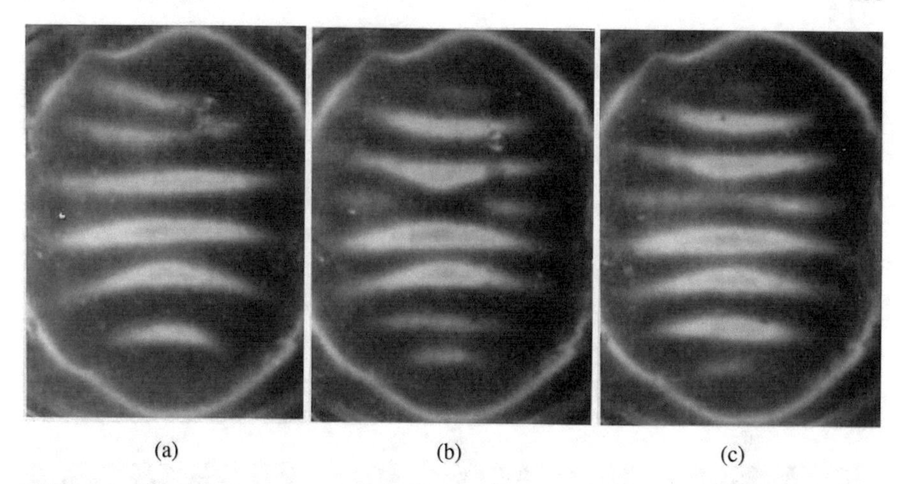

Fig. 6.30 Blinking state convection in a horizontal ferrofluid layer heated from below at $\Delta T/\Delta T_c = 2.5$ and placed in magnetic field $H = 17\,\mathrm{kA/m}$ parallel to the layer. The time intervals between snapshots (a)–(b) and (b)–(c) are 3 and 1 min, respectively.

The filled triangles in Figure 6.24(b) mark the regime where convection rolls aligned with the applied field propagate in the direction perpendicular to the field from the centre of the layer towards its edges, that is, up and down in Figure 6.30. Such a dynamics is caused by a climbing dislocation arising at the left and right edges of the layer (see Figure 6.30(a) and (b)) and resulting in an additional pair of rolls appearing near the centre of the layer and displacing the existing rolls in the direction perpendicular to their axes; see Figure 6.30(c). The formation of new convection rolls typically takes several minutes and leads to the roll displacement speed of the order of several millimetres per minute. The intensity of the displaced rolls changes as they approach the edges of the layer, and this creates an impression of blinking convection patterns. The total number of bright stripes in Figure 6.30 changes from 4 to 6 which corresponds to the convection roll number varying from 8 to 12. This indicates the simultaneous presence of two convection modes with different wavenumbers that leads to a blinking state instability [21].

A more complicated dynamics of convection patterns observed at the same values of physical parameters is illustrated in Figure 6.31. In the lower half of image (a), defects lead to the so-called pinning effect [99, 160] resulting in two fork-like structures. These defects annihilate leading to the formation of new rolls; see images (b) and (c). At the same time in the upper half of images (a)–(c), defects lead to alternating coalescence of neighbouring rolls and formation of new defects, which is similar to the dynamics observed in Hele-Shaw convection [150]. A relatively simple pattern illustrated in image (c) is subsequently destroyed by the birth of new defects seen in images (d)–(f). The three-branch structures are formed alternately in the lower and upper parts of the images. The appearance and disappearance of defects in the bulk of the layer occurs simultaneously with those of a roll near the lower edge of

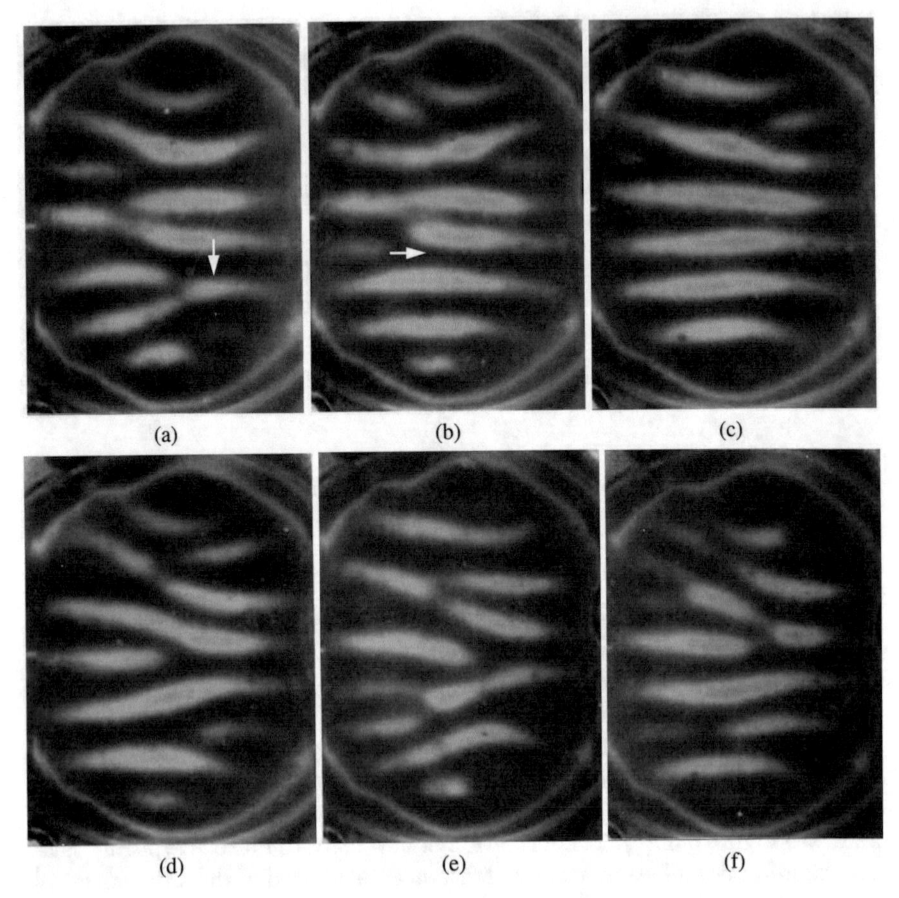

Fig. 6.31 Regime of propagating rolls in a horizontal ferrofluid layer heated from below at $\Delta T/\Delta T_c = 2.5$ and placed in magnetic field $H = 17\,\mathrm{kA/m}$ parallel to the layer. The vertical and horizontal arrows show sliding and climbing dislocations, respectively. The time intervals between the consecutive snapshots are 12, 5, 6, 30 and 10 min.

the images and "blinking" (variation of the intensity) of rolls near the upper edge.

In strong magnetic fields convection rolls aligned with the field pinch in the central part of the layer and break leading to the birth of dislocations. Such regimes are marked by empty triangles in Figure 6.24(b). The formed dislocations move along the rolls towards the left and right edges of the layer and disappear once they reach them. In contrast, new dislocations appear near the upper and lower edges of the image that collide and form new rolls aligned with the applied field.

To conclude, magnetic field parallel to the plane of a horizontal ferrofluid layer does not cause the appearance of convection but has the orienting influence on thermogravitational convection patterns. In the next section, we will consider the opposite situation where cross-layer convection arises due to the action of a magnetic field, while the fluid motion caused by the gravitational buoyancy plays the orienting role aligning thermomagnetic patterns with the direction of the fluid velocity.

6.3 Vertical Layer

6.3.1 Problem Overview

This chapter will focus on experimental studies of convection flows arising in a flat ferrofluid layer heated from a side and placed into a uniform magnetic field normal to the layer. In this case magnetic and gravitational forcing act in perpendicular directions and thus are easier to distinguish. Such flows have been studied in the past; see [20, 116, 250]. Yet despite the apparent simplicity of a setup, there was only limited success in detecting thermomagnetic effects in these early experiments. There were two main reasons for this: the relatively large thickness of the used experimental layers ($\sim 1\,\mathrm{cm}$) and small magnetisation of working fluids. These resulted in the small values of the magnetic Grashof (Rayleigh) number relative to those of the gravitational one. Thus the observed convection flows were dominated by gravitational buoyancy effects.

Earlier theoretical studies [121] were also unconvincing as they assumed from the outset that instability patterns in a vertical ferrofluid layer placed in a normal magnetic field will be similar to those observed in purely buoyancy-driven flows and form two-dimensional horizontal structures. Yet later experimental studies [36, 37] and a more comprehensive three-dimensional linear stability analysis [234] showed that the thermomagnetic convection in a vertical layer of ferrofluid placed in a normal magnetic field sets in the form of vertical rolls. The details of this analysis and the discussion of the interplay between gravitational and thermomagnetic mechanisms of convection for an infinite vertical layer have been given in Chapter 3. Here we discuss experimental results for finite ferrofluid layers and parametric regimes precluding the appearance of thermogravitational waves.

6.3.2 Thermomagnetic Convection Patterns

Here we will discuss the experimental results for ferrofluid flows in vertical differentially heated layers of various thicknesses h and heights l. The experiments were performed with kerosene-based ferrofluids with the saturation magnetisations $M_s = 43$ and $55\,\mathrm{kA/m}$. When the magnetic field applied normally to the layer is not sufficiently strong, the fluid in the layer rises along the hot wall and descends along the cold one as illustrated in Figure 6.32 (regime A, region [1] in Figure 3.21(a)). The colouring of a thermosensitive film remains horizontally uniform. The vertical colour variation is due to a vertical temperature stratification arising because of the finite height of the experimental cavity. When the strength of magnetic field increases, the magnetoconvection structures appear that are superposed onto the basic flow and are visualised by the emerging colour patterns on the thermosensitive film; see regime B in Figure 6.32 (regions [3] and [10] in Figure 3.21(a)). The first thermomagnetic convection mode assumes the form of vertical rolls aligned with the direction of the up-down basic flow. Each blue-brown stripe pair corresponds to a roll with the cross-layer velocity component directed from the hot to cold wall (blue) or vice versa (brown). Such colour variations have been used to characterise various spatial flow patterns developing in the layer in supercritical regimes.

The resulting approximate stability boundary for the fluid layer with thickness $d = 3.5\,\mathrm{mm}$ (see Figure 4.4) is shown by a line in Figure 6.33. As seen from this figure, the increase of the cross-layer temperature difference and of the magnetic field strength leads to the variation of wavenumber β of the magnetoconvection rolls: its value increases further away from the basic flow stability boundary in supercritical regimes. Similar behaviour of the instability structures is also observed in horizontal ferrofluid layers [38, 217]. The critical value of the nondimensional wavenumber (scaled using the half-thickness $d/2$ of the layer) $\beta_c = 2.0 \pm 0.1$ agrees well with the theoretical value reported in Table 3.1 (point D).

Figure 6.34 presents the values of a dominant convection roll wavenumber β in the horizontal z direction (see Figure 3.2) as a function of the supercriticality parameter. According to the analysis of Section 3.3, it grows with magnetic Grashof number.The available experimentally measured wavenumbers

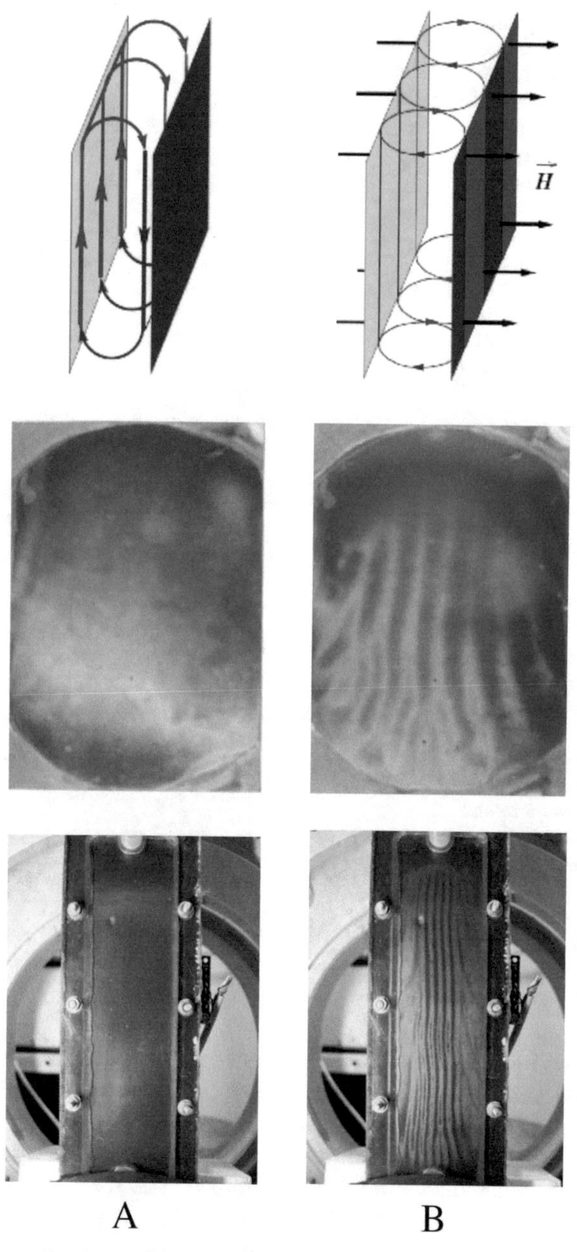

Fig. 6.32 Schematic view and experimental visualisation of the buoyancy-driven parallel up-down basic flow (A, $H = 0$ kA/m) and thermomagnetic rolls (B) in the disk-shaped ($d = 3.5\,\mathrm{mm}$, $l = 75\,\mathrm{mm}$) and rectangular ($d = 4.0\,\mathrm{mm}$, $l = 250\,\mathrm{mm}$) vertical layers. Experimental parameters for the disk-shaped and rectangular cavities are ($H = 19\,\mathrm{kA/m}$, $\Delta T = 20\,\mathrm{K}$, $M_s = 55\,\mathrm{kA/m}$) and ($H = 21\,\mathrm{kA/m}$, $\Delta T = 13.3\,\mathrm{K}$, $M_s = 43\,\mathrm{kA/m}$), respectively. Magnetic field is perpendicular to the plane of the photographs.

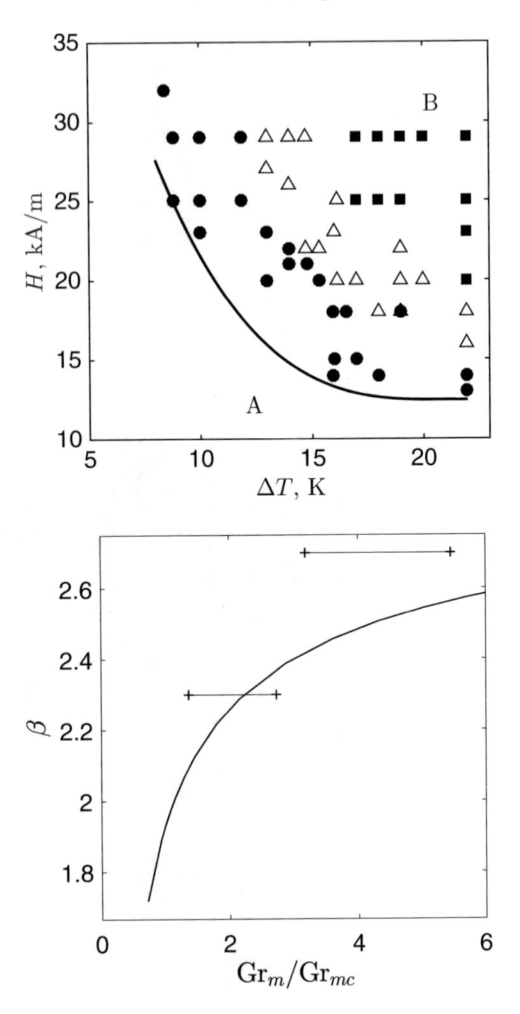

Fig. 6.33 Map of convection flows in the form of vertical rolls with nondimensional wavenumbers $\beta = 2.1$ (circles), 2.4 (triangles) and 2.7 (squares) in a vertical layer ($d = 3.5\,\mathrm{mm}$, $l = 75\,\mathrm{mm}$) of ferrofluid ($M_s = 55\,\mathrm{kA/m}$) heated from a side and placed in a uniform transverse magnetic field. Stable parallel basic flow exists in region A and an instability is observed in region B.

Fig. 6.34 The dominant wavenumber for thermomagnetic convection observed in a normal magnetic field and taking the form of vertical stationary roll as determined by the linear stability analysis of Section 3.3. Wavenumbers of convection roll experimentally observed in a 3.5 mm thick layer over a range of magnetic Grashof numbers are shown by the connected "+" symbols.

are also shown in this plot. The agreement is very good near the criticality, but the analytical values are somewhat lower than experimental in far supercritical regimes. This is expected since in reality the unstable wavenumbers spectrum widens away from a bifurcation point so that a spatial modulation of periodic instability patterns brought about by the enclosure boundaries occurs. This may lead to the deviation of the dominant wavenumber predicted for an infinite layer from that observed in finite geometry.

While the wavenumber of the observed magnetoconvection patterns is well defined, the total number of convection rolls across the width of the fluid layer changes in time as a result of their amplitude modulation and movement of dislocation defects. For example, consider series of snapshots in Figure 6.35. The left- and rightmost rolls in the top row series of snapshots (the disk-shaped layer) periodically disappear and reappear. A weaker modulation of

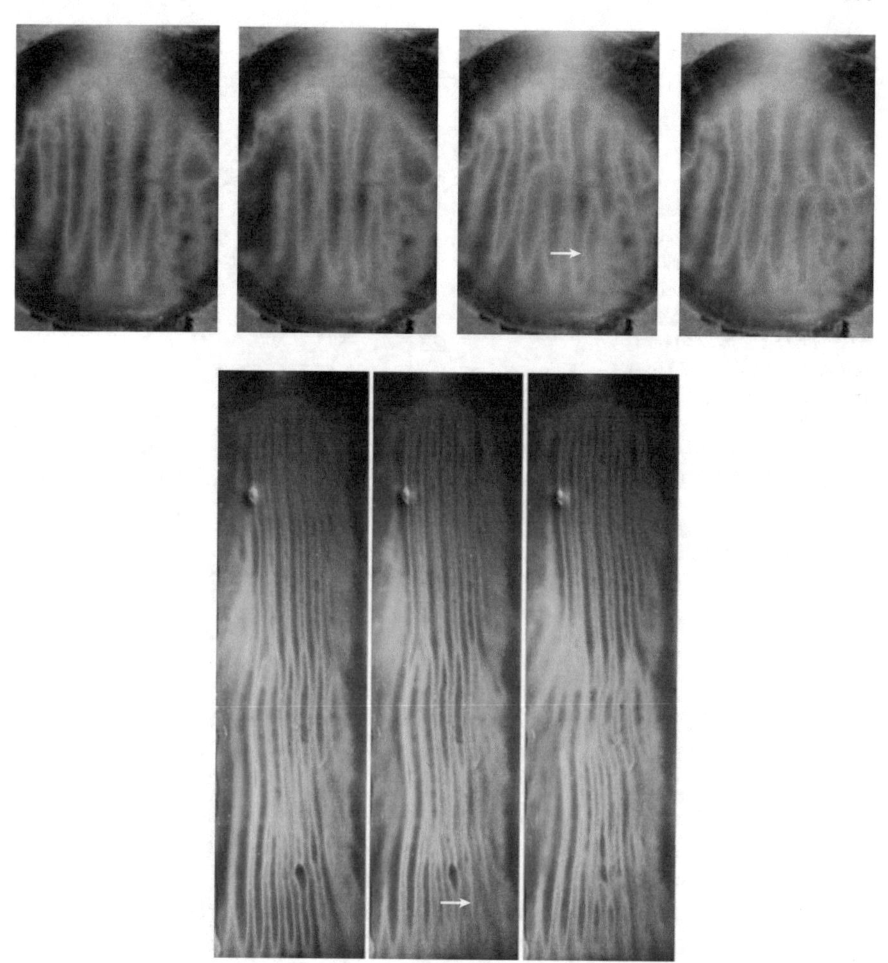

Fig. 6.35 Amplitude modulation of convection rolls in the vicinity of the side boundaries and the motion of dislocations observed at the cold wall of the disk-shaped (top row; d = 3.5 mm, l = 75 mm, M_s = 55 kA/m, ΔT = 15 K, H = 17 kA/m) and rectangular (bottom row; d = 4.0 mm, l = 250 mm, M_s = 43 kA/m, ΔT = 18 K, H = 21 kA/m) ferrofluid layers. The time between snapshots is 3 min in the top row and 1.4 min in the bottom row.

the near-edge rolls is also visible in the bottom row (rectangular layer). Similar spatio-temporal behaviour with variable intensity of convection rolls have been also observed in binary mixture convection [21, 129]. Moreover, the snapshot series also demonstrate the appearance of dislocation defects leading to the formation of additional convection rolls. Such dislocation can be climbing (moving along the roll axis) or gliding (moving perpendicular to the rolls) [99, 160]. The translation of gliding defects is shown by a white arrow in Figure 6.35. There a part of a green stripe shifts to the right perpendicular to the axis of the rolls. In addition, in the bottom row in Figure 6.35, the variation of a vertical length of a defect is detected signifying its climbing nature.

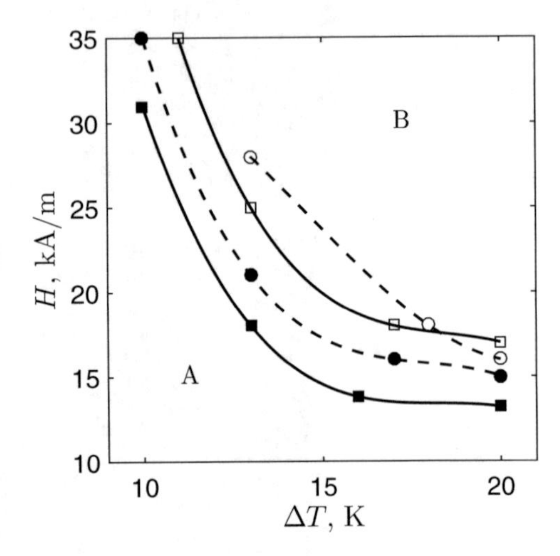

Fig. 6.36 Stability diagram of the parallel up-down basic flow in a transverse magnetic field in vertical ferrofluid (M_s = 43 kA/m) layers with height l = 250 mm (squares) and l = 70 mm (circles). Filled (empty) symbols correspond to the downward (upward) flow near the observation wall. Stable flow exists in region A and an instability is observed in region B.

The onset of thermomagnetic convection in a transverse magnetic field was investigated in layers of different heights and in two heating configurations: when the observation wall of the layer with a thermosensitive film attached was either heated or cooled. The experimental parallel basic flow stability boundaries are shown in Figure 6.36. It is seen there that the thermomagnetic instability sets in taller layers at smaller values of the applied magnetic field H and temperature difference ΔT. Therefore, the vertical temperature stratification, which is stronger pronounced in shorter layers, plays a stabilising role. Thus the theoretically obtained stability boundaries reported for the infinite layer in Section 3.3.4 is expected to provide the parametric lower bound for the convection onset in realistic layers. A similar dependence of the instability onset on the aspect ratio of the fluid layer is observed in purely gravitational convection [18, 147, e.g.]. A surprising fact is though that the onset of thermomagnetic convection depends on the direction of the applied temperature gradient stronger than on the height of the layer. Indeed, due to the symmetry of a problem, it is natural to expect that the onset of thermomagnetic convection should be independent of the sign of the applied cross-layer temperature gradient. However, as seen from Figure 6.36 the critical values of H and ΔT are noticeably larger when the layer is cooled from the observation side where a thermosensitive film is attached. The reason for this seemingly contradictory fact is that in order to make visualisation of convection structures possible, the temperature of a thermosensitive film has to be maintained in the 18–20° C range regardless of whether it is attached to the heated or cooled side of the layer. Therefore the average fluid temperature is larger when the observation side of the layer is cooler and fluid flows downwards along it. For example, when the temperature difference between the walls is maintained at $\Delta T = 20\,\mathrm{K}$ as in Figure 6.37, the difference between the average temperatures in the layer heated from the back and from the front is 10 K. The difference in average values of viscosity and magnetisation of a ferrofluid in this case reaches 10–20% and 1%, respectively. Since the value of magnetic Grashof number characterising thermomagnetic convection is $\mathrm{Ra}_m \sim (\Delta T/\eta)^2$, it sets for smaller values of the applied temperature difference in a less viscous fluid, that is, when the layer's front wall is cooler.

The evolution of convective structures in a layer heated from the back and from the front in an increasing magnetic field is shown in Figure 6.37. When no magnetic field is applied, the colour of the thermosensitive film is mostly uniform: brown (18° C, heating from the back) in image (a) and blue (22° C, heating from the front) in image (f). When a relatively weak magnetic field is applied, its distortion near the layer boundaries causes localised flows there that is seen as narrow green (image (b)) or brown (image (g)) stripes near the vertical edges; see the discussion in Section 4.4.1. When magnetic field is increased (images (c) and (h)), thermomagnetic rolls appear over the complete layer. Further increase of the applied field leads to the reduction of the area occupied by vertical rolls in favour of unsteady oblique rolls appearing near the vertical edges of the layer (images (d) and (i)). These rolls tend to align with the in-layer component of the distorted near-edge magnetic field and drift upwards. Their speed grows with the increase of a magnetic field. The comparison of images (e) and (j) corresponding to the identical values of the applied temperature difference and magnetic field but opposite directions of heating shows that the thermomagnetic motion near the observation wall is more intense when the fluid is heated from the back and the average temperature of the fluid is higher. In this configuration oblique near-edge rolls form cascades with multiple dislocations and intense fluid mixing. Such an evolution of convective structures is robust and is observed in layers with various aspect ratios and of different shapes (rectangular and circular).

The detailed parametric map of the observed convection regimes is given in Figure 6.38 for the fluid layer with thickness $d = 6\,\mathrm{mm}$ (see Figure 4.5). The onset boundary is shown by a solid line interpolating the experimental data. The measurement accuracy shown by error bars is determined primarily by the step size of discrete parameter variation: $\pm 0.7\,\mathrm{K}$ for the temperature difference at fixed magnetic field and $\pm 0.7\,\mathrm{kA/m}$ for the magnetic field at fixed temperature difference. Various patterns arising in the layer heated from the back are illustrated in Figure 6.39 for $\Delta T = 18.3\,\mathrm{K}$. Image (a) corresponds to the unperturbed up-down basic flow. The vertical colour gradient indicates the existence of a vertical temperature stratification with the maximum temperature difference (between blue and dark brown regions) of about 4 K.

As H and, subsequently, the value of magnetic Grashof number increase, various instability patterns arise on a background of basic up-down flow: stationary vertical thermomagnetic rolls (Figure 6.39(b), squares in Figure 6.38)

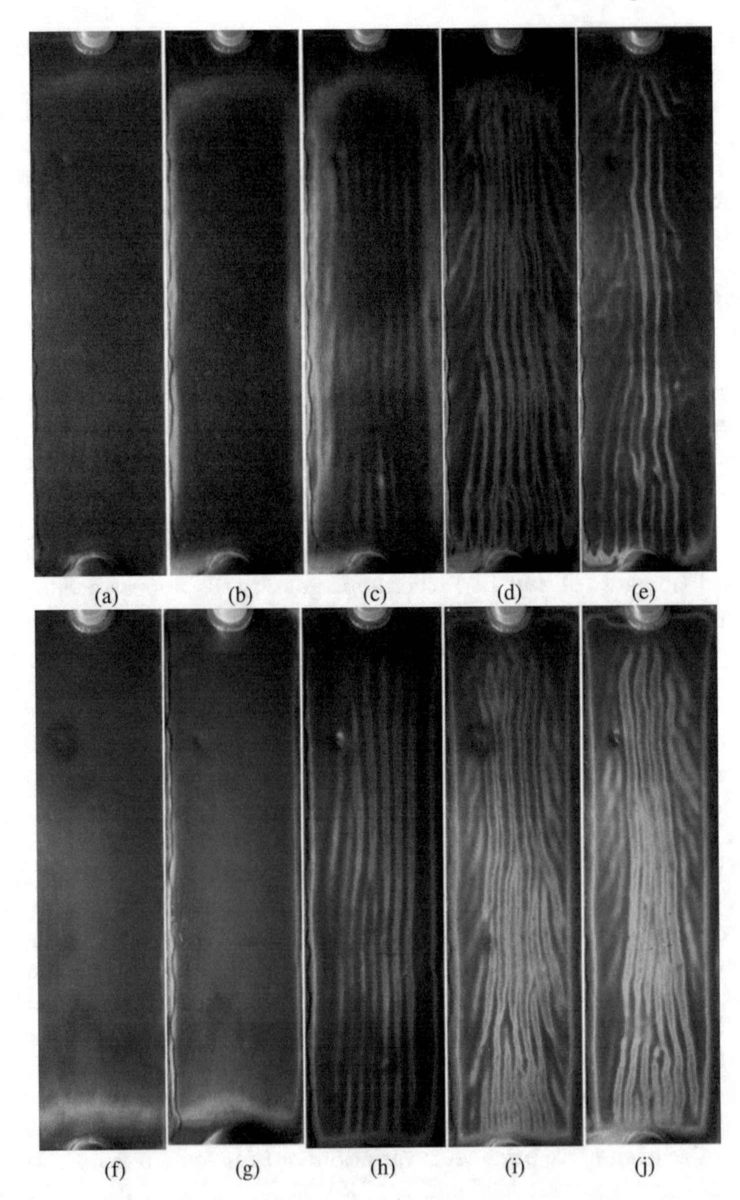

Fig. 6.37 Evolution of convective structures in the increasing uniform transverse magnetic field at $\Delta T = 20\,\mathrm{K}$ in a ferrofluid ($M_s = 43\,\mathrm{kA/m}$) layer with height $l = 250\,\mathrm{mm}$ and thickness $d = 4.0\,\mathrm{mm}$. Top row, heating from the back (a) $H = 0\,\mathrm{kA/m}$, (b) $H = 11\,\mathrm{kA/m}$, (c) $H = 15\,\mathrm{kA/m}$, (d) $H = 25\,\mathrm{kA/m}$, (e) $H = 35\,\mathrm{kA/m}$; bottom row, heating from the front (f) $H = 0\,\mathrm{kA/m}$, (g) $H = 13\,\mathrm{kA/m}$, (h) $H = 21\,\mathrm{kA/m}$, (i) $H = 31\,\mathrm{kA/m}$, (j) $H = 35\,\mathrm{kA/m}$.

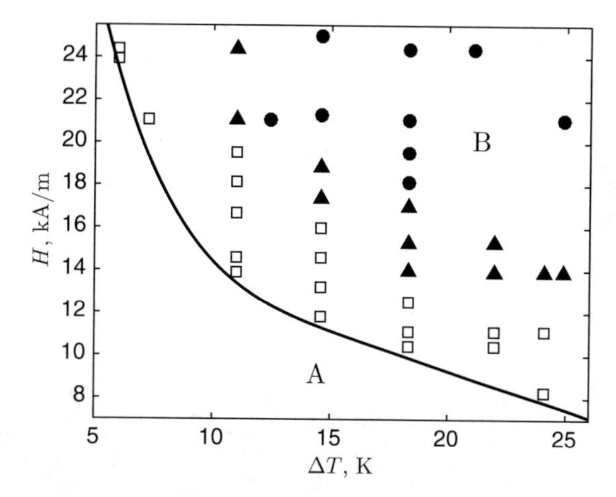

Fig. 6.38 Map of convective flows in a vertical layer ($d = 6\,\text{mm}$, $l = 250\,\text{mm}$) of ferrofluid ($M_s = 43\,\text{kA/m}$) heated from the back and placed in a uniform transverse magnetic field. Stable parallel basic flow exists in region A and an instability is observed in region B. Various flow patterns are detected in region B: squares, stationary vertical thermomagnetic rolls; triangles, superposition of stationary vertical rolls and propagating waves with a single frequency; and circles, superposition of stationary vertical rolls and propagating waves with two frequencies.

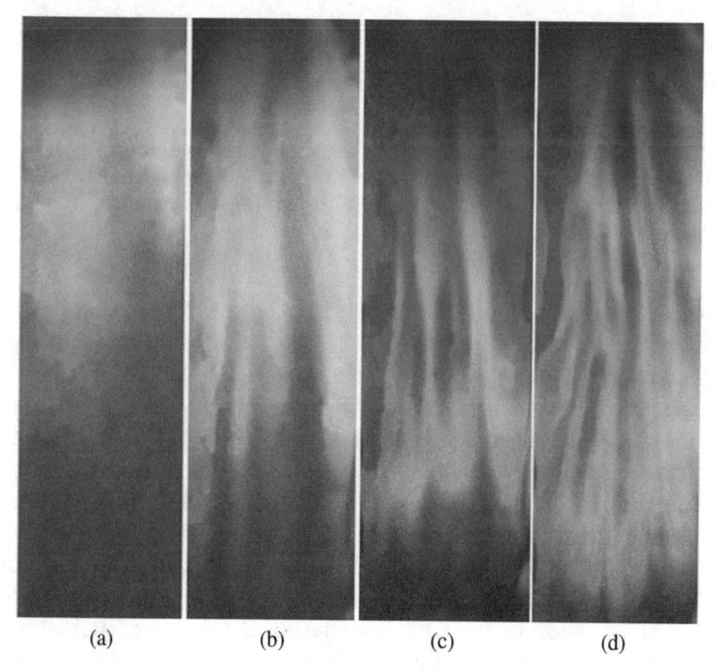

Fig. 6.39 Evolution of convective structures in the increasing uniform transverse magnetic field at $\Delta T = 18.3\,\text{K}$ in a ferrofluid ($M_s = 43\,\text{kA/m}$) layer with height $l = 250\,\text{mm}$ and thickness $d = 6.0\,\text{mm}$: (a) $H = 0\,\text{kA/m}$, (b) $H = 12\,\text{kA/m}$, (c) $H = 17\,\text{kA/m}$, (d) $H = 21\,\text{kA/m}$. Heating from the back.

and then a superposition of stationary rolls and propagating waves (Figure 6.39(c) and (d), triangles and circles in Figure 6.38). The total number of convection rolls increases further away from the onset of convection. According to the stability analysis of Section 3.3.4, the increasing complexity of flow patterns is due to the co-existence of stationary rolls and thermomagnetic waves with different wavenumbers. Images (b)–(d) in Figure 6.39 show that thermomagnetic convection patterns arising in a progressively stronger magnetic field mixes ferrofluid bringing warm fluid from the back of the layer to the front; see the domination of blue regions in images (c) and (d).

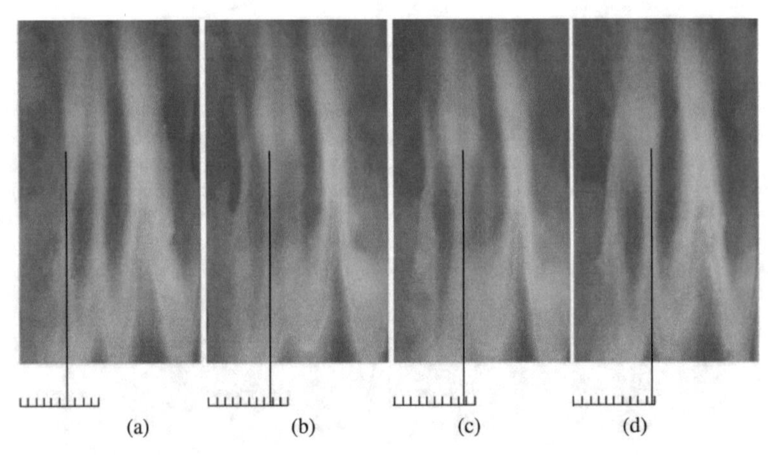

Fig. 6.40 Superposition of stationary rolls and a single thermomagnetic wave at $\Delta T = 22\,\mathrm{K}$ and $H = 14$ kA/m in a ferrofluid ($M_s = 43\,\mathrm{kA/m}$) layer with height $l = 250\,\mathrm{mm}$ and thickness $d = 6.0\,\mathrm{mm}$. The time delay between images (a) and (b) and (b), (c) and (d) is 30 and 20 s, respectively. Heating from the back.

The spatio-temporal evolution of thermomagnetic convection pattern consisting of vertical stationary rolls and a propagating wave and corresponding to the parametric region marked by triangles in Figure 6.38 is illustrated in Figure 6.40. The central part of the layer is shown. Each of the blue-green pairs in image (a) corresponds to a thermomagnetic roll. The location of the leftmost green stripe is marked by a vertical line in image (a) is traced in images (b)–(d). It is seen that it shifts to the right signifying the presence of horizontally propagating waves. Their experimentally measured speed is approximately 0.1 mm/s, which agrees with the computations reported in [236, 238]; see Figure 3.7(c). The superposition of waves and rolls leads to the temporal variation of the spatial period of the observed pattern. For example, the second from the right blue stripe in image (a) splits into two in image (c) and becomes a single stripe again in image (d).

Figure 6.41 shows a typical visualisation of flow regimes marked by circles in Figure 6.38. The two vertical rolls corresponding to blue stripes marked 1 and 3 in image (b) remain stationary and dominate the pattern throughout the observations. At the same time secondary blue stripes 2 and 4 travel from left to right. Their wavelength is 1.5–1.8 times smaller than that of stationary rolls, which is in good agreement with computational results presented in Figure 3.7(b).

Superficially, the images presented in Figures 6.40 and 6.41 may appear quite similar. However, they correspond to qualitatively different regimes. To establish this, a time series of thermocouple (see Figure 6.41(a)) readings corresponding to the difference between the temperature in the fluid layer and

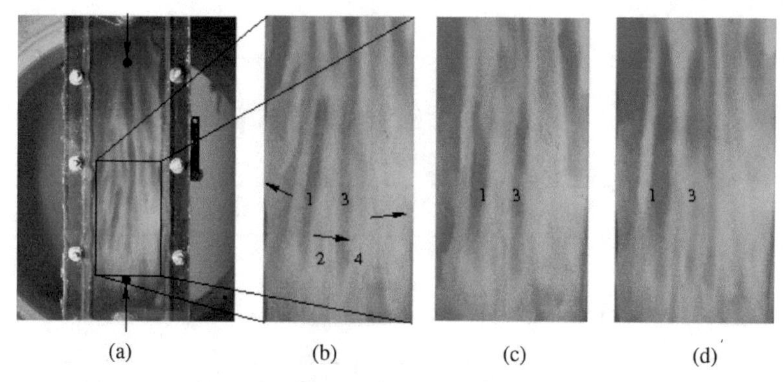

Fig. 6.41 Superposition of stationary rolls and two waves at $\Delta T = 18.3\,\mathrm{K}$ and $H = 21$ kA/m in a ferrofluid ($M_s = 43\,\mathrm{kA/m}$) layer with height $l = 250\,\mathrm{mm}$ and thickness $d =$ 6.0 mm: (a) full view of the experimental setup, (b)–(d) close-up snapshots. The time delay between images (b) and (c) and (c) and (d) is 30 and 50 s, respectively. Heating from the back. Black arrows in image (a) show the location of thermocouples; arrows in image (b) indicate the direction of motion of convection rolls.

at a reference point in the body of a copper heat exchanger—thermograms similar to those shown in Figure 6.42—were recorded and analysed.

When experimental runs are performed in a weak magnetic field that varies in a stepwise manner, the thermograms remain flat as in Figure 6.42(a). However, the temperature registered by thermocouples changes abruptly every time the magnetic field changes reflecting different modes of heat transfer across the layer. For $H = 0\,\mathrm{kA/m}$ only basic buoyancy-driven up-down flow exists in the cavity so that heat is transferred between the hot and cold walls of the layer mostly via conduction. When magnetic field is switched on but remains weak ($H = 7\,\mathrm{kA/m}$ in Figure 6.42(a)), the drop of the temperature is recorded. It is caused by the enhancement of heat transfer due to the appearance of toroidal flow near the edges of the layer that is caused by the distortion of magnetic field at the boundary between magnetic and non-magnetic media discussed in detail in Section 4.4.1. Further increase of the applied magnetic field (up to $H = 10\,\mathrm{kA/m}$ in Figure 6.42(a)) triggers the onset of stationary thermomagnetic rolls and related intensification of heat transfer leading to yet another drop in the recorded temperature.

When the magnetic field increases to 14 kA/m, temperature oscillations with the period of approximately 480 s occur; see Figure 6.42(b). The power spectrum of the corresponding time series is shown in the top plot in Figure 6.43(a). The fundamental harmonic has the frequency $\nu_0 \approx 0.0021$ Hz. Due to a nonlinear self-interaction, it leads to the appearance of superharmonics (bound modes) with multiple frequencies $2\nu_0$, $3\nu_0$ and $4\nu_0$ visible in the power spectrum. The self-interaction of the fundamental harmonic also leads to the modification of mean flow corresponding to a local maximum of the power spectrum near zero frequency. Thus the first oscillatory mode of

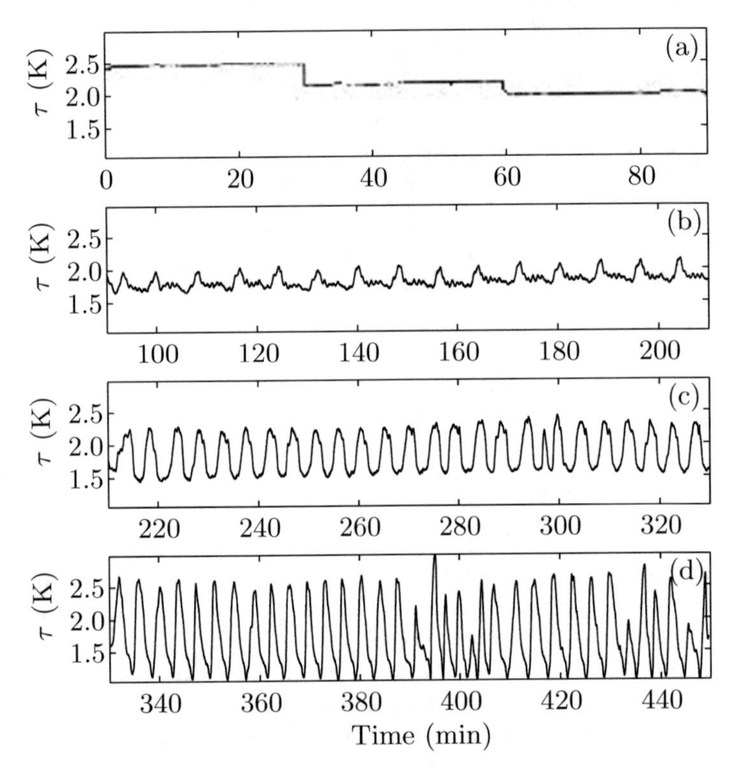

Fig. 6.42 Thermograms recorded by a lower thermocouple (see Figure 6.41(a)) for $\Delta T = 18.3$ K in a layer ($d = 6$ mm, $L = 250$ mm) of ferrofluid ($M_s = 43$ kA/m) placed in a uniform transverse magnetic field (a) $H = 0$, 7 and 10 kA/m, (b) $H = 14$ kA/m, (c) $H = 17$ kA/m and (d) $H = 21$ kA/m.

convection corresponds to a single thermomagnetic wave and the corresponding bound modes, which agrees with computational results of [236, 238].

A further increase of magnetic field leads to the decrease of the period of oscillations to 290 s (see Figure 6.42(c)) and to the qualitative change of the Fourier power spectrum (see Figure 6.43(b)): instead of the presence of equi-spaced superharmonics, the spectrum widens near the fundamental frequency $\nu_0 \approx 0.0035\,\mathrm{Hz}$. This is a result of sideband instability that narrows the spectrum towards the fundamental frequency.

At $H = 21\,\mathrm{kA/m}$ the basic periodic signal becomes visibly modulated; see Figure 6.42(d). This signifies the appearance of a second oscillatory mode (thermomagnetic wave) with a close frequency. The Fourier power spectrum in this regime has two closely located maxima $\nu_1 \approx 0.0041\,\mathrm{Hz}$ and $\nu_2 \approx 0.0047$ near the fundamental frequency $\nu_0 = (\nu_1 + \nu_2)/2 \approx 0.0044\,\mathrm{Hz}$ corresponding to the fundamental period 225 s; see Figure 6.43(c). The modulation fre-

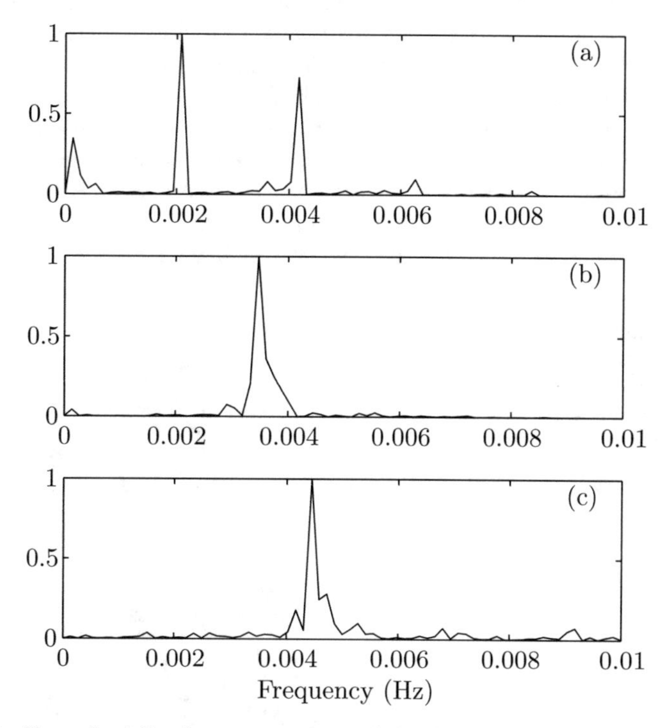

Fig. 6.43 Normalised Fourier power spectra of the thermograms shown in Figure 6.42 for $\Delta T = 18.3\,\mathrm{K}$ and (a) $H = 14\,\mathrm{kA/m}$, (b) $H = 17\,\mathrm{kA/m}$ and (c) $H = 21\,\mathrm{kA/m}$.

quency is $\nu_m = (\nu_2 - \nu_1)/2 \approx 0.0003\,\mathrm{Hz}$ corresponds to the modulation period that is approximately 15 times larger than the fundamental period.

6.3.3 Heat Transfer Characteristics

While a visualisation using a thermosensitive film and thermocouple readings of a local temperature provides valuable insight into the spatio-temporal structure of arising thermomagnetic convection patterns, they do not provide quantitative information on the heat transfer characteristics across the ferrofluid layer that is strongly influenced by the edge effects. In order to minimise them, a specially built chamber with a centrally located sensor described in Section 4.5 was used. The compact chamber design (the layer thickness $d = 2\,\mathrm{mm}$; see Figure 4.10) enabled placing it between the poles of a strong electromagnet creating a uniform magnetic field up to 220 kA/m. Along with the use of ferrofluid with a high degree of magnetisation up to $M_s = 55\,\mathrm{kA/m}$, this enabled reaching the magnetoconvection onset at significantly smaller cross-layer temperature differences. The value of Nusselt number in these experiments was calculated as described in Section 4.5, and its deviation from unity was used to define the onset. The typical Nu curves are shown in Figure 6.44. The comparison of lines 1 and 5 in Figure 6.44

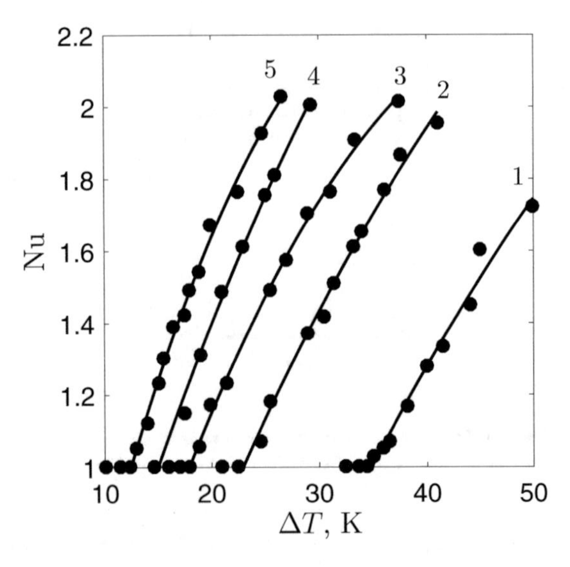

Fig. 6.44 Dimensionless heat transfer rate across a vertical layer ($d = 2\,\mathrm{mm}$, $l = 75\,\mathrm{mm}$) of ferrofluid ($M_s = 55\,\mathrm{kA/m}$) heated from a side and placed in a uniform transverse magnetic field 1: 9.1 kA/m, 2: 14.6 kA/m, 3: 20.0 kA/m, 4: 25.5 kA/m and 5: 35.0 kA/m.

shows that quadrupling the magnetic field reduces the critical value of the temperature difference by approximately a factor of 3.

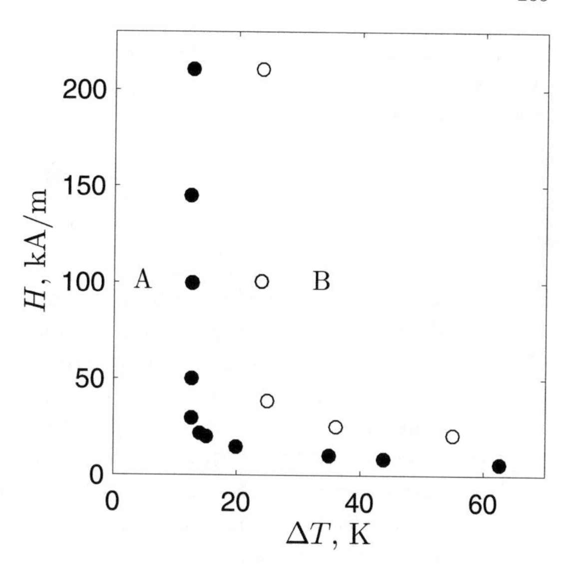

Fig. 6.45 Nusselt number for a vertical layer ($d = 2\,\text{mm}$, $L = 75\,\text{mm}$) of ferrofluid ($M_s = 55\,\text{kA/m}$) placed in a uniform transverse magnetic field: filled circles, Nu = 1; empty circles, Nu = 2. Stable parallel basic flow exists in region A and an instability is observed in region B.

The Nusselt number isolines are shown in Figure 6.45. The Nu = 1 line corresponds to the onset of thermomagnetic convection. The shape of the isolines demonstrates that while magnetoconvection leads to a rapid increase of the heat transfer rate in moderate fields up to about 30 kA/m, the application of stronger magnetic fields did not influence the heat transfer. Thus there exists a maximum magnetic field beyond which heat transfer across the layer remains constant for a given temperature difference. This is expected when the fluid magnetisation approaches saturation.

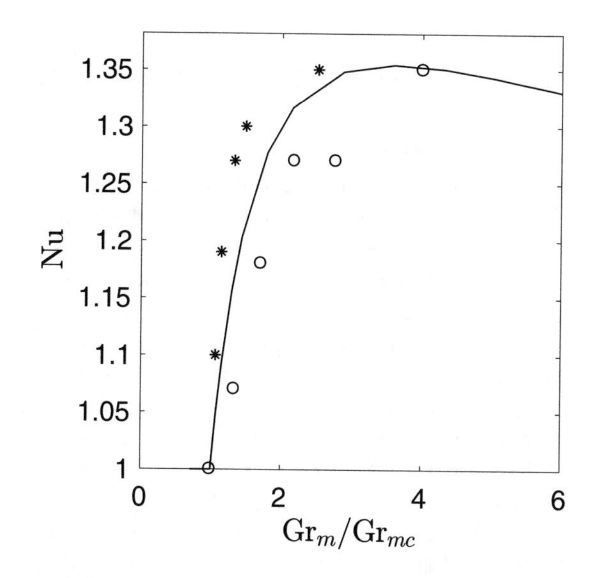

Fig. 6.46 Experimental (symbols) and analytical (line) values of the Nusselt number for thermomagnetic convection observed in a normal magnetic field and taking the form of vertical stationary rolls. The star and circle symbols show the experimental results for 2 mm and 5 mm thick layers, respectively.

In Figure 6.46 the experimentally measured and analytically estimated values of Nusselt number Nu are shown. The theoretical value of Nusselt number is computed as

$$\mathrm{Nu} = 1 - |A_e|^2 \left.\frac{d\theta_{20}}{dx}\right|_{x=-1}, \tag{6.3}$$

where $|A_e|$ is the equilibrium amplitude defined by (3.119) and θ_{20} is the averaged in the z direction (see Figure 3.2) deviation of the fluid temperature from that corresponding to a pure conduction state; see Section 3.5. Figure 6.46 demonstrates that the experimental and computed values of the Nusselt number are in reasonable agreement near the convection threshold. Yet the difference between the analytical and experimental values obtained for different enclosures is noticeable. While this discrepancy may be attributed to the difficulties with estimating the values of experimental parameters, which were discussed in Section 4.8, there appears to exist an experimentally observed trend: the Nusselt number values found for a thicker layer characterised by larger gravitational Grashof numbers are somewhat lower than those for a thin enclosure. This is consistent with the analytical conclusions of [234] and discussion in Section 3.3, where it was shown that the buoyancy effects characterised by the gravitational Grashof number tend to suppress magnetoconvection. It was also found in experiments that the effective heat flux across the layer rapidly increases once convection sets, reaches its maximum and then starts decreasing for larger supercritical values of the magnetic Grashof number. Similar behaviour is predicted by the analysis of Section 3.5; see the line in Figure 6.46.

6.3.4 Influence of Fluid Stratification

In the context of magnetoconvection, ferrofluids are typically assumed to be homogeneous. This is indeed a reasonable physical approximation when fluid is well mixed. However, there are several reasons why ferrofluid's uniformity can be disrupted. One of them, the fluid density stratification caused by the gravitational sedimentation of magnetic particles, was discussed in Section 5.3. Because convection is a very sensitive physical process such a stratification cannot be ignored in its context (see also estimations in Section 5.1). In this section we summarise the major effects a fluid stratification has on magnetoconvection flow patterns in a vertical layer. Such effects can be quite dramatic as Figures 6.47 and 6.48 demonstrate.

The full stratification regime (see Section 5.3.1) illustrated in Figure 6.47 establishes when the isothermal fluid layer rests horizontally on its wide side for a sufficiently long time (for a few weeks in the experiment illustrated in the figure) and then is rotated to its upright position just before the cross-layer temperature difference is applied. The corresponding distribution of magnetic particle concentration is shown in Figure 6.47(a). Cellular convection structures appear with horizontal boundaries slowly drifting downwards; see Figure 6.47(b). When a uniform external magnetic field normal to the layer is applied, vertical magnetoconvection rolls appear inside each cell as demonstrated in Figure 6.47(c). Such a complex flow pattern exists for a substantial time (for at least an hour), but eventually the cell boundaries erode due to convective mixing of the fluid, the concentrational density stratification disappears and a convection pattern with long vertical thermomagnetic rolls occupying the complete vertical extent of the layer similar to those shown in Figure 6.37 establish.

In the case of the partial stratification (see Section 5.3.2), the fluid layer remained vertical and isothermal for about a month. The established concentration profile of magnetic particles is shown in Figure 6.48(a). A uniform

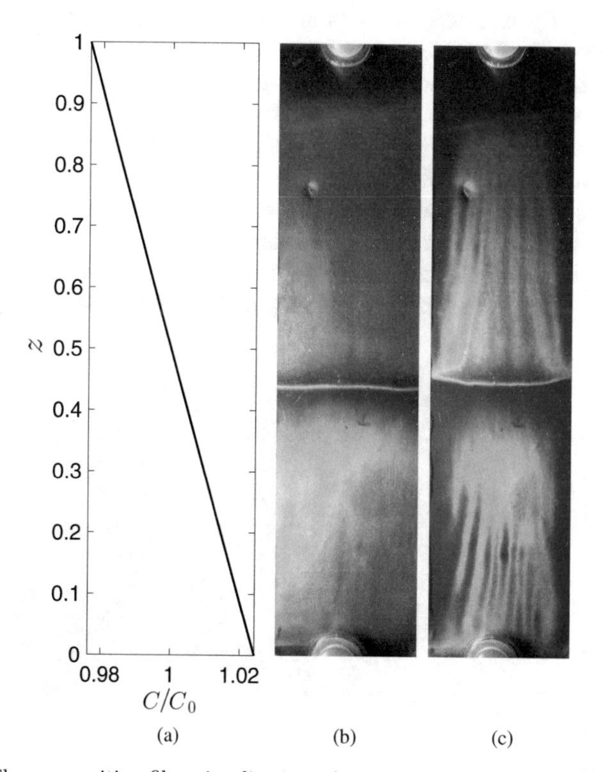

Fig. 6.47 Thermosensitive film visualisation of convection in a fully stratified vertical layer of magnetic fluid: (a) vertical distribution of the concentration of magnetic particles, (b) convection cells in the absence of magnetic field and (c) thermomagnetic patterns arising in a uniform magnetic field $H = 21\,\mathrm{kA/m}$ normal to the layer. Heating from the back with the temperature difference $\Delta T = 13\,\mathrm{K}$ applied between the walls. The layer is 4 mm thick and 250 mm high. The blue and brown regions correspond to warm and cold fluid, respectively.

magnetic field was applied normally to the plane of the fluid layer simultaneously with the cross-layer temperature difference. As a result a cellular flow pattern depicted in Figure 6.48(b) was formed. The locations of the horizontal convection cell boundaries agree well with those of the edges of the density-stratified regions predicted computationally by solving Equation (5.3). The regular vertical thermomagnetic roll pattern similar to that shown in Figure 6.37 is clearly visible in the middle part of the layer where the fluid remained homogeneous, but it is disrupted in the top and bottom stratified regions.

A well-defined cellular flow structure was established from the start and existed for 1.5 hours. The smaller top cell where the primary up-down thermogravitational flow remained relatively weak due to its small vertical extent contained inclined, wavy and spiral thermomagnetic rolls typically seen in the Rayleigh-Bénard configuration; see Figures 6.7 and 6.8. In contrast, in the larger middle cell with a strong up-down flow component predominantly vertical and slightly inclined, thermomagnetic rolls and waves [234, 237, 238] were observed. As time progressed the structures in a smaller top cell remained relatively stable, while the rolls in the larger middle cell underwent a series of secondary instabilities eventually leading to the formation of irregular unsteady patterns evidencing a higher effective supercriticality of the conditions in the larger cell.

These experiments also demonstrated that fluid mixing enhancement caused by thermomagnetic convection significantly accelerates the process of fluid homogenisation so that the cellular structures get washed out within a much shorter time in comparison with the situation discussed in Sections 5.3.1

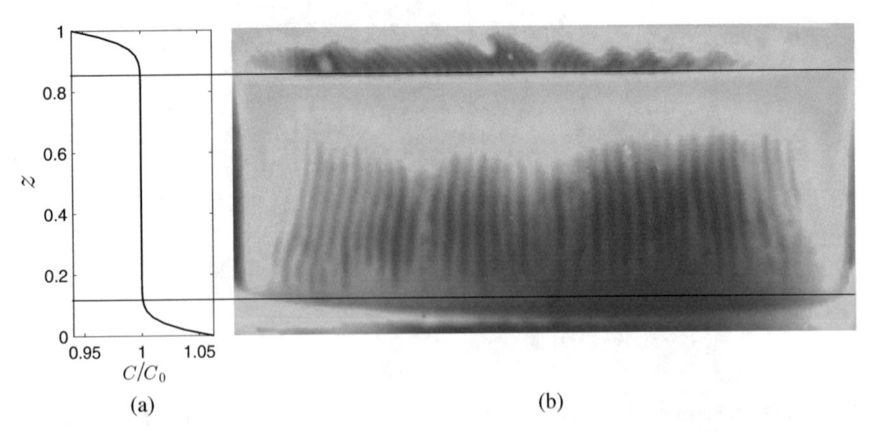

Fig. 6.48 Magnetoconvection in a partially stratified vertical layer of magnetic fluid: (a) vertical distribution of the concentration of magnetic particles and (b) infrared image of thermomagnetic patterns arising in a uniform magnetic field $H = 35\,\mathrm{kA/m}$ normal to the layer. Cooling from the back with the temperature difference $\Delta T = 23\,\mathrm{K}$ applied between the walls. The colours correspond to the temperature measured in degrees Celsius (colour scale is given in Figure 5.13). The layer is 6 mm thick and 180 mm high.

and 5.3.2 when no magnetic field was applied. In the experiments described above, the magnetic field was applied from the start of the observations, and the unicellular flow, which evidences fluid uniformity, took about 2 hours to establish compared to almost 7 hours in experiments in the absence of magnetic convection described in Section 5.3.

In addition to the fluid density stratification due to the gravitational sedimentation of solid phase in a finite vertical fluid layer heated from one vertical side and cooled from the other, a vertical temperature gradient is always established in addition to the applied horizontal temperature difference. It occurs because denser cool fluid tends to accumulate near the bottom of a finite-height layer, while warmer and less dense fluid collects near the top [140]. Such a vertical temperature stratification reduces as the height-to-width ratio of the layer increases, but it is only zero in the limit of the infinite aspect ratio. In experiments reported here, including those pictured in Figures 5.13, 6.47 and 6.48, the vertical temperature difference reaches 3–4 K. When external magnetic field is applied normally to the layer, this leads to a stronger fluid magnetisation near the bottom of the layer where the fluid is cooler. The fluid magnetisation in the bottom part of the vertical layer is strengthened further by the gravitational sedimentation of magnetic particles and their aggregates that leads to the increase of their concentration. Because of such a vertical stratification of the fluid magnetisation, the magnetoconvection patterns are stronger pronounced in the bottom part of each primary convection cell; see Figures 6.47(c) and 6.48(b). The number of visible magnetoconvection rolls in the bottom part of the vertical ferrofluid layer increases, which leads to the formation of a fan-like roll patterns (also seen in Figure 6.32). This occurs either due to the increase of the spatial period of the rolls or the interaction of thermomagnetic waves and stationary rolls. Both of these effects are traced back to the local increase of the magnetic Grashof number due to a stronger magnetisation of the fluid [40, 43, 238].

If the layer of ferrofluid is bounded by inclined edges as in a disk-shaped cavity (see Figure 6.35), the concentration isolines in the vertically stratified ferrofluids deviate from horizontal in the boundary regions. This breaks a mechanical equilibrium and can lead to additional flow there as discussed, for example, in [58, 182].

There are further factors that can lead to the stratification of non-isothermal ferrofluids placed into a magnetic field. According to experiments reported in [23, 25, 75, 197] thermo- and magnetophoresis can lead to a significant variation of particle concentration and, consequently, fluid magnetisation. For example, if positive thermophoresis is observed [253], then particles tend to drift towards a cold wall of the layer, where they are swept down by the basic flow further increasing the fluid magnetisation near the bottom edge of the layer. At the same time due to magnetophoresis, the particles tend to drift towards the regions of stronger magnetic field, that is, in the direction opposite to that of thermophoresis. The two processes compete with the net result depending on the specifics of experiment.

6.3.5 *Other Factors Influencing Experimental Flow Patterns*

As discussed in Section 4.4.1, the distortion of magnetic field near the edges of a ferrofluid layer leads to the appearance of a near-edge vortex flow. Such a flow interacts with the nearby thermomagnetic rolls forcing them to drift towards the edges (see similar observations in [56, 97]). As noted in Section 6.3.4, the number of such rolls is greater at the bottom of a vertical layer, and the thermomagnetic rolls there are closer to the edges. As a result the near-edge vortex essentially "sucks" them in from the bottom and subsequently "rolls" them up. As a result thermomagnetic rolls near the vertical edges bend towards them and then move up. There are two other factors favouring the orientation of the near-edge rolls perpendicular to the wall. The first is the appearance of the in-layer magnetic field component normal to the edge discussed in Section 4.4.1. As noted in [13] and shown computationally in [202], the preferred orientation of thermomagnetic rolls is when they are aligned with the applied magnetic field. The second is that rolls perpendicular to the solid wall minimise friction losses in the system [99].

The flow patterns are also influenced by the variation of the fluid viscosity. It is caused by three main factors. The first is the variation of the average fluid temperature associated with the need to maintain the thermosensitive film within its working thermal range that is discussed in Section 6.3.2. The second is magnetoviscous effect [170]: when the flow shear is small (in the experiments discussed here, it is of the order $0.1\,\mathrm{s}^{-1}$), the increase of fluid viscosity in strong magnetic fields may reach up to 100%. The third is the formation of particle aggregates in ferrofluid that have a strong effect on both rotational and magnetic viscosity of the fluid. Their presence also influences the intensity of fluid stratification due to the gravitational sedimentation as well as the values of magnetic susceptibility and thermal conductivity of the fluid [100, 119, 142, 151, 170]. Among other factors the number and size of the aggregates depends on the material of solid particles, the used surfactant, the strength of the applied magnetic field and history of experiment [8, 187, 193, 199]. Unfortunately, it is virtually impossible to accurately quantify these factors in practice.

6.4 Inclined Layer

6.4.1 *Convection in an Inclined Layer Placed in a Normal Magnetic Field*

Thermogravitational convection in an inclined layer of ferrocolloid has been discussed in detail in Section 5.4. Here we will consider convection arising in an inclined fluid layer placed in a magnetic field $\mathbf{H}$ that is perpendicular to the plane of the layer and parallel to the applied temperature gradient ∇T as shown in Figure 6.49. In this case the thermally induced non-uniformity of fluid magnetisation results in the appearance of the magnetic field gradient ∇H and the ponderomotive force. It enhances the effect of gravitational buoyancy caused by the fluid density gradient $\nabla \rho_T$ and leads to the onset of convection at smaller cross-layer temperature differences. An additional effect promoting the onset of convection is the migration of solid particles and their aggregates due to their thermodiffusion that also leads to the unstable density stratification $\nabla \rho_{\mathrm{TD}}$. In contrast, the gravitational sedimentation of particles delays the onset of convection resulting in a stable density stratification $\nabla \rho_{\mathrm{GS}}$. The longitudinal temperature gradient $\nabla T'$ arising due to the finite extent of an experimental layer also has a stabilising effect on convection and so does the magnetoviscosity of the fluid which is most pronounced when the applied magnetic field is perpendicular to the direction of basic flow vorticity as is indeed the case in Figure 6.49.

Experiments with rectangular enclosures with small aspect ratios showed that magnetic field in corner regions becomes essentially non-uniform, and, similar to cubic enclosures [123, 133], magnetoconvection in them starts immediately after the magnetic field is switched on. To avoid a strong distortion of the externally applied uniform magnetic field, enclosures in the shape of thin discs with the diameter 75 mm and thicknesses $d = 2.0$ and 3.5 mm were used. They were heated and cooled at their circular faces; see Figures 4.4 and 4.10. The enclosures were filled with a kerosene-based ferrofluid with $M_s = 55\,\mathrm{kA/m}$.

Similar to experiments with horizontal layers, the critical cross-layer temperature differences at which convection rolls were first detected at fixed layer inclination angles were determined using the integral heat flux sensor shown in Figure 4.10 as described in Section 6.2.2. The dependences of Nusselt number on the applied cross-layer temperature difference ΔT and the magnitude H of the applied normal magnetic field for the layer inclined at 60° to the horizontal are shown in Figure 6.50. The extrapolation of experimental data shown in Figure 6.50(a) for $H = 7\,\mathrm{kA/m}$ indicates that at $\Delta T = 28 \pm 1\,\mathrm{K}$ the cross-layer thermomagnetic convection arises (crosses in Figure 6.50(a)) and superposes onto the base flow that is parallel to wide faces of the inclined experimental chamber and that is caused by gravitational buoyancy. When the magnitude of the applied magnetic field is doubled, thermomagnetic convection sets at a smaller cross-layer temperature difference $\Delta T = 16 \pm 1\,\mathrm{K}$ (stars in Figure 6.50(a)). In contrast, in the absence of a magnetic field, the cross-layer bulk convection does not set even at $\Delta T \approx 50\,\mathrm{K}$ (circles in Figure 6.50(a)). However, the nondimensional heat flux measured in the zero field is still found to increase gradually up to Nu ≈ 1.18 from the initial value of approximately 1.1 (see also Figure 6.50(b)). Such an increase occurs because of the intensification of the base flow in an inclined layer of finite extent: a warm (cool) fluid rises (descends) along the lower heated (upper

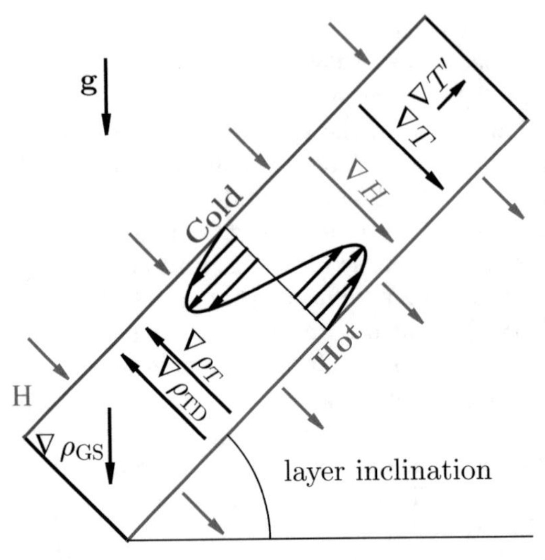

Fig. 6.49 Schematic view of an experimental arrangement with magnetic field applied normally to an inclined fluid layer.

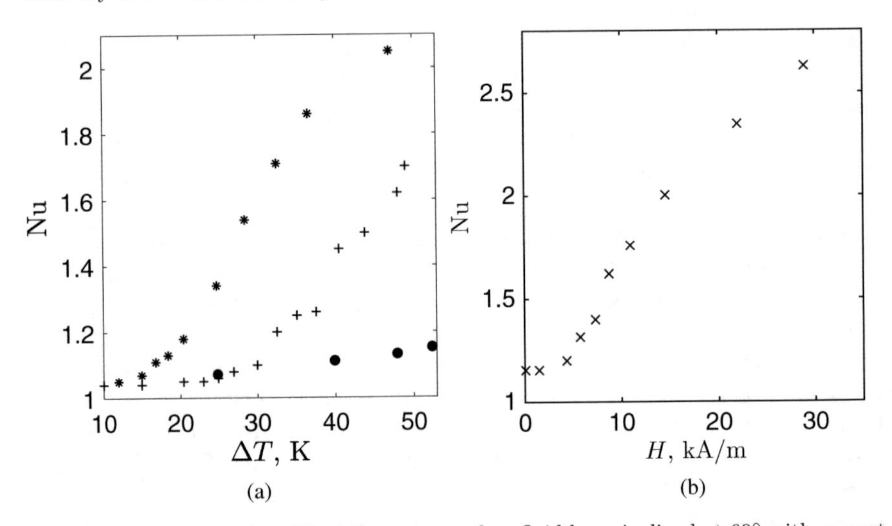

Fig. 6.50 Nondimensional heat flux across a ferrofluid layer inclined at 60° with respect to the horizontal: (a) Nu(ΔT) for fixed values of $H = 0$ (circles), 7 (crosses) and 14 (stars) kA/m; (b) Nu(H) at $\Delta T \approx 50\,\mathrm{K}$ ($d = 2\,\mathrm{mm}$).

cooled) face of the layer, hits the upper (lower) edge of the experimental chamber, turns and starts descending (ascending) along the upper cooled (lower heated) face. Because of this, near the upper (lower) edge of the layer, the warm (cool) fluid gets in contact with the cooled (heated) wall increasing the overall heat transfer rate across the layer. Such an effect reduces when layers with a larger aspect ratios are used or when the layer inclination angle with respect to the horizontal is reduced. In a magnetic field such a base-flow induced enhancement of heat transfer could be somewhat weakened or enforced by the interaction of the base flow with toroidal structures that appear near the layer edges and that have been described in Section 4.4.1.

It is more convenient to detect the onset of magnetoconvection in weak fields from the Nu(H) dependence obtained experimentally for large values of the applied temperature difference ΔT as illustrated in Figure 6.50(b). This figure shows that at $\Delta T \approx 50\,\mathrm{K}$, thermomagnetic convection sets at $H \approx 4.4\,\mathrm{kA/m}$. The experiments show that for large layer inclination angles greater than 50°, the cross-layer heat transfer intensity can only be increased using a thermomagnetic convection mechanism by increasing the magnitude of the applied magnetic field. Typical threshold values of the cross-layer temperature difference in various magnetic fields are given in Table 6.2.

Table 6.2 The dependence of the critical cross-layer temperature difference on the magnitude of the normal magnetic field applied to the ferrofluid layer inclined at 60° with respect to the horizontal.

H (kA/m)	4.4	7	14
ΔT (K)	50	28	16

Figures 6.51 and 6.52 demonstrate the dependence of cross-layer heat transfer rate on the applied temperature difference at different layer inclinations at $H = 7$ and $14\,\mathrm{kA/m}$. In small fields ($H = 7\,\mathrm{kA/m}$) and at small inclinations (0° and 30°), the values of the nondimensional heat flux coincide within the measurement error (filled circles and diamonds in Figure 6.51). This indicates that the base flow and the associated longitudinal temperature stratification arising due to the finite aspect ratio of the layer have no significant influence on the cross-layer heat flux for small layer inclinations: heat is transferred primarily by conduction. This observation is consistent with the data presented for the zero field in Figure 5.16(a). The onset of cross-layer convection at 60° inclination occurs at a significantly larger temperature difference and for 90° does not occur at all in the investigated thermal regimes. When the strength of the applied magnetic field is doubled, the values of the total heat flux approach each other in a wider range of the inclination angles between 0° and 60°; see filled circles and stars in Figure 6.52. This means that magnetoconvection arising in stronger fields becomes the dominant heat transfer mechanism that does not depend on the layer inclination. Such type of convection can occur even in layers with an inverted heating; see the data points marked by the empty circles and diamonds in Figure 6.52.

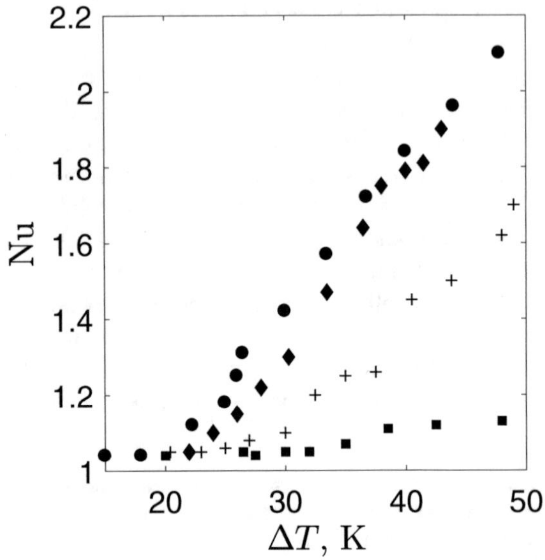

Fig. 6.51 Nondimensional heat flux across a ferrofluid layer inclined at 0° (circles, heating from below), 30° (diamonds), 60° (crosses) and 90° (squares, heating from a side) with respect to the horizontal and placed in a perpendicular magnetic field $H = 7\,\mathrm{kA/m}$ as a function of the cross-layer temperature difference ($d = 2\,\mathrm{mm}$).

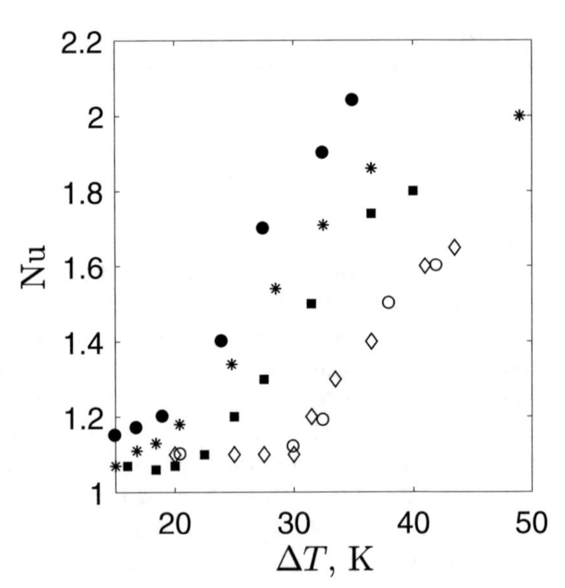

Fig. 6.52 Nondimensional heat flux across a ferrofluid layer inclined at 0° (circles, heating from below), 60° (stars), 90° (squares, heating from a side), 150° (circles) and 180° (diamonds, heating from top) with respect to the horizontal and placed in a perpendicular magnetic field $H = 14\,\mathrm{kA/m}$ as a function of the cross-layer temperature difference ($d = 2\,\mathrm{mm}$).

Figure 6.53 shows the values of the nondimensional heat flux measured at the maximum cross-layer temperature difference achieved in experiments. The data demonstrates that thermomagnetic convection sets in a layer inclined at 60° (crosses) at a somewhat weaker magnetic field than for larger inclinations. However, for fields stronger than about $20\,\mathrm{kA/m}$, the nondimensional heat flux becomes independent of the magnitude of the applied field.

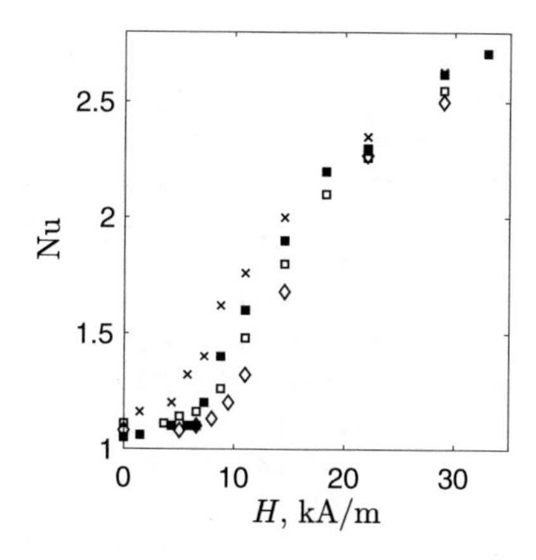

Fig. 6.53 Nondimensional heat flux across a ferrofluid layer inclined at 60° (crosses), 90° (filled squares), 120° (empty squares) and 165° (diamonds) with respect to the horizontal as a function of the applied magnetic field at $\Delta T \approx 50\,\mathrm{K}$ ($d = 2\,\mathrm{mm}$).

Thus, in strong normal magnetic fields, the spatial orientation of the layer has virtually no influence on the heat transfer rate across the layer.

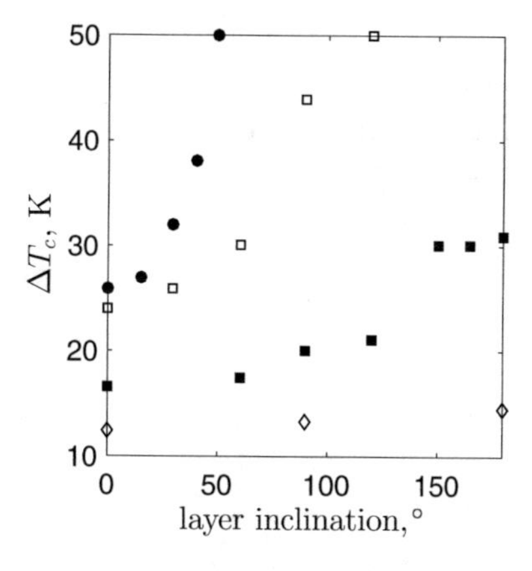

Fig. 6.54 Critical cross-layer temperature difference for the onset of convection as a function of the layer inclination angle in fixed magnetic field $H = 0$ (circles), 7.3 (empty squares), 14 (filled squares) and 29 (diamonds) kA/m ($d = 2\,\mathrm{mm}$).

The dependence of the critical cross-layer temperature difference at which the bulk convection sets on the layer inclination angle in the applied magnetic field of a fixed magnitude is shown in Figure 6.54. In the absence of the magnetic field, the critical temperature difference sharply increases starting from about 30° (filled circles); see a similar detailed stability map for the layer with thickness $d = 3.5\,\mathrm{mm}$ in Figure 5.16(a). The appearance of an additional thermomagnetic convection mechanism in relatively weak ($H = 7.3\,\mathrm{kA/m}$) fields does not change the qualitative behaviour (approximately parabolic; see empty squares) of the threshold temperature difference, but reduces its magnitude and expands the range of the layer inclination angles at which convection is detected by a factor of two compared to pure thermogravitational convection. In stronger fields ($H = 14\,\mathrm{kA/m}$) the dependence of the critical temperature on the layer inclination angle becomes approximately linear (filled squares) and in even stronger fields ($H = 29\,\text{kA/m}$) it is almost constant: ΔT_c changes only by about $2\,\text{K}$ over the complete range of layer orientations from heated from below to heated from above. Thus, in strong fields thermomagnetic convection mechanism that is insensitive to the spatial orientation of the fluid layer with respect to gravity dominates the heat transfer across the layer.

The overall stability diagram in the three-parameter space is shown in Figure 6.55. The value of $\Delta T_c = 25\,\text{K}$ corresponding to the onset of convection in a horizontal fluid layer heated from below in the absence of a magnetic field is used. The fluid is motionless below the shown surface in a horizontal layer, while a parallel basic flow exists away from the edges of the layer when it is inclined below this surface. Above the shown surface, a bulk convection is detected. The cases of horizontal and vertical layers have been discussed in detail in Sections 6.2 and 6.3. In layers inclined at small angles (up to 60°) and placed in weak magnetic fields, convection arising above the gently sloping part of the marginal stability surface has a mixed thermogravitational and thermomagnetic character. For large inclination angles (above 60° and up to 180°) only thermomagnetic convection is possible, which corresponds to the horizontal section of the surface. In strong magnetic fields the onset of convection is almost independent of the layer inclination or the applied temperature difference. This corresponds to a vertical section of the stability boundary.

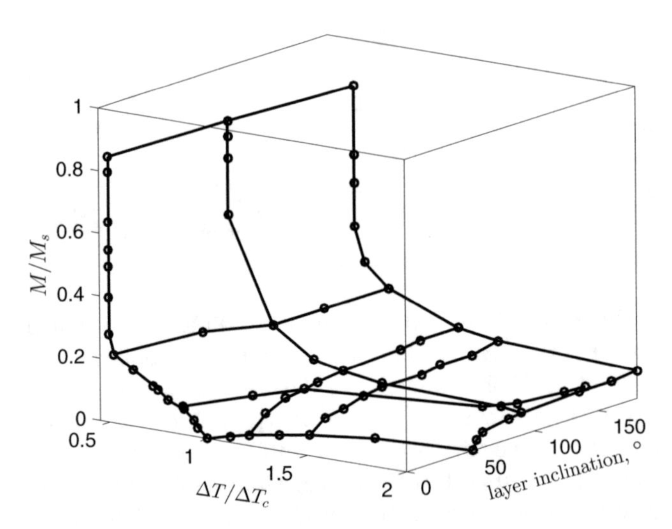

Fig. 6.55 Stability diagram for an inclined layer of ferrofluid placed in a uniform external transverse magnetic field ($M_s = 55\,\text{kA/m}$, $\Delta T_c = 25\,\text{K}$, $d = 2.0\,\text{mm}$).

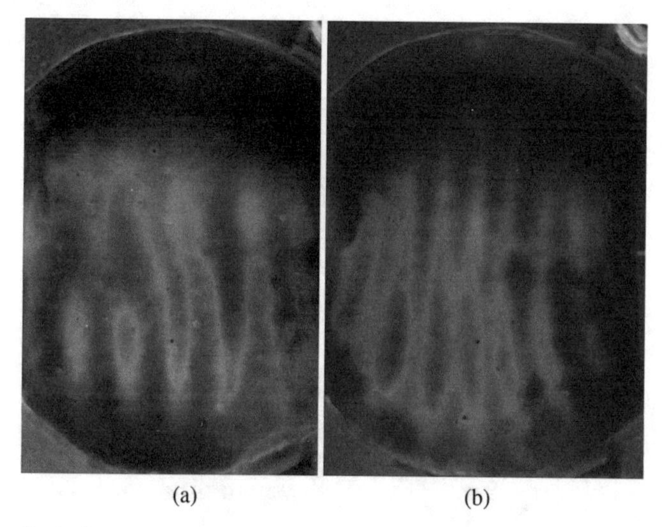

(a) (b)

Fig. 6.56 Typical convection patterns in a ferrofluid layer inclined at 15° with respect to horizontal at $\Delta T/\Delta T_c = 2.5$ and (a) $H = 0$ and (b) 22 kA/m ($M_s = 55$ kA/m, $\Delta T_c =$ 5.1 K, $d = 3.5$ mm).

Finally, in Figure 6.56 the thermal imprints of typical convection patterns are shown for a 3.5 mm thick layer of ferrofluid inclined at 15° to the horizontal. The comparison of images obtained in the absence of the field and at $H = 22$ kA/m shows that when the field is applied the wavelength of convection patterns reduces: 10 convection rolls can be distinguished in image (a) and 14 in image (b). Another observation is that the longitudinal thermal stratification clearly seen in image (a) in the absence of a magnetic field (cold brown region along the lower edge of the layer and hot blue region near the top) is greatly reduced when magnetic field is applied. This is because of the appearance of a toroidal flow structure along the perimeter of a layer placed in a normal field discussed in Section 4.4.1.

6.4.2 Convection in an Inclined Layer Placed in a Magnetic Field Parallel to the Layer

It was discussed in Sections 5.4 and 6.2.3 that the basic flow driven by the gravitational buoyancy in an inclined layer and external magnetic field parallel to the plane of the layer has the orienting influence on the arising cross-layer thermogravitational convection rolls: in the former case, they align with the direction of the basic flow velocity, in the latter with the in-layer component of the vector of the applied magnetic field. Linear stability results discussed in Section 3.4.2 also support these observations. In this section the

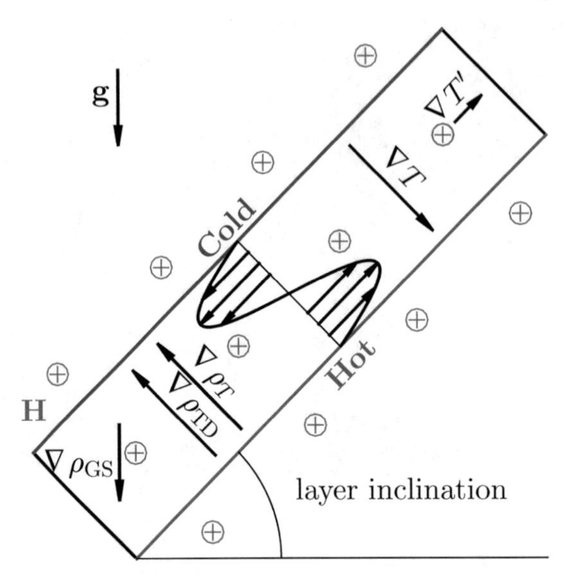

Fig. 6.57 Schematic view of an experimental arrangement with magnetic field applied horizontally in the plane of an inclined fluid layer.

situation when the vectors of basic flow velocity and the applied magnetic field are perpendicular will be discussed so that the two aligning influences (see Figures 5.15 and 6.21) compete with each other. Experimental chambers shown in Figures 4.4, 4.5 and 4.10 inclined with respect to the horizontal, heated from below and placed in a horizontal magnetic field parallel to their wide faces as shown schematically in Figure 6.57 were used for flow observations and heat flux measurements. In contrast to the situation illustrated in Figure 6.49, when the magnetic field is applied in the direction parallel to the layer and perpendicular to the temperature gradient ∇T, the magnetic field remains uniform and a ponderomotive force enhancing thermogravitational convection does not arise.

As discussed in Section 5.4, when the fluid layer is inclined, convection of Rayleigh-Bénard type sets at larger values of the cross-layer temperature differences and thus of the gravitational Rayleigh number than in a horizontal layer. The application of the magnetic field parallel to the plane of a horizontal layer does not shift the convection threshold but removes the spatial degeneracy of the arising patterns aligning them with the field; see Section 6.2.3. A similar effect is known to exist when a shear flow is superposed perpendicular to the applied temperature gradient [56]. In contrast, the application of a magnetic field perpendicular to the direction of the basic flow in an inclined layer suppresses thermogravitational convection rolls aligned with the slope of the layer as schematically shown in Figure 5.15(b). The mechanism of such a suppression is illustrated in Figures 6.22 and 6.23. A similar effect was previously found in thermally stratified Couette and

Poiseuille flows with flow velocity perpendicular to the axes of the convection rolls [56, 72, 92, 93, 97].

The suppression of Rayleigh-Bénard-type convection in an inclined layer when the magnitude H of the applied magnetic field is increased at fixed values of the cross-layer temperature difference ΔT is demonstrated in Figure 6.58. To the right of symbols, only basic flow with cubic velocity profile exists, while to the left of them, Rayleigh-Bénard-type convection develops on its background. At large layer inclinations and in weak magnetic fields such

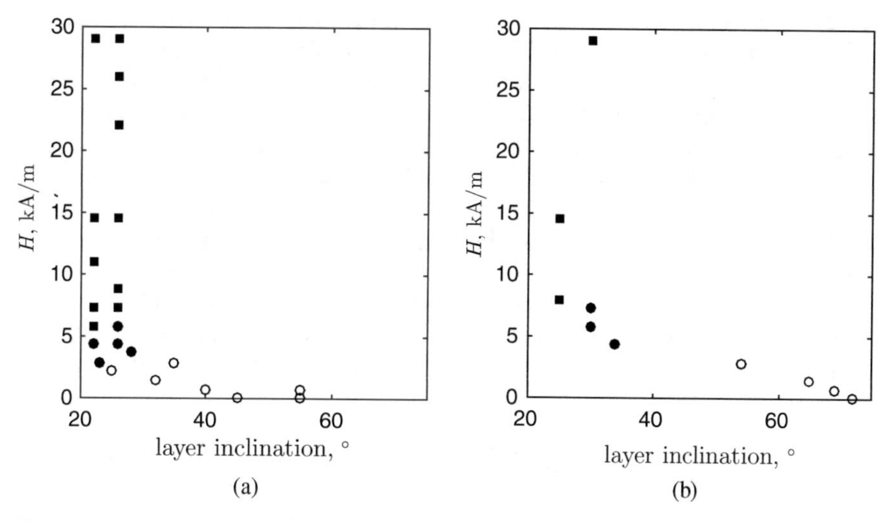

Fig. 6.58 The experimental boundary for the onset of Rayleigh-Bénard type convection in an inclined ferrofluid layer placed in a horizontal magnetic field parallel to the plane of the layer for (a) $\Delta T = 6.0$ (the left set of symbols) and 9.0 K (the right set of symbols) and (b) $\Delta T = 26.0$ K. Convection exists to the left of the symbols. The convection sets in the form of rolls aligned with the basic flow velocity (empty circles), aligned with magnetic field (squares) and of superposition of rolls aligned with the flow and the field (filled circles). $M_s = 55$ kA/m, $d = 3.5$ mm, $l/d = 21$.

convection assumes the shape of longitudinal rolls aligned with the slope of the layer (the empty circles). The vertical lines of symbols (squares) correspond to transverse convection rolls aligned with the magnetic field (across the layer slope). Filled circles denote regimes where a combination of mutually perpendicular rolls are observed that result in the formation of localised convection cells. The comparison of Figure 6.58(a) and (b) reveals that the range of the layer inclination angles in which the rolls with perpendicular orientations are observed simultaneously increases with the magnitude of the applied cross-layer temperature difference, and this is generally true for pure longitudinal and transverse rolls. However, such an enlargement of the inclination angle range for transverse rolls observed in fields of moderate strengths is somewhat retarded; see Figure 6.58(a).

The parametric map of various flow patterns observed in the inclined layer of ferrofluid heated from below and placed in a uniform horizontal magnetic field parallel to the layer is shown in Figure 6.59. The value $\Delta T_c = 5.1\,\mathrm{K}$ corresponding to the convection onset in a horizontal layer heated from below in the absence of a magnetic field is chosen to present the relative temperature difference across the layer. The vertical section of the diagram at zero inclination corresponds to the case when the magnetic field is applied along a horizontal layer. In this case the convection onset is independent of the strength of the field, which corresponds to a straight line in the vertical back plane of the diagram. The horizontal plane in the diagram corresponding to $H = 0$ contains line shown in Figures 5.16(a), 5.17(a), and 5.20(a). In region **A** in Figure 6.59, either motionless or parallel basic flow (see Figure 5.15(a)) states exist corresponding to the conduction regime. When magnetic field is applied in the direction parallel to the layer, these states are disturbed near the layer boundaries due to the edge effects discussed in Section 4.4.1; see the left image in Figure 4.6. The boot-shaped surface in Figure 6.59 represents the stability boundary for the basic flow state behind which Rayleigh-Bénard-type convection patterns develop. At small magnetic fields (region **B**), the pattern-orienting influence of the basic flow prevails and convection rolls are aligned with the slope of the layer. In strong magnetic fields (region **C**), the rolls are aligned with the field. The peculiar boot-like shape of

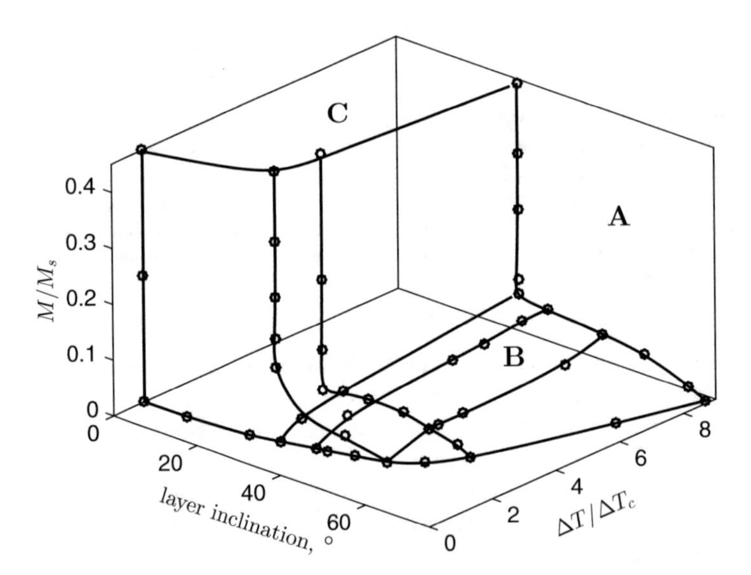

Fig. 6.59 Stability diagram for an inclined layer of ferrofluid placed in a uniform horizontal external magnetic field parallel to the plane of the layer: **A**, basic parallel flow; **B**, convection rolls aligned with the basic flow direction; and **C**, convection rolls and plumes aligned with the applied magnetic field ($M_s = 55\,\mathrm{kA/m}$, $\Delta T_c = 5.1\,\mathrm{K}$, $d = 3.5\,\mathrm{mm}$, $l/d = 21$).

the stability surface is due to the fact that when gravitational buoyancy and magnetic effects characterised by the ratios $\Delta T/\Delta T_c$ and M/M_s, respectively, are strong, they suppress convection rolls in both longitudinal and transverse directions enhancing the stability of the basic state. A wide variety of convection patterns is observed in a relatively narrow region near the "rise of the boot" where transition from one preferred roll orientation to the other occurs. The patterns there include cells (Figure 6.60(b)), travelling modulated oblique (Figure 6.60(c)) and undulated (Figures 6.61, 6.62, and 6.65(b)) rolls and modulated stationary (Figure 6.63) and travelling plumes (Figures 6.60(d), 6.65(c,d), 6.67, and 6.68).

The variation of oscillatory convection patterns arising in a layer inclined at 15° with respect to the horizontal for parametric values within the bound-

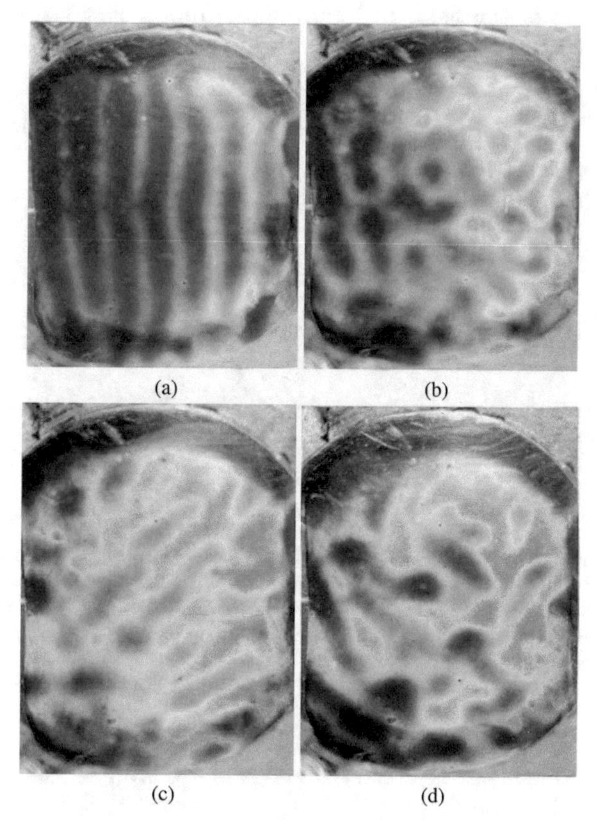

Fig. 6.60 The evolution of convection patterns in a ferrofluid layer inclined at 15° to the horizontal, heated from below with $\Delta T/\Delta T_c = 2$ and placed in a uniform horizontal magnetic field H parallel to the layer (left to right in the photographs): (a) $H = 0$, (b) $H = 1.0\,\mathrm{kA/m}$, (c) $H = 1.5\,\mathrm{kA/m}$ and (d) $H = 2.0\,\mathrm{kA/m}$ ($M_s = 55\,\mathrm{kA/m}$, $d = 3.5\,\mathrm{mm}$, $l/d = 21$).

ary surface shown in Figure 6.59 as the strength of the applied magnetic field increases is illustrated in Figures 6.60, 6.61, 6.62, and 6.63. In the absence of a magnetic field, convection pattern consists of regular rolls aligned with the direction of the basic flow; see Figure 6.60(a). When a weak magnetic field perpendicular to the basic flow velocity is applied, these rolls break into cells illustrated in Figure 6.60(b). An increase of magnetic field leads to the formation of rolls and elongated plumes making the angle of about 45° with the direction of **H**; see Figure 6.60(c,d). This is indicative of comparable orienting influences of the basic flow and magnetic field in these regimes. A similar re-orientation of convection rolls was demonstrated computationally in [42] and analytically in Section 3.4.

A further increase of a magnetic field leads to the appearance of non-stationary modulated rolls and horseshoe-shaped plumes arranged into lines in the direction of the field. In Figure 6.61(a) such flow structures have a

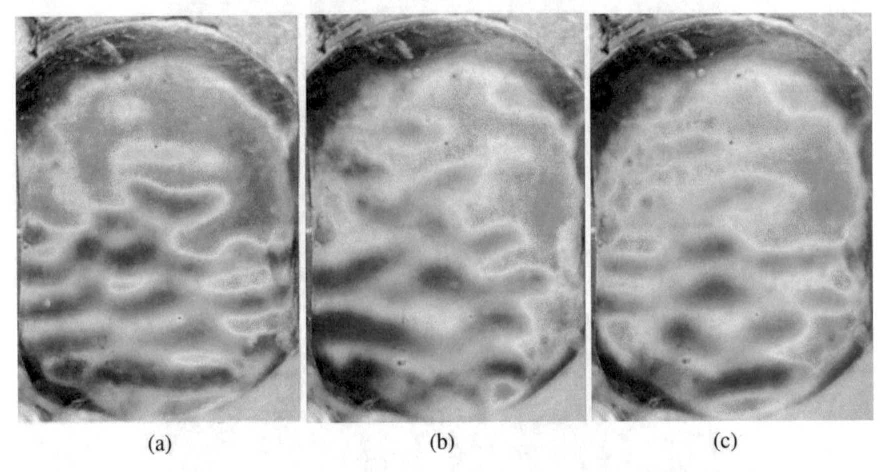

Fig. 6.61 The spatio-temporal evolution of convection patterns in a ferrofluid layer inclined at 15° to the horizontal, heated from below with $\Delta T/\Delta T_c = 2$ and placed in a uniform horizontal magnetic field $H = 2.5\,\text{kA/m}$ parallel to the layer. The time interval between snapshots (a) and (b) is 4 min and between (b) and (c) is 2 min ($M_s = 55\,\text{kA/m}$, $d = 3.5\,\text{mm}$, $l/d = 21$).

uniform intensity over the most of the layer. However, some time later convection slightly weakens near the right edge of the layer (Figure 6.61(b)) and then near the left one (Figure 6.61(c)). Consequently, the application of a magnetic field not only influences the orientation of convection patterns but also their intensity. Strengthening magnetic field also leads to the overall reduction of the area occupied by convection rolls and limits it to the lower part of the layer, where on average the fluid is somewhat cooler due to the longitudinal temperature stratification associated with a finite vertical extent of the layer.

In even stronger magnetic field, its orienting influence becomes dominant; see Figure 6.62. However, the interaction with basic flow is still felt: convection patterns aligned with the field are not stationary. For example, in Figure 6.62(a) convection rolls occupy the complete layer, but a minute later they decay in the middle part of the layer as seen from Figure 6.62(b). Subsequently, the rolls reappear in the centre but weaken near the bottom of the layer. As a result convection rolls form a chaotic pattern seen in Figure 6.62(b,c). Similar localised states were also observed in an inclined layer of ferrofluid heated from below in the absence of magnetic field (see Figures 5.18 and 5.19) and in a horizontal layer placed in a horizontal field (see Figure 6.29).

Spatio-temporal evolution of standing "blinking" plumes arising near the vertical section of the surface shown in Figure 6.59 is illustrated in Figure 6.63. Image (a) shows the remaining part of a decaying convection roll and two small plumes near the lower edge of the layer. As time progresses these plumes grow, see image (b), and then decay, see image (c). Thus in sufficiently strong fields localised convection plumes appear and evolve changing their sizes and shapes.

As has been shown in Figure 6.58(b), at $\Delta T = 26.0\,\mathrm{K}$ and layer inclinations between 25° and 30°, a stabilisation of basic conduction state is observed associated with the mutual suppression of convection rolls of different orientations promoted by basic flow velocity and the applied magnetic field. A different view of a parametric region where this occurs is given in Figure 6.64 for the layer inclined at 27° with respect to the horizontal. The stabilisation region **A** appears as a gap between regions **B** and **C** correspond-

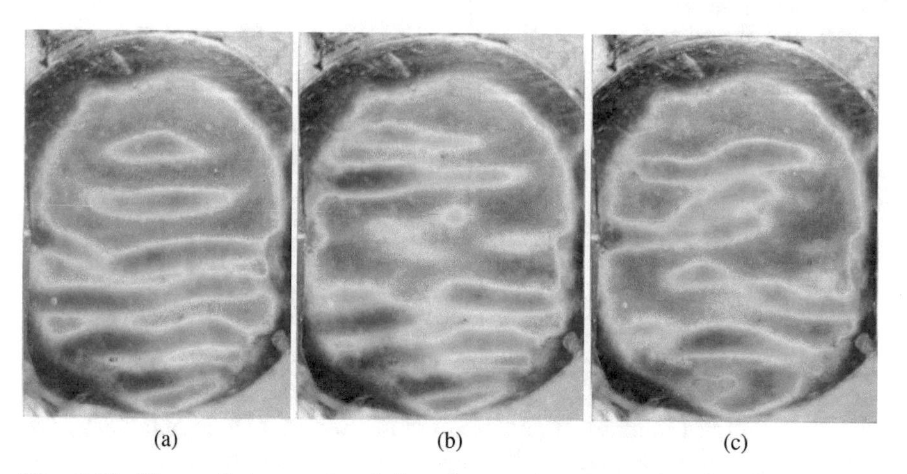

Fig. 6.62 The spatio-temporal evolution of convection patterns in a ferrofluid layer inclined at 15° to the horizontal, heated from below with $\Delta T/\Delta T_c = 2$ and placed in a uniform horizontal magnetic field $H = 8.0\,\mathrm{kA/m}$ parallel to the layer. The time interval between snapshots (a) and (b) is 1 min and between (b) and (c) is 2 min ($M_s = 55\,\mathrm{kA/m}$, $d = 3.5\,\mathrm{mm}$, $l/d = 21$).

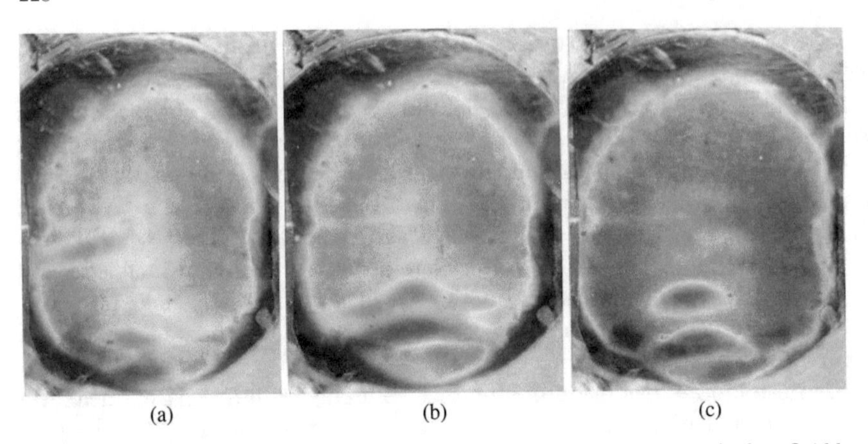

Fig. 6.63 Standing blinking convection rolls existing in the lower part of a ferrofluid layer inclined at 15° to the horizontal, heated from below with $\Delta T/\Delta T_c = 2$ and placed in a uniform horizontal magnetic field $H = 15.0\,\text{kA/m}$ parallel to the layer. The time interval between snapshots (a) and (b) is 5 min and between (b) and (c) is 1 min ($M_s = 55\,\text{kA/m}$, $d = 3.5\,\text{mm}$, $l/d = 21$).

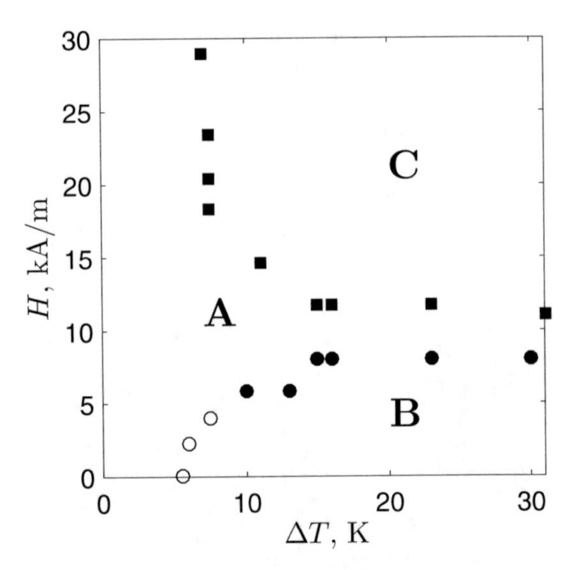

Fig. 6.64 Experimental boundaries separating the regions of existence of various convection patterns in a ferrofluid layer inclined at 27° to the horizontal, heated from below and placed in a uniform horizontal magnetic field: empty circles, rolls aligned with the slope of the layer; filled circles, cells and localised vortices; and squares, rolls aligned with the magnetic field ($M_s = 55\,\text{kA/m}$, $d = 3.5\,\text{mm}$, $l/d = 21$). Regions **A**, **B** and **C** correspond to similar regions marked in Figure 6.59.

ing to the existence regions of rolls of different orientations. Namely, at small values of the cross-layer temperature difference ΔT and magnetic field H, the empty circles separate conduction regime and regime dominated by the rolls aligned with the slope of the layer. At larger values of ΔT, the filled circles signify the appearance of localised convection plumes drifting up the layer. Above the squares the rolls aligned with the magnetic field dominate the observed convection patterns. When magnetic field is increased, the location of horizontal rolls shifts towards the lower edge of the layer and their

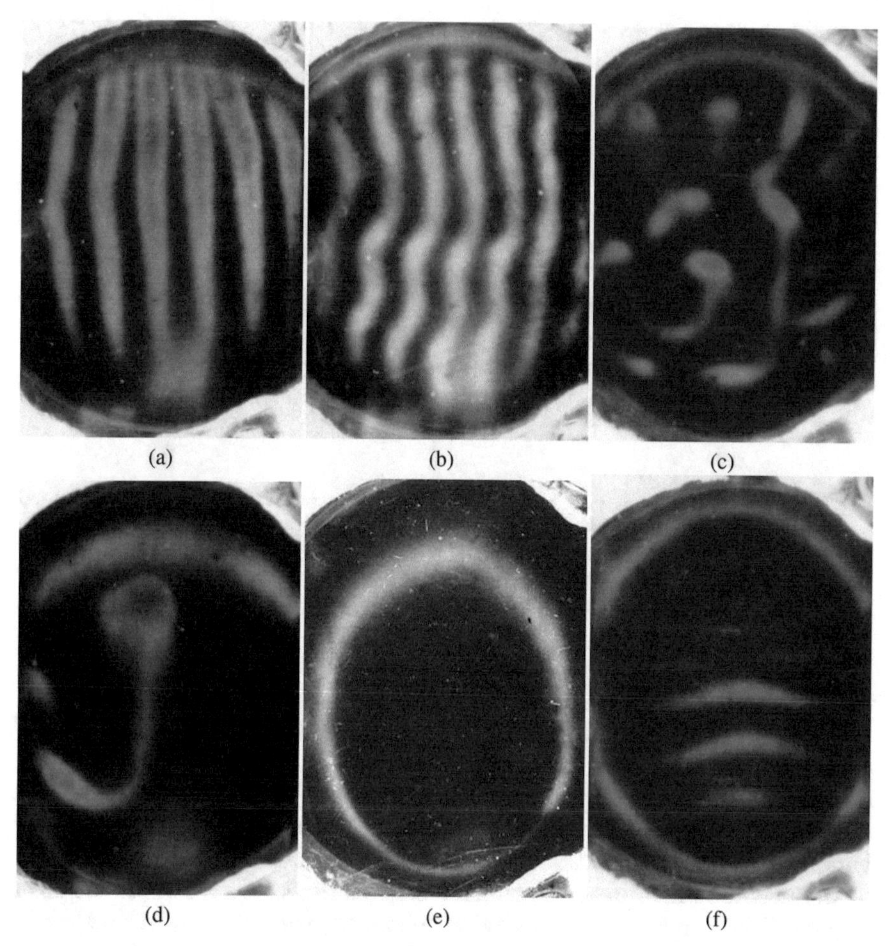

Fig. 6.65 Snapshots of convection patterns observed in a ferrofluid layer inclined at 27° to the horizontal, heated from below with $\Delta T/\Delta T_c = 3$ and placed in a uniform horizontal magnetic field parallel to the layer: (a) $H = 0$, (b) $H = 2.0\,\mathrm{kA/m}$, (c) $H = 4.0\,\mathrm{kA/m}$, (d) $H = 6.0\,\mathrm{kA/m}$, (e) $H = 9.0\,\mathrm{kA/m}$ and (f) $H = 12.0\,\mathrm{kA/m}$ ($M_s = 55\,\mathrm{kA/m}$, $d = 3.5\,\mathrm{mm}$, $l/d = 21$).

number decreases. Convection is fully suppressed between squares and filled circles in Figure 6.64.

The sequence of various convection structure observed in the layer inclined at 27° with respect to the horizontal as the magnitude of the applied horizontal magnetic field increases at the fixed cross-layer temperature difference $\Delta T = 16\,\mathrm{K}$ (corresponding to a vertical cut through Figure 6.64) is illustrated in Figure 6.65. At $H = 0$ convection rolls are aligned with the basic flow (vertical in the photograph (a)). When a weak magnetic field (up to $H = 3\,\mathrm{kA/m}$) is applied perpendicularly to these rolls they deform assuming a characteristic wave shape seen in photograph (b). At $H = 4\,\mathrm{kA/m}$ convec-

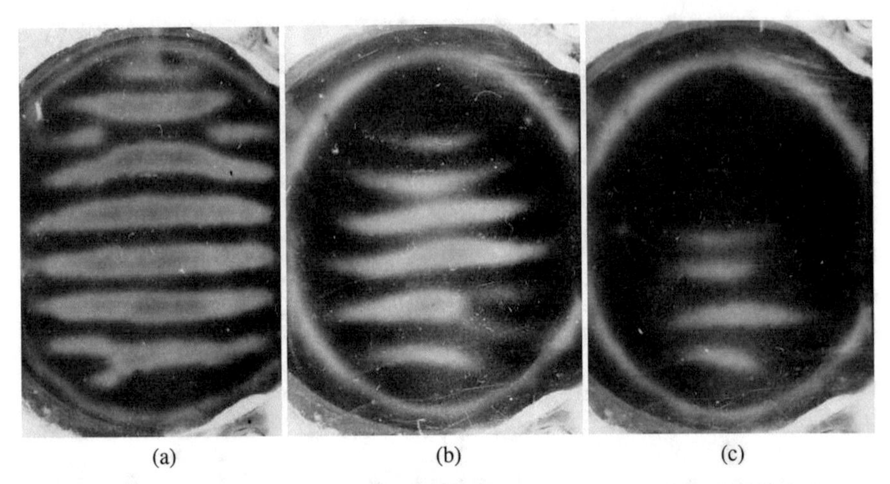

Fig. 6.66 The reduction of a convection region in an inclined ferrofluid layer heated from below with $\Delta T/\Delta T_c = 3$ and placed in a uniform horizontal magnetic field with the increase of the inclination angle from (a) 10° to (b) 15° in the fixed field $H = 7\,\mathrm{kA/m}$ and with the increase of the field from (b) $H = 7\,\mathrm{kA/m}$ to (c) $H = 18\,\mathrm{kA/m}$ at the fixed inclination angle of 15° ($M_s = 55\,\mathrm{kA/m}$, $d = 3.5\,\mathrm{mm}$, $l/d = 21$).

tion cells and plumes appear as shown in image (c). Plumes drifting upwards appear near the horizontal section of a line connecting solid circles in Figure 6.65. They are shown in photograph (d). Only a toroidal convection roll is visible in image (e) that corresponds to the gap between filled circles and squares in Figure 6.65. Convection in the bulk of the layer is fully suppressed here, and only the boundary roll survives, that is, caused by the distortion of the magnetic field near the circular edge of the layer. For $H > 10\,\mathrm{kA/m}$ (region **C** in Figure 6.64), convection rolls align with the applied magnetic field with stronger patterns located in the lower part of the inclined layer; see Figure 6.65(f).

The suppression of convection rolls in the upper part of the layer is illustrated in Figure 6.66. Two factors are responsible for such an effect. First is the layer inclination. When it increases the longitudinal (vertical in Figure 6.66) temperature stratification is established in a layer of a finite extent. This reduces the effective cross-layer temperature gradient responsible for the onset of convection so that the overall convection pattern weakens and the number of convection rolls decreases; compare images (a) and (b). The second factor leading to the suppression of convection is the increasing magnitude of a horizontal magnetic field; compare photos (b) and (c). It results in shortening of the rolls and the associated reduction of their amplitude due to the intensification of a toroidal fluid motion. It is caused by the distortion of a magnetic field at the boundaries of the experimental chamber, which is proportional to the strength of the applied field.

In the layer inclination angle range between 23° and 33° ("the rise of the boot" in Figure 6.59), a peculiar regime has been observed: convection sets in the form of moving localised plumes or vortices. Most commonly a pair of plumes appears spontaneously. Since the fluid inside them has a higher temperature near the upper cooled observation face of the experimental chamber, the plumes drift up the inclined layer; see Figure 6.67. The shape and size of plumes change as they move. In image (a) a larger plume is located at the centre of the layer, and a smaller one is below it. As the plumes drift up, the first one becomes smaller, while the second grows; see image (b). Once the first plume reaches the upper edge of the layer, it disappears, but a new plume forms at the bottom of the layer as seen from image (c) and starts drifting up the slope; see image (d).

Another scenario involving the formation of a localised vortex is illustrated in Figure 6.68. A localised plume seen in image (a) is initially elongated along the horizontal magnetic field lines and stretched vertically by the background flow resulting in a vortex shaped as shown in image (b). Subsequently, it breaks into three oblique localised vortices shown in image (c). Once they drift up the layer and disappear at the upper edge of the layer, similar to Figure 6.67, a new plume is formed near the bottom edge and starts moving up; see image (d).

It is of interest to note that localised structures similar to those shown in Figures 6.67 and 6.68 have also been observed in different convection systems: in a vertical differentially heated layer oscillating in its own plane [263], in an inclined air layer heated from below [71] and in electroconvection of nematic liquid crystals [76].

The discussed above convection patterns observed in a disk-shaped inclined enclosure also exist in a rectangular inclined layer with a larger aspect ratio ($l/d = 42$) shown in Figure 4.5 in the cross-layer temperature difference range $\Delta T/\Delta T_c = 5 - 15$. When such a long layer is inclined at small angles, topologically different structures can occur along the slope at fixed values of ΔT and H. For example, at 10° inclination oblique rolls exist in the middle part of the layer (see image (a) in Figure 6.69) that break into several shorter plumes as time progresses; see image (b). Subsequently, most of them recombine forming rolls aligned with the basic flow; see image (c). Similar recombination of individual convection cells into rolls is observed in the upper part of the layer. This is referred as the zipper state in literature [21]. In the lower part of the layer, convection patterns remain in the shape of cells or plumes of various sizes.

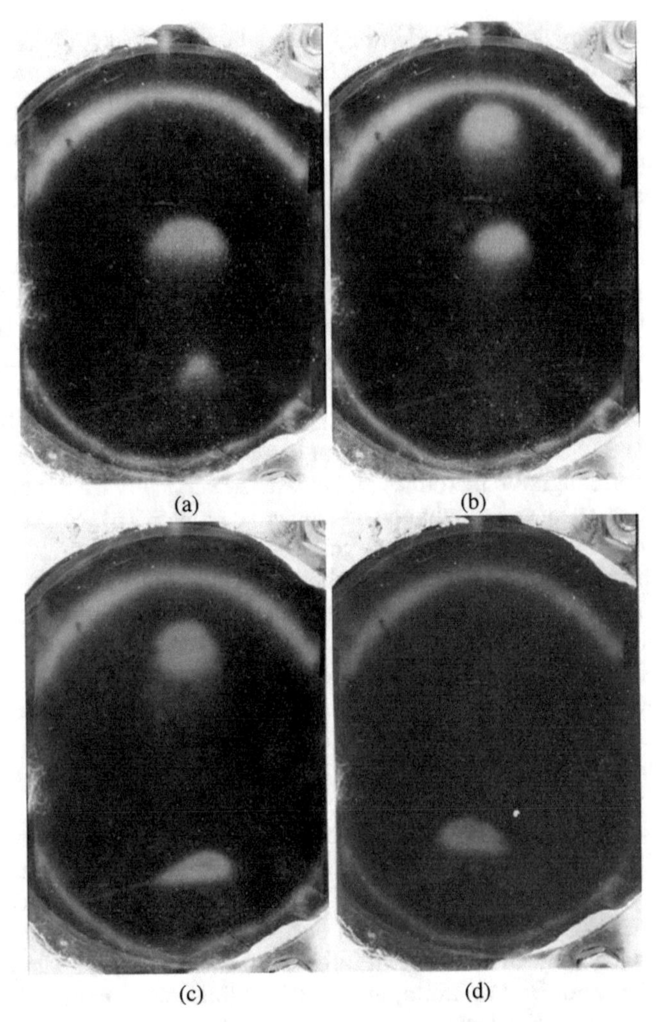

Fig. 6.67 Displacement of localised convection plumes in a ferrofluid layer inclined at 25° to the horizontal, heated from below with $\Delta T/\Delta T_c = 2$ and placed in a uniform horizontal magnetic field $H = 6.0\,\mathrm{kA/m}$ parallel to the layer. The time interval between the snapshots is 2 min ($M_s = 55\,\mathrm{kA/m}$, $d = 3.5\,\mathrm{mm}$, $l/d = 21$).

At larger inclination angles, a suppression of convection patterns is observed in a long layer as the strength of the applied horizontal magnetic field increases. This is similar to the pattern evolution observed in the disk-shaped enclosure of a smaller aspect ratio $l/d = 21$ (region **A** in Figure 6.59). For example, image (a) in Figure 6.70 shows that in the absence of the field, the complete flow region in the layer inclined at 44° is occupied by convection rolls aligned with the basic flow (ten rolls corresponding to five blue stripes are visible). The application of a magnetic field perpendicular to the rolls leads to the reduction of the total number of the rolls (see images (b) and

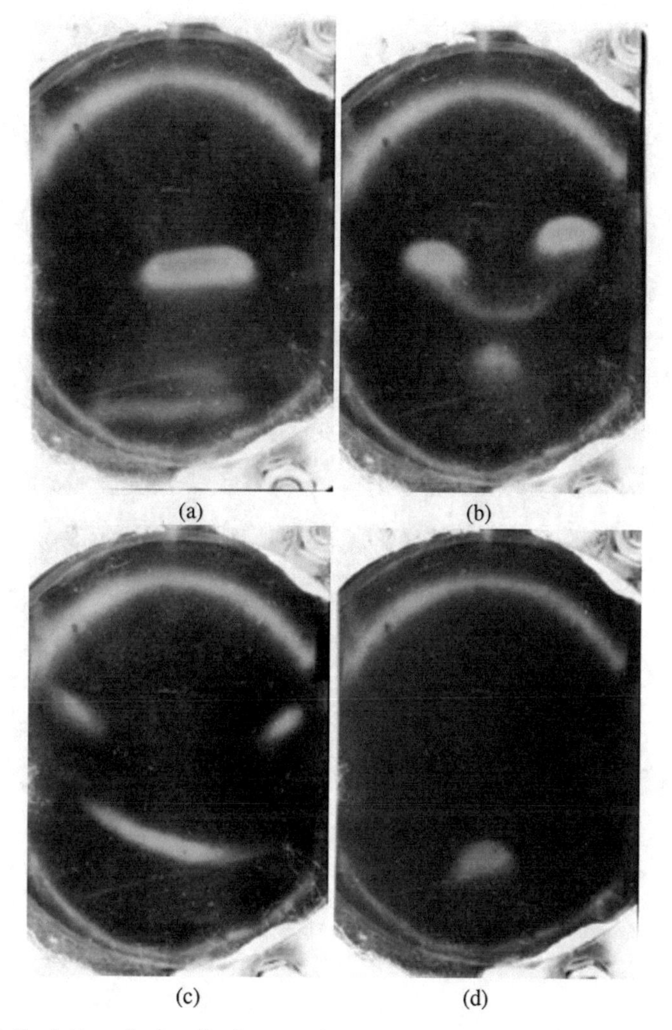

Fig. 6.68 Evolution of a localised convection vortex in a ferrofluid layer inclined at 25° to the horizontal, heated from below with $\Delta T/\Delta T_c = 2$ and placed in a uniform horizontal magnetic field $H = 6\,\mathrm{kA/m}$ parallel to the layer. The time interval between the snapshots (a) and (b) and (b) and (c) is 2 min, between (c) and (d) 1 min ($M_s = 55\,\mathrm{kA/m}$, $d = 3.5\,\mathrm{mm}$, $l/d = 21$).

(c)) and then to the completion of their suppression (see image (d)): only rolls near the left and right edges survive (blue regions in image (d)), but they are associated with the local distortion of the magnetic field at the boundary between two media rather than by thermogravitational buoyancy effects. The comparison of images (b)–(d) shows that the width of the region affected by such edge effects discussed in Section 4.4.1 increases with the strength of the applied magnetic field.

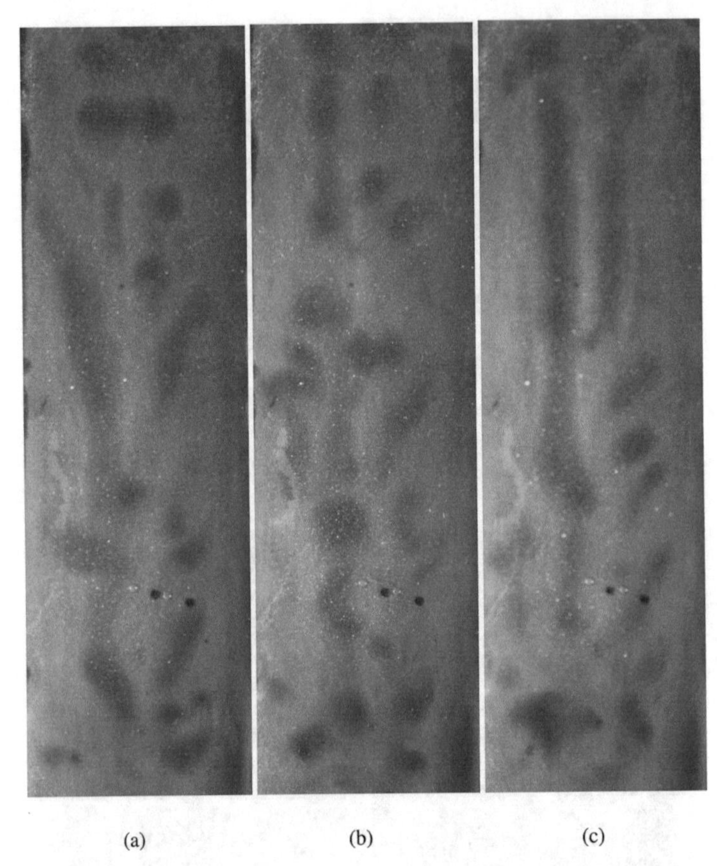

Fig. 6.69 Spatio-temporal convection patterns in a ferrofluid layer inclined at 10° with respect to the horizontal heated from below with $\Delta T = 12\,\mathrm{K}$ and placed in a uniform horizontal magnetic field $H = 5.0\,\mathrm{kA/m}$. The time interval between snapshots (a) and (b) is 2 min and between (b) and (c) 6 min ($M_s = 43\,\mathrm{kA/m}$, $d = 6.0\,\mathrm{mm}$, $l/d = 42$).

The heat flux measurements results across an inclined disk-shaped ferrofluid layer confined by two copper heat exchangers (see Figure 4.10) are summarised in Figures 6.71, 6.72, 6.73, and 6.74. Since the 17 mm integral heat flux sensor was positioned at the centre of the layer, it was not able to detect variations of heat flux due to the localised plumes and vortices appearing and moving near the edges of the layer and only recorded the changes due to flow structures immediately above it. Therefore its readings could only represent the true heat flux at small layer inclination angles when convection rolls occupied the complete layer. Such results for a layer inclined at 10° to the horizontal are presented in Figures 6.71 and 6.72. In weak magnetic fields $H < 1\,\mathrm{kA/m}$ heat flux is reduced compared with that in the absence of the

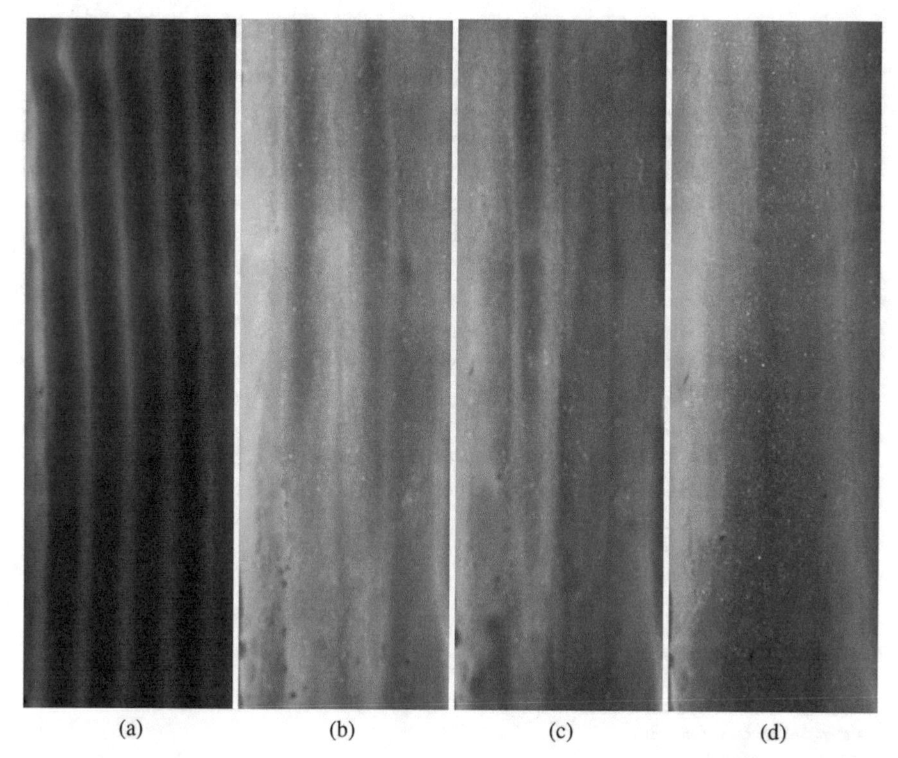

Fig. 6.70 Suppression of convection rolls aligned with the basic flow in a rectangular ferrofluid layer inclined at 44° with respect to the horizontal and heated from below with $\Delta T = 14\,\mathrm{K}$ by the increasing horizontal magnetic field perpendicular to the rolls: (a) $H = 0$, (b) 5.5, (c) 7.0 and (d) $9.7\,\mathrm{kA/m}$ ($M_s = 43\,\mathrm{kA/m}$, $d = 6.0\,\mathrm{mm}$, $l/d = 42$).

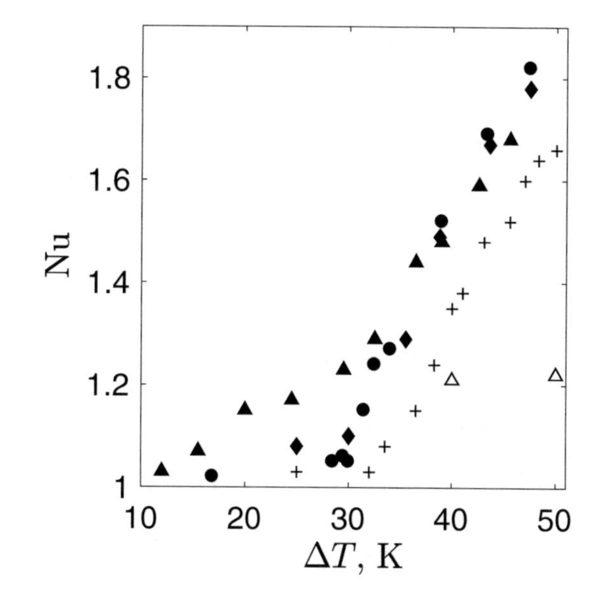

Fig. 6.71 Dependence of the nondimensional heat flux on the temperature difference applied across a ferrofluid layer inclined at 10° with respect to the horizontal and placed in a horizontal magnetic field H parallel to the layer: circles, $H = 0$; crosses, $H = 0.7\,\mathrm{kA/m}$; diamonds, $H = 7\,\mathrm{kA/m}$; and triangles, $H = 14\,\mathrm{kA/m}$. Heating from below except empty triangles ($M_s = 55\,\mathrm{kA/m}$, $d = 2.0\,\mathrm{mm}$, $l/d = 35$).

field (see filled circles and crosses in Figure 6.71). This is likely to be associated with the suppression of convection rolls aligned with the basic flow by the perpendicular field. In stronger fields the heat flux reduction is not detected (diamonds and filled triangles in Figure 6.71) indicating that in this case magnetic field fully suppresses perpendicular convection rolls but has no influence on rolls that are now aligned with the field. Another conclusion that follows from Figure 6.71 is that in strong fields the Nusselt number increase from the zero-field level occurs as soon as such a field is applied with no obvious convection threshold. This occurs because of the appearance of the toroidal near-edge vortex (see Figure 6.65(e)) caused by the distortion of magnetic field lines there. Such a flow results in the Nusselt increase up to 1.2 for $\Delta T = 30\,\mathrm{K}$ without the onset of convection in the bulk of the layer. The hypothesis that such a gradual increase of Nusselt number is due to the near-edge flow perturbations is confirmed in experiments when the layer was heated from above (empty triangles in Figure 6.71). In this arrangement thermogravitational convection rolls do not occur in the bulk of the layer, yet the values of Nusselt number larger than 1 were measured.

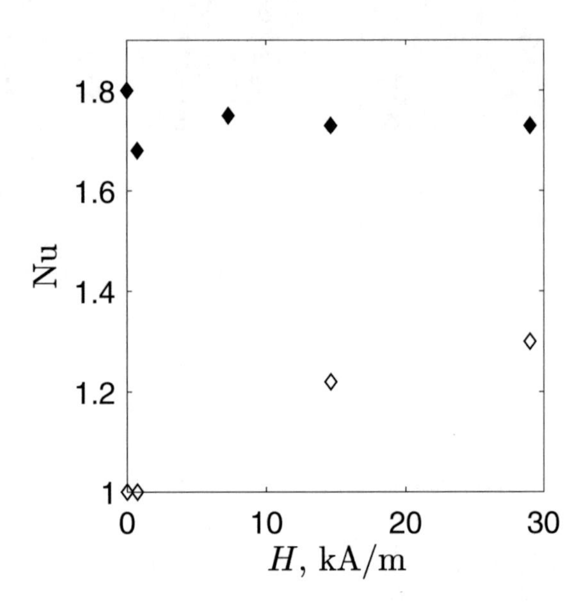

Fig. 6.72 Dependence of the nondimensional heat flux on the magnitude of the horizontal magnetic field parallel to a ferrofluid layer inclined at 10° with respect to the horizontal at $\Delta T = 50\,\mathrm{K}$. Filled and empty symbols correspond to heating from below and above, respectively ($M_s = 55\,\mathrm{kA/m}$, $d = 2.0\,\mathrm{mm}$, $l/d = 35$).

The slight suppression of heat transfer across the layer placed in a weak magnetic field is also seen in Figure 6.72 (the filled diamonds). At the same time when the layer is heated from above, the value of Nusselt number gradually increases with the strength of the applied magnetic field, which once again confirms that the heat flux across the layer is strongly influenced by the near-edge toroidal flow. The Nusselt number associated with such a vortex reaches the value of 1.3 at $H = 30\,\mathrm{kA/m}$.

The variation of heat flux across a ferrofluid layer inclined at 30° with respect to the horizontal is illustrated in Figures 6.73 and 6.74. In the absence

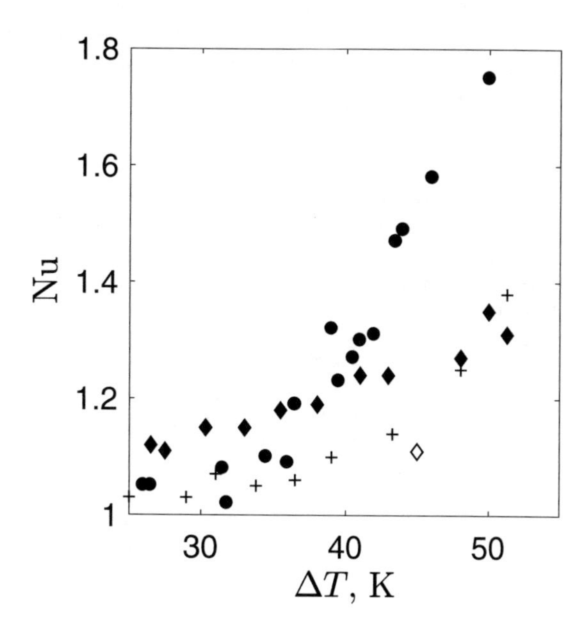

Fig. 6.73 Dependence of the nondimensional heat flux on the temperature difference applied across a ferrofluid layer inclined at 30° with respect to the horizontal and placed in a horizontal magnetic field H parallel to the layer: circles, $H = 0$; crosses, $H = 0.7\,\mathrm{kA/m}$; and diamonds, $H = 7\,\mathrm{kA/m}$. Heating from below except empty diamonds ($M_s = 55\,\mathrm{kA/m}$, $d = 2.0\,\mathrm{mm}$, $l/d = 35$).

of the magnetic field at cross-layer temperature differences close to the critical value $\Delta T_c \approx 30\,\mathrm{K}$, the measured values of Nusselt number are scattered (filled circles in Figure 6.73; see also Section 5.4). The scatter is associated with the temporal oscillations of heat flux near the onset of convection in ferrofluid. After magnetic field is switched on, two trends are detected. First, the Nusselt number increases from 1.02 at $H = 0.7\,\mathrm{kA/m}$ to 1.15 at $H = 7\,\mathrm{kA/m}$ because of the intensification of the near-edge toroidal motion. Second, convection in the bulk of the layer weakens starting from $\Delta T \approx 37\,\mathrm{K}$; see crosses and filled circles in Figure 6.73. At $\Delta T \approx 50\,\mathrm{K}$ the decrease of the heat flux reaches almost 30% of the heat flux measured in the absence of the field even though convection rolls aligned with the applied magnetic field (see Figure 6.66) still exist: they are responsible for the heat flux difference in the layers heated from below (filled diamonds) and above (empty diamonds) in Figure 6.73.

Finally, Figure 6.74 demonstrates that the strongest suppression of heat transfer across the ferrofluid layer inclined at 30° with respect to the horizontal occurs for $0 < H < 5\,\mathrm{kA/m}$ when Nusselt number decreases to about 1.1, the value corresponding to the heat flux due to the near-edge vortex. This means that bulk convection in these regimes is fully suppressed by the application of a horizontal magnetic field parallel to the layer. The other observation from Figure 6.74 is that in strong magnetic fields the values of Nusselt number measured for the layers heated from below (filled diamonds) and above (empty diamonds) are almost identical. The likely explanation of this peculiar fact is that in such regimes the heat flux due to the near-edge toroidal flow, which is independent of the direction of heating, is much larger than that of bulk Rayleigh-Bénard-type convection.

Thus, a horizontal magnetic field parallel to a ferrofluid layer tends to suppress Rayleigh-Bénard-type convection when the layer inclination increases. The inclination angle at which the full suppressions occurs depends on the magnitude of the applied field, the cross-layer temperature difference and the aspect ratio of the layer. A partial suppression of convection rolls aligned with the basic flow is associated with the re-orienting influence of the magnetic field perpendicular to them. An intriguing feature of such intermediate regimes is the formation of travelling localised thermal plumes and solitary convection vortices.

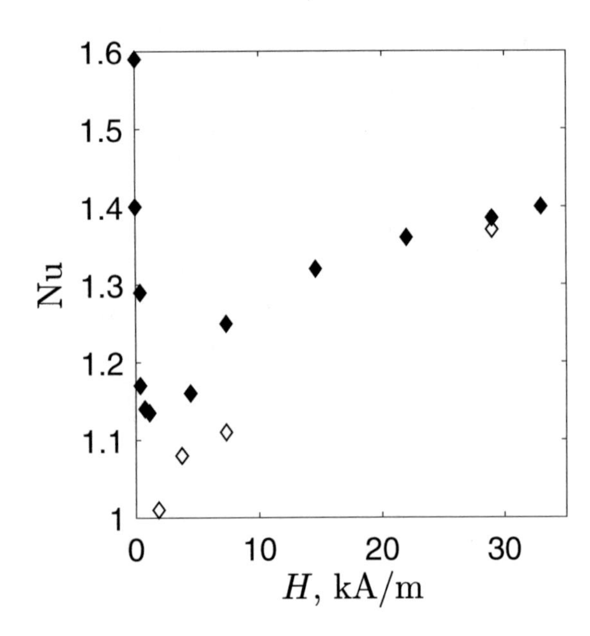

Fig. 6.74 Dependence of the nondimensional heat flux on the magnitude of the horizontal magnetic field parallel to a ferrofluid layer inclined at 30° with respect to the horizontal at $\Delta T = 44\,\mathrm{K}$. Filled and empty symbols correspond to heating from below and above, respectively ($M_s = 55\,\mathrm{kA/m}$, $d = 2.0\,\mathrm{mm}$, $l/d = 35$).

6.5 Sphere

6.5.1 Problem Overview

The choice of a spherical geometry for studies of magnetoconvection is dictated by the fact that the externally applied uniform magnetic field remains so within a spherical cavity filled with uniformly magnetised ferrofluid [144]. This is also so for ellipsoidal cavities and their limiting cases—infinitely long cylinders and infinite flat layers. However, as discussed in Section 6.2.2 in finite flat layers, the distortion of magnetic field near their edges is unavoidable, and it increases with the strength of the applied magnetic field and with the thickness of the layer relative to its lateral dimensions. This restricts the choice of geometrical characteristics of flat layers. In contrast the uniformity of a magnetic field is preserved in spherical cavities of an arbitrary size making them a geometry of choice when the influence of a uniform magnetic field is the main focus of study. Yet, to make sure that in experiments conducted in normal gravity (ground-based experiments as contrasted to experiments performed onboard spacecrafts) the magnetic effects are not obscured by those of

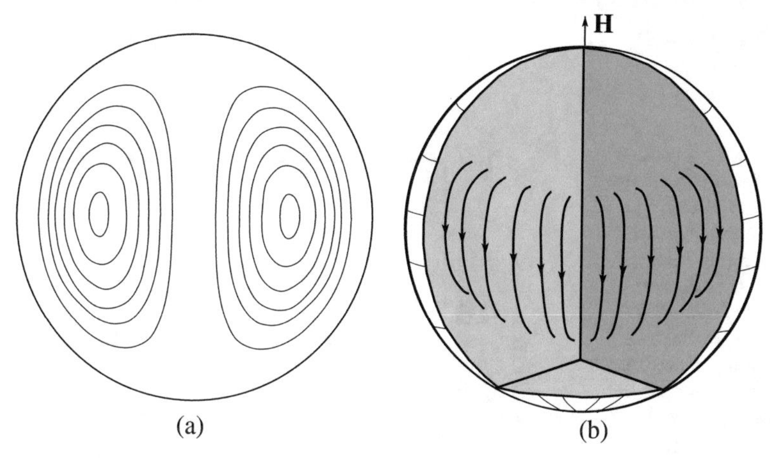

Fig. 6.75 Schematic view of weak thermomagnetic motion arising in a sphere heated from above, filled with a ferrofluid and placed in a uniform vertical external magnetic field: (a) streamlines, (b) overall view of the flow.

gravitational buoyancy, sufficiently small spherical cavities need to be used; see Equation (4.1) and the relevant discussion in Section 4.2.

While the uniform external magnetic field inside a cavity filled with an isothermal ferrofluid remains uniform, the analysis and experimental results reported in [34] for a sphere heated from above show that magnetic field inside a thermally stratified sphere becomes non-uniform and a positive vertical gradient of its intensity is established. This leads to the appearance of the ponderomotive force driving a cooler and stronger magnetised ferrofluid up along the vertical axis of a sphere, which results in a toroidal fluid motion schematically shown in Figure 6.75. Such a motion occurs as soon as the vertical magnetic field is applied. Similar thresholdless convective motions have been detected in various physical systems with broken symmetries, for example, in a cylinder with a slightly inclined with respect to the vertical axis [179] and in a slightly deformed sphere [46], carved inside large non-uniformly heated solid blocks. In the context of magnetoconvection, thresholdless fluid motion was reported previously for ferrofluid [133] and air [123] filling a cubic cavity. In this case such a motion arises by the non-uniformity of a magnetic field induced by the geometry of the cavity. The existence of weak magnetically induced toroidal motions in spherical cavities somewhat obscures experimental investigation of the onset of the main convection mode taking the form of a single vortex with a horizontal axis located in an equatorial plane shown in Figure 4.13 as will be demonstrated in Section 6.5.2.

6.5.2 Thermomagnetic Convection in a Sphere Heated from Top

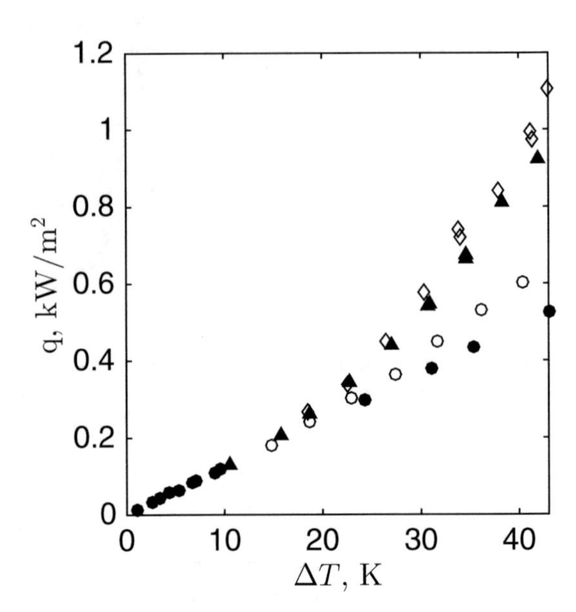

Fig. 6.76 Heat flux as a function of the temperature difference between the poles of a sphere carved inside a Plexiglas block, filled with a transformer-oil-based ferrofluid with $M_s = 44.9\,\mathrm{kA/m}$ heated from above and placed in a uniform vertical magnetic field with magnitude $H = 0$ (filled circles), 15.9 (empty circles), 35 (triangles) and 107 (diamonds) kA/m.

In order to focus completely on the thermomagnetic mechanism of convection, in this section we will discuss experimental results obtained when the spherical cavity shown in Figure 4.12 was heated from above so that the occurrence of thermogravitational convection was precluded. In this case the influence of a uniform magnetic field is maximised if it is applied vertically along the temperature gradient. Figure 6.76 shows the dependence of the measured heat flux q on the temperature difference ΔT between the poles of the sphere filled with a transformer-oil-based ferrofluid. The physical properties of such a ferrofluid are listed in Table 4.1. As seen from this figure in the absence of the magnetic field, the heat flux increases linearly with the applied temperature difference as expected in pure conduction regime. When magnetic field is applied, the values of heat flux start to increase faster from about $\Delta T = 15\,\text{K}$ indicating the appearance of magnetoconvection. The heat flux detected in a magnetic field at the maximum experimental temperature difference $\Delta T = 43\,\text{K}$ is about 20% larger than that in the absence of the field at $H = 15.9\,\text{kA/m}$, but this increase reaches 100% for $H = 107\,\text{kA/m}$. The difference between the values of heat flux detected for magnetic fields in the range between 35 and 107 kA/m is insignificant indicating that once magnetoconvection sets in, the average strength of the magnetic field plays only a weak role in defining the structure of the flow, which is consistent with theoretical findings of [89] and [234] for flat ferrofluid layers placed in a normal magnetic field; see also the discussion in Section 3.3. Because of this the onsets of main convection mode schematically shown in Figure 4.13 in this interval of magnetic field strength is hard to distinguish experimentally. The Nusselt number dependences on the applied temperature difference and

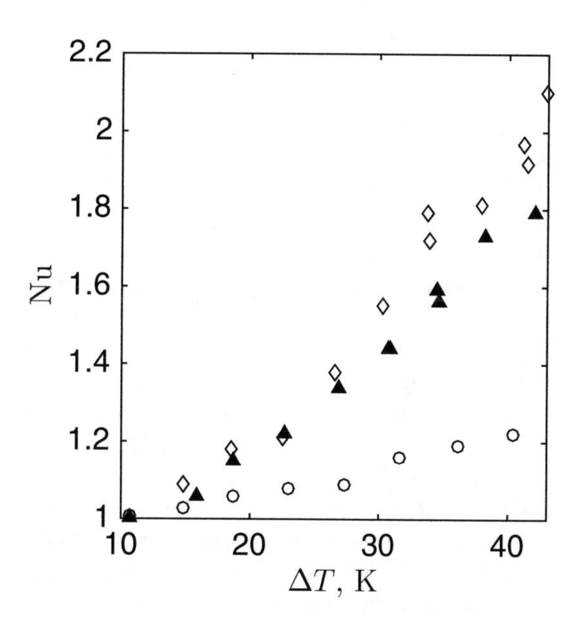

Fig. 6.77 Nondimensional heat flux as a function of the temperature difference between the poles of a sphere for the same parameters as in Figure 6.76.

magnetic field shown in Figure 6.77 are rather smooth. However, the rate of increase of Nu with ΔT changes at $\Delta T \approx 23\,\text{K}$ in strong and at $\Delta T \approx 27\,\text{K}$ in weak magnetic fields. The slow growth of Nu up to the value of approximately 1.2 in strong and 1.1 in weak fields is likely to be caused by weak convection flows illustrated in Figure 6.75.

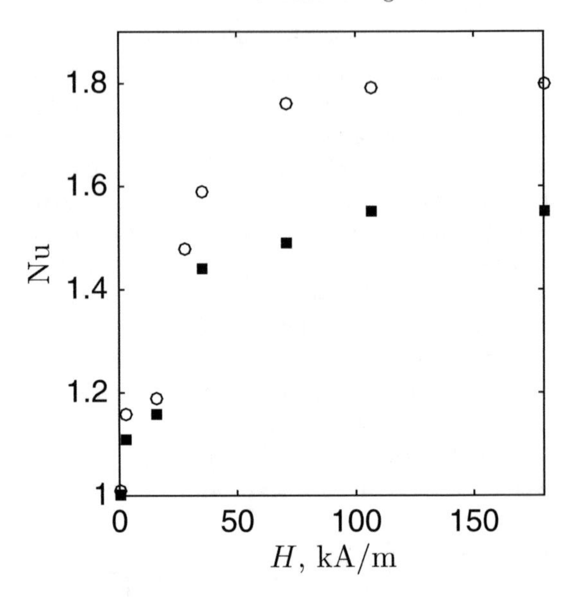

Fig. 6.78 Nondimensional heat flux as a function of the magnitude of the applied uniform vertical magnetic field at fixed temperature differences $\Delta T = 41.8$ (circles) and 36.6 (squares) K between the poles of a sphere carved inside a Plexiglas block, filled with a transformer-oil-based ferrofluid with $M_s = 44.9\,\text{kA/m}$ and heated from above.

As seen from Figure 6.78 the nondimensional heat flux reaches its maximum at about 100 kA/m and then remains constant. Such a behaviour is consistent with the variation of the magnetic parameter (see Figure 6.3) responsible for the strength of convection. Since the strength of convection also depends on the applied temperature difference ($\mathrm{Ra}_m \sim \Delta T^2$), its increase by about 5 K leads to the increase of the saturated Nusselt number by 16% in Figure 6.78. Note also from Figure 6.78 that for small fields $H < 25\,\text{kA/m}$ the heat flux in the considered transformer-oil-based ferrofluid increases with the field insignificantly. In a ferrofluid based on polyethylsiloxane with a 5 times larger viscosity (see Table 4.1), such an increase was not possible at all since magnetoconvection could not be initiated even at $H = 107\,\text{kA/m}$. On the other hand, in a less viscous kerosene-based ferrofluid, thermomagnetic convection sets at smaller temperature differences; see Figure 6.79. Similar to a ferrofluid based on transformer oil in a kerosene-based ferrofluid, initially the nondimensional heat flux increases slightly with ΔT up to $\mathrm{Nu} \approx 1.1$ due to a weak thresholdless toroidal fluid motion. However, starting from some critical temperature difference ΔT_c that depends on the strength of the applied field, the heat flux starts rapidly growing. Such a transition to a strong convection flow occurs at $\Delta T_c \approx 21\,\text{K}$ for $H = 17\,\text{kA/m}$ and at $\Delta T_c \approx 9\,\text{K}$ for $H = 100\,\text{kA/m}$. As seen from Figure 6.79, the intensity of heat exchange in less viscous ferrofluids can be increased by a factor of up to 3 by the application of a sufficiently strong magnetic field.

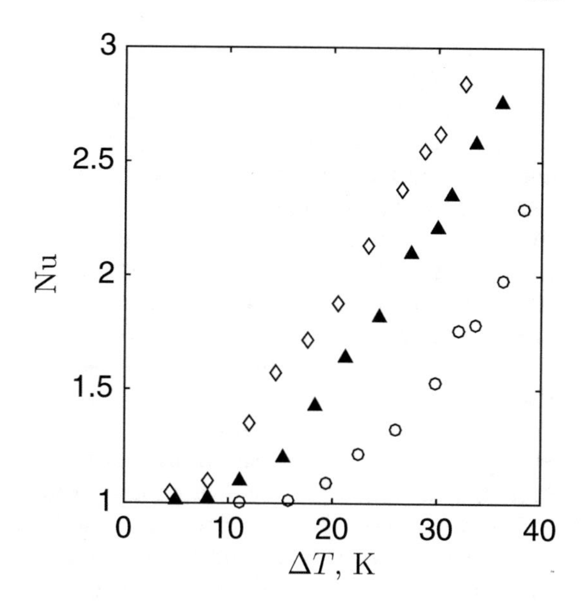

Fig. 6.79 Nondimensional heat flux as a function of the temperature difference between the poles of a sphere carved inside a Plexiglas block, filled with a kerosene-based ferrofluid with $M_s = 55\,\mathrm{kA/m}$ heated from above and placed in a uniform vertical magnetic field with magnitude $H =$ 17 (circles), 44 (triangles) and 100 (diamonds) kA/m.

The heat transfer results presented in Figure 6.79 were measured in a well-mixed ferrofluid with a uniform density. The comparison of heat fluxes measured in the same fluid after it was left at rest for several days and developed the density stratification due to the gravitational sedimentation of a solid phase is given in Figure 6.80. Convection in a stratified fluid sets at a larger temperature difference and heat flux remains 10–30% smaller than in a well-mixed fluid.

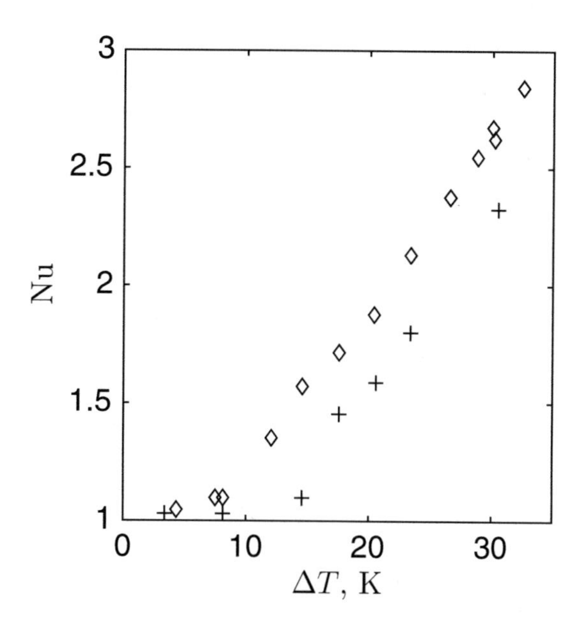

Fig. 6.80 Same as Figure 6.79 but for well-mixed (diamonds) and stratified (crosses) ferrofluids at $H = 100\,\mathrm{kA/m}$.

The destabilising influence of a uniform vertical external magnetic field on the mechanical equilibrium of ferrofluid in a spherical cavity heated from above is similar to that in a horizontal flat layer discussed in Section 6.2.2. It is important to note that thermally induced non-uniformity of the fluid magnetisation that sometimes is not accounted for in studies using the so-called inductionless approximation [133, e.g.] can have a strong influence both on the type of the arising thermomagnetic flows and the corresponding heat transfer. Moreover, the experimental studies reported in this section demonstrate that a significant thermomagnetic enhancement of heat transfer can be achieved in fluid volumes with a characteristic size of tens of millimetres, that is, an order of magnitude larger than initially thought [89]. However, the temperature difference required for the onset of magnetoconvection in a gravitationally stably stratified fluid (i.e. fluid heated from above) is an order of magnitude larger than that for the onset of thermogravitational convection in an unstably stratified fluid (fluid heated from below).

The direction of the applied magnetic field with respect to the temperature gradient also has a strong influence on the arising convection flows. When the fluid is heated from above, applying a vertical field is the optimal way to enhance heat transfer through the sphere. However, when the fluid is heated from below, the application of a vertical or horizontal field leads to a suppression of gravitational convection and thus to a decrease of the overall heat flux [35, 135]. The main reasons for such a stabilisation of buoyancy-driven convection by a magnetic field are the increase of the effective fluid viscosity due to the magnetoviscous effect [170] and the intensification of gravitational sedimentation in a magnetic field due to the formation magnetic particle aggregates caused by their dipole-dipole interaction [53, 187, 198, 199].

Chapter 7
Concluding Remarks

Abstract A brief summary of the book is presented in this chapter.

This book contains a comprehensive discussion of theoretical and experimental studies of thermogravitational and thermomagnetic convection in ferro-nanofluids. The reported numerical and analytical results are generally in line with experimentally observed convection flows. However, we avoid claiming that the overall "theoretical results are in good agreement with experiment" because present model descriptions of processes taking place in ferrofluids are not always sufficiently complete to warrant such a claim. This is so for objective reasons discussed in the book. Despite this, the reported theoretical and experimental results are mutually complementary. For example, the existence of vertical thermomagnetic rolls was first observed experimentally and then quantified theoretically. On the other hand, thermomagnetic waves were discovered via the flow stability analysis, which guided subsequent dedicated experiments that confirmed their existence.

The reported experimental investigations demonstrate that when ferrofluids are well mixed as a result of sufficiently strong thermogravitational or thermomagnetic convection, their flow behaviour is often similar to that of single-component media. However, there are a number of physical situations where the spatial non-uniformity of solid-phase concentration that can be caused by the gravitational sedimentation, thermodiffusion or magnetophoresis of magnetic particles and their aggregates plays a major role in determining fluid behaviour both in a magnetic field and in its absence. The presence of solid magnetic particles is also responsible for a stabilisation of convectionless state due to the increase of the effective fluid viscosity attributed to its rotational and magnetic components. Another demonstrated feature of ferrofluids is the dependence of its transport coefficients on the microstructure of a colloid that can be strongly influenced by the conditions of storage and the his-

A. A. Bozhko, S. A. Suslov, *Convection in Ferro-Nanofluids: Experiments and Theory*, Advances in Mechanics and Mathematics 40,
https://doi.org/10.1007/978-3-319-94427-2_7

tory of a ferrofluid use. Systematic accounting of all these factors is frequently impossible for objective reasons such as the lack of experimental methods of registering local shear rate or determining the instantaneous composition of the fluid given that it is dictated by the unmeasurable local thermal and magnetic fields. Despite these objective difficulties, the presented experimental and analytical results provide practically important information on the ways of controlling various flow instabilities by means of externally applied magnetic field that in turn can be used for a significant enhancement or suppression of heat transfer in ferrofluids. The presented experiments also show that in certain conditions, heat flux through a non-isothermal ferrofluid can vary spontaneously and irregularly. The possibility of this has to be taken into account when designing practical heat management applications involving ferrofluids. This book attempted to disclose physical reasons for such a complex behaviour and formulate further recommendations for the practical use of non-isothermal ferrofluids. It also outlined the current open questions in theoretical modelling of magneto-polarisable media as a whole and in specific geometries and thermal conditions.

Given that up to a half of world energy consumption is currently associated with heat management, the study of a thermomagnetic mechanism of heat transfer capable of a manyfold enhancement of heat exchange rate reported here makes it very timely. Since ferrofluids belong to a wider class of nanofluids, the study of their behaviour is also a step forward in research on their synthesis and potential applications and ultimately in addressing global problems of energy saving and reducing the effects of greenhouse gas emission and global warming.

Appendix A
Brief Summary of the Used Numerical Approximation

In order to obtain sufficiently accurate numerical results, the governing basic flow and stability equations have been solved using the Chebyshev pseudo-spectral collocation method initially proposed in [110, 138] and implemented in [239, 240]. In this method the solution is represented by a linear combination of Chebyshev polynomials of degrees ranging from 0 to the pre-set values N_r and N_z that define the total number of Chebyshev modes used for approximating the solutions in the cross-layer direction x, where $x \in [-1, 1]$. A rectangular Gauss-Lobatto collocation point grid $[x_k]$ is introduced, where $x_k = \cos[\pi(k-1)/(N-1)]$, $k = 1, 2, \ldots, N$. The required solution, say, $f(x)$, is then evaluated at these nodes. As shown in [110, 138], these values uniquely define the coefficients in the linear combinations of Chebyshev polynomials used in the spectral approximation of the solution. Therefore, the function values at any point within a computational domain can be found since the explicit expressions for Chebyshev polynomials are readily available. Thus, the spectral accuracy of the method is ensured. At the same time as discussed in [110, 138], the collocation formulation enables one to reduce the integration and differentiation procedures to a simple matrix-vector multiplication. Namely, introducing the vector $\mathbf{f} = [f_1, f_2, \ldots, f_N]^T$ of function values at points x_l in a grid, one can write

$$\frac{\mathrm{d}\mathbf{f}}{\mathrm{d}x} = \hat{\mathsf{G}}_1^N \mathbf{f}\,, \quad \frac{\mathrm{d}^2\mathbf{f}}{\mathrm{d}x^2} = \hat{\mathsf{G}}_2^N \mathbf{f}\,, \quad \mathbf{F} = \mathsf{W}^N \mathbf{f}\,, \tag{A.1}$$

where $F = \int_{-1}^{x} f(\tilde{x})\mathrm{d}\tilde{x}$, $\hat{\mathsf{G}}_2^N = \hat{\mathsf{G}}_1^N \times \hat{\mathsf{G}}_1^N$ and the standard $N \times N$ matrices $\hat{\mathsf{G}}_1^{N\times N}$ and $\mathsf{W}^{N\times N}$ are defined in [110, 138].

A. A. Bozhko, S. A. Suslov, *Convection in Ferro-Nanofluids: Experiments and Theory*, Advances in Mechanics and Mathematics 40,
https://doi.org/10.1007/978-3-319-94427-2

Appendix B
Copyright Permissions

The following is the list of the materials previously published by the authors and reused here with permission.

Chapter 2

- Figure 2.1 and the related discussion are reproduced from
  H. Rahman and S. A. Suslov: Magneto-gravitational convection in a vertical layer of ferrofluid in a uniform oblique magnetic field. J. Fluid Mech. 795, 847–875 (2016),
  `http://dx.doi.org/doi:10.1017/jfm.2014.709` with permission of Cambridge University Press.

Chapter 3

- Figures 3.2, 3.39–3.47 and the related discussions are reproduced from
  H. Rahman and S. A. Suslov: Magneto-gravitational convection in a vertical layer of ferrofluid in a uniform oblique magnetic field. J. Fluid Mech. 795, 847–875 (2016),
  `http://dx.doi.org/doi:10.1017/jfm.2016.231` with permission of Cambridge University Press.

- Figures 3.3–3.5, 3.27, 3.29–3.36 and the related discussions are reproduced from

A. A. Bozhko, S. A. Suslov, *Convection in Ferro-Nanofluids: Experiments and Theory*, Advances in Mechanics and Mathematics 40,
https://doi.org/10.1007/978-3-319-94427-2

H. Rahman and S. A. Suslov: Thermomagnetic convection in a layer of ferrofluid placed in a uniform oblique external magnetic field. J. Fluid Mech. 764, 316–348 (2015),
http://dx.doi.org/doi:10.1017/jfm.2014.709 with permission of Cambridge University Press.

- Figures 3.6, 3.8–3.14, 3.20–3.26 and the related discussions are reproduced from
  S. A. Suslov: Thermomagnetic convection in a vertical layer of ferromagnetic fluid. Phys. Fluids 20, 084101 (2008),
  http://dx.doi.org/doi:0.1063/1.2952596 with permission of AIP Publishing.

- Figures 3.15, 3.18–3.19 and the related discussions are reprinted with permission from
  S. A. Suslov, A. A. Bozhko, A. S. Sidorov and G. F. Putin: Thermomagnetic convective flows in a vertical layer of ferrocolloid: Perturbation energy analysis and experimental study. Phys. Rev. E 86, 016301 (2012),
  http://dx.doi.org/doi:10.1103/PhysRevE.86.016301.
  Copyright (2012) by the American Physical Society.

- Figures 3.49 and 3.50 and the related discussions are reproduced from
  P. Dey and S. A. Suslov: Thermomagnetic instabilities in a vertical layer of ferrofluid: nonlinear analysis away from a critical point. Fluid Dyn. Res. 48, 061404 (2016),
  http://dx.doi.org/doi:10.1088/0169-5983/48/6/061404
  with permission of IOP Publishing.

Chapter 4

- Figures 4.5, 4.14 and the related discussions are reprinted with permission from
  S. A. Suslov, A. A. Bozhko, A. S. Sidorov and G. F. Putin: Thermomagnetic convective flows in a vertical layer of ferrocolloid: Perturbation energy analysis and experimental study. Phys. Rev. E 86, 016301 (2012),
  http://dx.doi.org/doi:10.1103/PhysRevE.86.016301.
  Copyright (2012) by the American Physical Society.

- Figure 4.12 and the related discussion are reproduced from
  M. T. Krauzina, A. A. Bozhko, P. V. Krauzin, S. A. Suslov: Complex behaviour of a nanofluid near thermal convection onset: Its nature and features. Int. J. Heat Mass Transfer 104, 688–692 (2017),
  http://dx.doi.org/10.1016/j.ijheatmasstransfer.2016.08.106
  with permission of Pergamon.

- Figure 4.13 and the related discussion are reprinted with permission from S. A. Suslov, A. A. Bozhko, A. S. Sidorov and G. F. Putin: Intermittent flow regimes near the convection threshold in ferromagnetic nanofluids. Phys. Rev. E 91, 013010 (2015),
  http://dx.doi.org/doi:10.1103/PhysRevE.91.013010.
  Copyright (2015) by the American Physical Society.

Chapter 5

- Figures 5.22–5.31 and the related discussions are reprinted with permission from
  S. A. Suslov, A. A. Bozhko, A. S. Sidorov and G. F. Putin: Intermittent flow regimes near the convection threshold in ferromagnetic nanofluids. Phys. Rev. E 91, 013010 (2015),
  http://dx.doi.org/doi:10.1103/PhysRevE.91.013010.
  Copyright (2015) by the American Physical Society.

- Figures 5.32, 5.33 and the related discussions are reproduced from
  M. T. Krauzina, A. A. Bozhko, P. V. Krauzin, S. A. Suslov: Complex behaviour of a nanofluid near thermal convection onset: Its nature and features. Int. J. Heat Mass Transfer 104, 688–692 (2017),
  http://dx.doi.org/10.1016/j.ijheatmasstransfer.2016.08.106
  with permission of Pergamon.

Chapter 6

- Figures 3.15, 3.19–3.18 and the related discussions are reprinted with permission from
  S. A. Suslov, A. A. Bozhko, A. S. Sidorov and G. F. Putin: Thermomagnetic convective flows in a vertical layer of ferrocolloid: Perturbation energy analysis and experimental study. Phys. Rev. E 86, 016301 (2012),
  http://dx.doi.org/doi:10.1103/PhysRevE.86.016301.
  Copyright (2012) by the American Physical Society.

References

1. Ageev, V.A., Balyberdin, V.V., Veprik, I.Y., Ievlev, I.I., Legejda, V.I., Fedonenko, A.I.: Convection of magnetic fluids in non-uniform magnetic field (in Russian). Magnetohydrodynamics **2**, 61–65 (1990)
2. Ageikin, D.I.: Determination of heat emission by means of thermomagnetic convection (in Russian). Proc. Acad. Sci. USSR **74**, 229–232 (1950)
3. Ahlers, G., Behringer, R.P.: Evolution of turbulence from the Rayleigh-Bénard instability. Phys. Rev. Lett. **40**(11), 712–716 (1978)
4. Ahlers, G., Lerman, K., Cannell, D.S.: Different convection dynamics in mixtures with the same separation ratio. Phys. Rev. E. **53**(3), 2041–2044 (1996)
5. Altan, C.L., Elkatmis, A., Yuksel, M., Aslan, N., Bucak, S.: Enhancement of thermal conductivity upon application of magnetic field to Fe_3O_4 nanofluids. J. Appl. Phys. **110**, 093917 (2011)
6. Arefyev, I.M., Arefyeva, T.A., Kazakov, Y.B.: Patent 2517704 of the russian federation (2014)
7. Assenheimer, M., Steinberg, V.: Rayleigh-Bénard convection near the gas-liquid critical point. Phys. Rev. Lett. **70**(25), 3888–3891 (1993)
8. Avdeev, M.V., Aksenov, V.L.: Small-angle neutron scattering in structure studies of magnetic fluids (in Russian). Phys. Usp. **180**(10), 1009–1034 (2010)
9. Avdeev, M.V., Bica, D., Vekas, L., Aksenov, V.L.: Comparative structure analysis of non-polar organic ferrofluids stabilised by saturated monocarboxylic acids. J. Colloid Interface Sci. **334**, 37–41 (2009)
10. Bajaj, K., Cannell, D.S., Ahlers, G.: Competition between spiral-defect chaos and rolls in Rayleigh-Benard convection. Phys. Rev. E. **55**(5), 4869–4872 (1997)

A. A. Bozhko, S. A. Suslov, *Convection in Ferro-Nanofluids: Experiments and Theory*, Advances in Mechanics and Mathematics 40,
https://doi.org/10.1007/978-3-319-94427-2

11. Bashirnezhad, K., Bazri, S., Safaei, M.R., Goodarzi, M., Dahari, M., Mahian, O., Dalklca, A.S., Wongwises, S.: Viscosity of nanofluids: a review of recent experimental studies. Int. Commun. Heat Mass Trans. **73**, 114–123 (2016)
12. Bashtovoy, V.G., Berkovsky, B.M., Vislovich, A.N.: Introduction to thermomechanics of magnetic fluids (in Russian). Institute of High Temperatures of the Russian Academy of Sciences, Moscow (1985)
13. Bashtovoy, V.G., Berkovsky, B.M., Vislovich, A.N.: Introduction to Thermomechanics of Magnetic Fluids. Hemisphere, Washington (1988)
14. Batchelor, G.K.: Heat transfer by free convection across a closed cavity between vertical boundaries at different temperatures. Q. Appl. Math. **12**, 209–233 (1954)
15. Bednarz, T., Patterson, J.C., Lei, C., Ozoe, H.: Enhancing natural convection in a cube using a strong magnetic field—Experimental heat transfer rate measurements and flow vizualizations. Int. Commun. Heat Mass Transfer **36**, 781–786 (2009)
16. Behringer, R.P., Ahlers, G.: Heat transport and temporal evolution of fluid flow near the Rayleigh-Bénard instability in cylindrical containers. J. Fluid Mech. **125**, 219–258 (1982)
17. Belyaev, A.V., Smorodin, B.L.: The stability of ferrofluid flow in a vertical layer subject to lateral heating and horizontal magnetic field. J. Magn. Magn. Mater. **322**, 2596–2606 (2010)
18. Bergholz, R.F.: Instability of steady natural convection in a vertical fluid layer. J. Fluid Mech. **84**, 743–768 (1978)
19. Berkovsky, B.M., Medvedev, V.F., Krakov, M.S.: Magnetic Fluids: Engineering Application. Oxford University Press, Oxford (1993)
20. Berkovsky, B.V., Fertman, V.E., Polevikov, V.K., Isaev, S.V.: Heat transfer across vertical ferrofluid layers. Int. J. Heat Mass Transfer **19**, 981–986 (1976)
21. Bestehorn, M., Friedrich, R., Haken, H.: Pattern formation in convective instabilities. J. Modern Phys. B **4**(3), 365–400 (1990)
22. Bibik, E.E., Lavrov, I.S.: Preparing ferrofluids (in Russian). USSR Patent 467666 (1975)
23. Blums, E., Cebers, A.O., Maiorov, M.M.: Magnetic Fluids. Walter de Gruyter, Berlin (1997)
24. Blums, E., Mezulis, A., Maiorov, M., Kronkalns, G.: Thermal diffusion of magnetic nanoparticles in ferrocolloids: experiments on particle separation in vertical columns. J. Magn. Magn. Mater. **169**, 220–228 (1997)
25. Blums, E., Odenbach, S., Mezulis, A., Maiorov, M.: Soret coefficient of nanoparticles in ferrofluids in the presence of magnetic field. Phys. Fluids **10**(9), 2155–2163 (1998)
26. Blums, E.Y., Maiorov, M.M., Tsebers, A.O.: Magnetic Fluids (in Russian). Zinatne, Riga, Latvia (1989)

27. Bodenschatz, E., de Bruyn, J.R., Ahlers, G., Cannell, D.S.: Transitions between patterns in thermal convection. Phys. Rev. Lett. **67**(22), 3078–3081 (1991)
28. Bodenschatz, E., Pesch, W., Ahlers, G.: Recent developments in Rayleigh-Bénard convection. Annu. Rev. Fluid Mech. **32**, 709–778 (2000)
29. Bogatyrev, G.P., Gilev, V.G.: Concentration dependence of the viscosity of a magnetic liquid in an external field. Magnetohydrodynamics **20**, 249–252 (1984)
30. Bogatyrev, G.P., Shaidurov, G.F.: Convection stability of a horizontal ferrofluid layer in the presence of homogeneous magnetic field. Magnetohydrodynamics **12**(3), 374–383 (1976)
31. Bozhko, A., Bulychev, P.V., Putin, G.F., Tynjälä, T.: Spatio-temporal chaos in colloid convection. Fluid Dyn. **42**(1), 24–32 (2007)
32. Bozhko, A.A.: Onset of convection in magnetic fluids. Physics Procedia **9**, 176–180 (2010)
33. Bozhko, A.A.: Thermal convection in magnetic fluids in gravitational and magnetic fields (in Russian). Doctor of Sciences Dissertation. Perm State University, Perm (2011). dslib.net
34. Bozhko, A.A., Bratukhin, Y.K., Putin, G.F.: Experiments on ferrofluid convection in spherical cavity. In: Proceedings of the Joint 15th Riga and 6th PAMIR International Conference on Fundamental and Applied MHD, Riga, Latvia, vol. 1, pp. 333–336 (2005)
35. Bozhko, A.A., Kuchukova, M.T., Putin, G.F.: The influence of external uniform magnetic field on convection in magnetic fluid filling a spherical cavity. Magnetohydrodynamics **49**(1–2), 3–10 (2013)
36. Bozhko, A.A., Pilugina, T.V., Putin, G.F., Shupenik, D.V., Suhanovsky, A.N.: On instability of thermogravitational flow in a ferrofluid vertical layer in the transversal magnetic field. In: Y.Y. Schelykalov (ed.) Proceedings of the 8th International Conference on Magnetic Fluids, Plyos, Russia, pp. 75–78 (1998)
37. Bozhko, A.A., Putin, G.F.: Experimental investigation of thermomagnetic convection in uniform external field. Bull. Acad. Sci. USSR Phys. Ser. **55**, 1149–1156 (1991)
38. Bozhko, A.A., Putin, G.F.: Heat transfer and flow patterns in ferrofluid convection. Magnetohydrodynamics **39**(2), 147–169 (2003)
39. Bozhko, A.A., Putin, G.F.: Magnetic action on convection and heat transfer in ferrofluid. Indian J. Eng. Mater. Sci. **11**, 309–314 (2004)
40. Bozhko, A.A., Putin, G.F.: Thermomagnetic convection as a tool for heat and mass transfer control in nanosize materials under microgravity conditions. Microgravity Sci. Technol. **21**, 89–93 (2009)
41. Bozhko, A.A., Putin, G.F., Sidorov, A.S., Suslov, S.A.: Convection in a vertical layer of stratified magnetic fluid. Magnetohydrodynamics **49**(1–2), 143–152 (2013)

42. Bozhko, A.A., Putin, G.F., Tynjälä, T.: Magneto-hydrodynamic interaction in an inclined layer of ferrocolloid heated from below. In: Magnetism and Magnetic Materials. Solid State Phenomena, vol. 152, pp. 159–162. Trans Tech Publications, Zürich (2009). https://doi.org/10.4028/www.scientific.net/SSP.152-153.159
43. Bozhko, A.A., Tynjälä, T.: Influence of gravitational sedimentation of magnetic particles on ferrofluid convection in experiments and numerical simulations. J. Magn. Magn. Mater. **289**, 281–285 (2005)
44. Braithwaite, D., Beaugnon, E., Tournier, R.: Magnetically controlled convection in a paramagnetic fluid. Nature **354**, 134–137 (1991)
45. Bratsun, D.A., Zyuzgin, A.V., Putin, G.F.: Nonlinear dynamics and pattern formation in a vertical fluid layer heated from the side. Int. J. Heat Fluid Flow **24**, 835–852 (2003)
46. Bratukhin, Y.K., Maurin, L.N.: On convective motions of fluid in a nearly spherical cavity heated from below (in Russian). J. Appl. Mech. Tech. Phys. **3**, 69–72 (1983)
47. Brown, W.F.: Micromagnetics. Wiley, New York (1963)
48. Brown, W.F.: Thermal fluctuations of a single-domain particle. Phys. Rev. **130**(5), 1677–1686 (1963)
49. Büscher, K., Helm, C.A., Gross, C., Glöckl, G., Romanus, E., Weitschies, W.: Nanoparticle composition of a ferrofluid and its effects on the magnetic properties. Langmuir **20**(6), 2435–2444 (2004)
50. Busse, F.H.: Fundamentals of thermal convection. In: Mantle Convection: Plate Tectonics and Global Dynamics, pp. 23–95. Montreux, Gordon and Breach (1989)
51. Busse, F.H., Clever, R.M.: Stability of convection rolls in the presence of a horizontal magnetic filed. J. Mécanique Théorique et Appliquée **2**(4), 495–502 (1983)
52. Busse, F.H., Clever, R.N.: Three-dimensional convection in an inclined layer heated from below. J. Eng. Math. **26**, 1–19 (1992)
53. Buzmakov, V.M., Pshenichnikov, A.F.: On the structure of microaggregates in magnetite colloids. J. Colloid Interface Sci. **182**, 63–70 (1996)
54. Carruthers, J.R., Wolfe, R.: Magnetothermal convection in insulating paramagnetic fluids. J. Appl. Phys. **39**, 5718–5722 (1968)
55. Chait, A., Korpela, S.A.: The secondary flow and its stability for natural convection in a tall vertical enclosure. J. Fluid Mech. **200**, 189–216 (1989)
56. Chandra, K.: Instability of fluids heated from below. Proc. R. Soc. A **164**, 231–242 (1938)
57. Chandrasekhar, S.: Hydrodynamic and hydromagnetic stability. International Series of Monographs on Physics. Clarendon Press, Oxford (1961)
58. Chashechkin, Y.D., Levitskiy, V.V.: Pattern of flow around a sphere oscillating at neutrally buoyancy horizon in a continuously stratified fluid. J. Vis. **6**(1), 59–65 (2003)

59. Chashechkin, Y.D., Popov, V.A.: Methods of laboratory modelling of processes in inhomogeneous systems in normal and reduced gravity conditions. In: Hydrodynamics and Heat and Mass Transfer in Weightless Environment (in Russian), pp. 119–146. Nauka, Moscow (1982)
60. Chekanov, V.V.: On the interaction of particles in magnetic colloids. In: E. Blums (ed.) Hydrodynamics and Thermal Physics of Magnetic Fluids (in Russian), Salaspils, Latvia, pp. 69–76 (1980)
61. Chekanov, V.V.: Formation of aggregates as a result of phase transition in magnetic colloids. In: Physical Properties of Magnetic Fluids (in Russian), pp. 42–49. Ural Branch of the Academy of Sciences of the USSR, Sverdlovsk (1983)
62. Chen, C.F., Briggs, D.G., Wirtz, R.A.: Stability of thermal convection in a salinity gradient due to lateral heating. Int. J. Heat Mass Transfer **14**, 57–65 (1971)
63. Chikazumi, S., Taketomi, S., Ukita, M., Mizukami, M., Miyajima, H., Setogava, M., Kurihara, Y.: Physics of magnetic fluid. J. Magn. Magn. Mater. **65**, 245–251 (1987)
64. Choi, S.U.S.: Nanofluids: from vision to reality through research. J. Heat Transfer **131**, 033106 (2009)
65. Croquette, V.: Convective pattern dynamics at low Prandtl number: part I. Contemp. Phys. **30**(2), 113–133 (1989)
66. Croquette, V., Williams, H.: Nonlinear waves of the oscillatory instability on finite convective rolls. Phys. D **37**(1–3), 300–314 (1989)
67. Cross, M.C., Hohenberg, P.C.: Pattern formation out of equilibrium. Rev. Mod. Phys. **65**(3), 851–1112 (1993)
68. Cross, M.C., Hohenberg, P.C.: Spatiotemporal chaos. Science **263**(5153), 1569–1570 (1994)
69. Curtis, R.A.: Flow and wave propagation in ferrofluids. Phys. Fluids. **14**(10), 2096–2102 (1971)
70. Daniels, K.E., Plapp, B.B., Bodenschatz, E.: Pattern formation in inclined layer convection. Phys. Rev. Let. **84**(23), 5320–5323 (2000)
71. Daniels, K.E., Richard, J.W., Bodenschatz, E.: Localized transverse bursts in inclined layer convection. Phys. Rev. Let. **91**(11), 114,501 (2003)
72. Deardoriff, J.W.: Gravitational instability between horizontal plates with shear. Phys. Fluids **8**(6), 1027–1030 (1965)
73. Decker, W., Pesch, W., Weber, A.: Spiral defect chaos in Rayleigh-Bénard convection. Phys. Rev. Lett. **73**(5), 648–651 (1994)
74. DeLucas, L.J., Tillotson, B.J.: Diamagnetic control of convection during crystal growth. In: Proceedings of the Joint 12th European and the 6th Russian Symposium on Physical Science in Microgravity, St. Petersburg, Russia, vol. 2, pp. 162–169 (1997)
75. Demouchy, G., Mezulis, A., Bee, A., Talbot, D., Bacri, J.C., Bourdon, A.: Diffusion and thermodiffusion studies in ferrofluids with a new two-dimensional forced Rayleigh-scattering technique. J. Phys. D: Appl. Phys. **37**, 1417–1428 (2004)

76. Dennin, M., Ahlers, G., Cannell, D.S.: Chaotic localized states near the onset of electroconvection. Phys. Rev. Lett. **77**(12), 2475–2478 (1996)
77. Devendiran, D.K., Amirtham, V.A.: A review on preparation, characterization, properties and applications of nanofluids. Renew. Sust. Energ. Rev. **60**, 21–40 (2016)
78. Dey, P., Suslov, S.A.: Thermomagnetic instabilities in a vertical layer of ferrofluid: nonlinear analysis away from a critical point. Fluid Dyn. Res. **48**, 061404 (2016)
79. Donzelli, G., Cerbino, R., Vailati, A.: Bistable heat transfer in a nanofluid. Phys. Rev. Lett. **102**(10), 104503 (2009)
80. Dvorchik, S.E., Nagorny, M.M., Reutskiy, S.Y., Rykov, V.G.: Study of a transformer cooled by magnetic fluid. In: Abstract Book of the VI all-USSR Conference on Magnetic Fluids (in Russian), Plyos, Russia, pp. 114–115 (1991)
81. Edwards, B.F., Gray, D.D., Hang, J.: Magnetothermal convection in nonconducting diamagnetic and paramagnetic fluids. In: Proceedings of 3rd International Microgravity Fluid Physics Conference, pp. 711–716. NASA, Cleveland (1996)
82. Einstein, A.: On the movement of small particles suspended in a stationary liquid demanded by the molecular kinetic theory of heat. Ann. Phys. **17**, 549–560 (1905)
83. Einstein, A.: Eine neue Bestimmung der moleküldimensionen. Ann. Phys. **19**, 298–306 (1906)
84. Elfimova, E.A., Ivanov, A.O., Lakhtina, E.V., Pshenichnikov, A.F., Camp, P.J.: Sedimentation equilibria in polydisperse ferrofluids: critical comparisons between experiment, theory, and computer simulationlaminar free convection in a slot. Soft Matter **12**, 4103–4112 (2016)
85. Elmore, W.C.: The magnetisation of ferromagnetic colloids. Phys. Rev. **54**, 1092–1095 (1938)
86. Engler, H., Borin, D., Odenbach, S.: Thermomagnetic convection influenced by the magnetoviscous effect. J. Phys. Conf. Ser. **149**, 012105 (2009)
87. Fauve, S., Laroch, C., Libchaber, A., Perrin, B.: Spatial instabilities and temporal chaos. In: J.E. Wesfreid, S. Zaleski (eds.) Lecture Notes in Physics. Cellular Structures in Instabilities, pp. 278–284. Springer, Berlin (1984)
88. Fertman, V.E.: Magnetic fluids—Natural convection and heat transfer. Izdatel'stvo Nauka i Tekhnika, Minsk (1978)
89. Finlayson, B.A.: Convective instability of ferromagnetic fluids. J. Fluid Mech. **40**, 753–767 (1970)
90. Fornalik, E., Filar, P., Tagawa, T., Ozoe, H., Szmyd, J.S.: Experimental study on the magnetic convection in a vertical cylinder. Exp. Therm. Fluid Sci. **29**, 971–980 (2005)

91. Fujita, T., Namiya, M., Jeyadevan, B.: Basic study of heat pipe using the developed temperature-sensetive magnetic fluid. In: Abstracts of the 5th International Conference on Magnetic Fluids, pp. 187–188. USSR, Riga (1989)
92. Gage, K.S., Reid, W.H.: The stability of thermally stratified plane Poiseuille flow. J. Fluid Mech. **33**, 21–32 (1968)
93. Gallagher, A.P., Mercer, A.M.: On the behaviour of small disturbances in plane Couette flow with a temperature gradient. Proc. R. Soc. A **268**, 117–128 (1965)
94. Gavili, A., Zabihi, F., Isfahani, T.D., Sabbaghzadeh, J.: The thermal conductivity of water base ferrofluids under magnetic field. Exp. Therm. Fluid Sci. **41**, 94–98 (2012)
95. Gebhart, B., Jaluria, Y., Mahajan, R., Sammakia, B.: Buoyancy-Induced Flow and Transpot. Hemisphere, Washington (1988)
96. Gershuni, G.Z., Zhukhovitsky, E.M.: On the stability of plane convective motion of a fluid. Zh. Tekh. Fiz **23**, 1838–1844 (1953)
97. Gershuni, G.Z., Zhukhovitsky, E.M.: Convective Stability of Incompressible Fluid. Keter, Jerusalem (1976)
98. Gershuni, G.Z., Zhukhovitsky, E.M., Nepomnjashchy, A.A.: Stability of Convective Flows (in Russian). Science, Moscow (1989)
99. Getling, A.V.: Rayleigh-Bénard Convection: Structures and Dynamics. World Scientific, Singapore (1998)
100. Glukhov, A.F.: Experimental investigation of thermal convection in mixtures in conditions of gravitational separation. Ph.D. Thesis, Perm State University, Perm (1995)
101. Glukhov, A.F., Putin, G.F.: Attainment of the equilibrium barometric distribution of magnetic fluid particles. In: Lyubimov, D.V. (ed.) Hydrodynamics: Collection of Papers (in Russian), vol. 12, pp. 92–103. Perm State University, Perm (1999)
102. Glukhov, A.F., Putin, G.F.: Convection of magnetic fluids in connected channels heated from below. Fluid Dyn. **45**(5), 713–718 (2010)
103. Godson, L., Raja, B., Lal, D.M., Wongwises, S.: Enhancement of heat transfer using nanofluids—an overview. Renew. Sust. Energ. Rev. **14**, 629–641 (2010)
104. Goldina, O.A., Lebedev, A.V., Ivanov, A.O., Elfmova, E.A.: Themperature dependence of initial magnetic susceptibility of polydisperse ferrofluids: a critical comparison between experiment and theory. Magnetohydrodynamics **52**, 35–42 (2016)
105. Gotoh, K., Yamada, M.: Thermal convection in a horizontal layer of magnetic fluids. J. Phys. Soc. Jpn. **51**, 3042–3048 (1982)
106. Groh, C., Richter, R., Rehberg, I., Busse, F.H.: Reorientation of a hexagonal pattern under broken symmetry: the hexagon flip. Phys. Rev. E **76**, 055301 (2007)

107. Gubin, S.P., Koksharov, Y.A., Khomutov, G.B., Yurkov, G.Y.: Magnetic nanoparticles: preparation, structure and properties. Russ. Chem. Rev. **74**(6), 489–520 (2005)
108. Hall, W.F., Busenberg, S.N.: Viscosity of magnetic suspensions. J. Chem. Phys. **51**, 137–144 (1969)
109. Hart, J.E.: Stability of the flow in a differentially heated inclined box. J. Fluid Mech. **47**, 547–576 (1971)
110. Hatziavramidis, D., Ku, H.C.: An integral Chebyshev expansion method for boundary-value problems of O.D.E. type. Comput. Maths Appl. **11**(6), 581–586 (1985)
111. Hennenberg, M., Wessow, B., Slavtchev, S., Desaive, T., Scheild, B.: Steady flows of laterally heated ferrofluid layer: influence of inclined strong magnetic field and gravity level. Phys. Fluids **18**, 093602 (2006)
112. Hirschberg, A.: Role of asphaltenes in compositional grading of a reservoirs fluid column. J. Petrol. Technol. **40**(1), 89–94 (1988)
113. Huang, J., Edwards, B.F., Gray, D.D.: Thermoconvective instability of paramagnetic fluids in a uniform magnetic field. Phys. Fluids **9**(6), 1819–1825 (1997)
114. Huang, J., Gray, D.D., Edwards, B.F.: Thermoconvective instability of paramagnetic fluids in a nonuniform magnetic field. Phys. Rev. E **57**(5), 5564–5571 (1998)
115. Huke, B., Lucke, M.: Roll, square, and cross-roll convection in ferrofluids. J. Magn. Magn. Mater. **289**, 264–267 (2005)
116. Isaev, S.V., Fertman, V.E.: Physical properties and hydrodynamics of dispersive ferromagnetics, chap. Investigation of oscillatory instability of convective motion in a vertical layer of ferromagnetic fluid, pp. 71–75 (in Russian). Ural Research Center of Academy of Science of USSR, Sverdlovsk (1977)
117. Ivanov, A.O., Kantorovich, S.S., Reznikov, E.N., Holm, C., Pshenichnikov, A.F., Lebedev, A., Chremos, A., Camp, P.J.: Magnetic properties of polydisperse ferrofluids: a critical comparison between experiment, theory, and computer simulation. Phys. Rev. E **75**, 061405 (2007)
118. Ivanov, A.S.: Magnetophoresis and diffusion colloid particles in a thin layer of magnetic fluid. Ph.D. Thesis, Institute of Continuous Media Mechanics, Ural Branch of the Russian Academy of Sciences, Perm (2011)
119. Ivanov, A.S., Pshenichnikov, A.F.: On natural solutal convection in magnetic fluids. Phys. Fluids **27**, 092001 (2015)
120. Jakobs, I.S., Bean, C.P.: Fine particles, thin films and exchange anisotropy. In: Rado, G.T., Suhl, H. (eds.) Magnetism, vol. 3, pp. 271–350. Academic Press, New York (1963)
121. Kamiyama, S., Sekhar, G.N., Ruraiah, N.: Mixed convection of magnetic fluid in a vertical slot. Rep. Inst. High Speed Mech. Tohoku Univ. **56**, 1–16 (1988)

122. Kandelousi, S.M., Ganji, D.D.: External Magnetic Field Effects on Hydrothermal Treatment of Nanofluid. William Andrew, Amsterdam (2016)
123. Kaneda, M., Tagawa, T., Ozoe, H.: Convection induced by a cuspshaped magnetic field for air in a cube heated from above and cooled from below. J. Heat Transfer **124**, 17–25 (2002)
124. Khaldi, F.: Removal of gravity buoyancy effects on diffusion flames by magnetic fields. In: Abstracts of the First International Seminar on Fluid Dynamics and Material Processing, Algiers, Algeria, pp. 57–58 (2007)
125. Khaldi, F., Noudem, J., Gillon, P.: On the stability between gravity and magneto-gravity convection within a non-electroconducting fluid in a differentially heated rectangular cavity. Int. J. Heat Mass Transfer **48**, 1350–1360 (2005)
126. Kirdyashkin, A.G., Leont'ev, A.I., Mukhina, N.V.: Stability of a laminar flow of fluid in vertical layers with free convection. Fluid Dyn. **6**, 884–888 (1971)
127. Koji, F., Hideaki, Y., Masahiro, I.: A mini heat transport device based on thermosensitive magnetic fluid. Nanoscale Microscale Thermophys. Eng. **11**(1), 201–210 (2007)
128. Kolodner, P., Bensimon, D., Surko, C.M.: Traveling-wave convection in annulus. Phys. Rev. Lett. **60**(17), 1723–1726 (1988)
129. Kolodner, P., Surko, C.M.: Weakly nonlinear traveling-wave convection. Phys. Rev. Lett. **61**(7), 842–845 (1988)
130. Koronovsky, A.A., Hramov, A.E.: Continuous Wavelet Analysis and its Applications (in Russian). Fizmatlit, Moscow (2003)
131. Korpela, S.A., Gözüm, D., Baxi, C.B.: On the stability of the conduction regime of natural convection in a vertical slot. Int. J. Heat Mass Transfer **16**, 1683–1690 (1973)
132. Koschmieder, E.L.: Bénard Cells and Taylor Vortices. Cambridge University Press, Cambridge (1993)
133. Krakov, M.S., Nikiforov, I.V., Reks, A.G.: Influence of the uniform magnetric field on natural convection in cubic enclosure: experiment and numerical simulation. J. Magn. Magn. Mater. **289**, 272–274 (2005)
134. Krauzina, M.T., Bozhko, A.A., Krauzin, P.V., Suslov, S.A.: Oscillatory instability of convection in ferromagnetic nanofluid and in transformer oil. Fluid Dyn. Res. **48**, 061407 (2016)
135. Krauzina, M.T., Bozhko, A.A., Krauzin, P.V., Suslov, S.A.: The use of ferrofluids for heat removal: advantage or disadvantage? J. Magn. Magn. Mater. **431**, 241–244 (2017)
136. Krauzina, M.T., Bozhko, A.A., Putin, G.F., Suslov, S.A.: Intermittent flow regimes near the convection threshold in ferromagnetic nanofluids. Phys. Rev. E **91**(1), 013010 (2015)

137. Kronkalns, G.E.: Measurements of coefficients of thermal and electric conductivity of a ferrofluid in magnetic field (in Russian). Magnetohydrodynamics **31**, 138–140 (1977)
138. Ku, H.C., Hatziavramidis, D.: Chebyshev expansion methods for the solution of the extended Graetz problem. J. Comput. Phys. **56**, 495–512 (1984)
139. Kumar, A., Subudhi, S.: Preparation, characteristics, convection and applications of magnetic nanofluids: a review. Heat Mass Transfer **54**, 241–265 (2018)
140. Kutateladze, S.S., Berdnikov, V.S.: Structure of thermogravitational convection in a flat variously oriented layers of liquid and on a vertical wall. Int. J. Heat Mass Transfer **27**(9), 1595–1611 (1984)
141. Lakhtina, E.V.: Centrifugation of dilute ferrofluids. Phys. Procedia **9**, 221–223 (2010)
142. Lakhtina, E.V., Pshenichnikov, A.F.: Dispersion of magnetic susceptibility and the microstructure of magnetic fluid. Colloid J. **68**(3), 327–337 (2006)
143. Lalas, D.P., Carmi, S.: Thermoconvective stability of ferrofluids. Phys. Fluids **14**(2), 436–437 (1971)
144. Landau, L.D., Lifshitz, E.M.: Electrodynamics of Continuous Media. Pergamon Press, Oxford (1960)
145. Lebedev, A.V., Lysenko, S.N.: Magnetic fluids stabilized by polypropylene glycol. J. Magn. Magn. Mater. **323**, 1198–1202 (2011)
146. Lee, J., Hyun, M.T., Kang, Y.S.: Confined natural convection due to lateral heating in a stably stratified solution. Int. J. Heat Mass Transfer **33**, 869–875 (1990)
147. Lee, Y., Korpela, S.A.: Multicellular natural convection in a vertical slot. J. Fluid Mech. **126**, 91–121 (1983)
148. Leshe, A.: Nuclear Induction. Foreign Literature (Russian translation), Moscow (1963)
149. Li, Q., Xuan, Y., Wang, J.: Experimental investigations on transport properties of magnetic fluids. Exp. Therm. Fluid Sci. **30**, 109–116 (2005)
150. Lubimov, D.V., Putin, G.F., Chernatavskiy, V.I.: On convective motions in a Hele-Shaw cell (in Russian). Akademiia Nauk SSSR, Doklady **235**, 554–557 (1977)
151. Lukashevich, M.V., Naletova, B.A., Tsurikov, C.N.: Redistribution of concentration of magnetic fluid in a nonuniform magnetic field. Magnetohydrodynamics **3**, 64–69 (1988)
152. Magnetic colloids manufactured in Ferrohydrodynamics laboratory at Ivanovo State Energy University, Ivanovo, Russian Federation and at "Ferrohydrodynamika" Scientific and Industrial Laboratory, Nikolaev, Ukraine (trademark "Blesk") were used in experiments.
153. Malomed, B.A., Nepomnyashchy, A.A., Tribelsky, M.I.: Domain boundaries in convection patterns. Phys. Rev. A **42**(12), 7244–7263 (1990)

154. Mamiya, H., Nakatani, I., Furubayshy, T.: Phase transitions of iron-nitride magnetic fluids. Phys. Rev. Lett. **84**, 6106–6109 (2000)
155. Martsenyuk, M.A.: Thermal conductivity of a suspension of ellipsoidal particles in a magnetic field. In: Proceedings of the 8th Riga MHD Conference, Riga, Latvia, vol. 1, pp. 108–109 (1975)
156. Matsuki, H., Murakami, K.: Performance of an automatic cooling device using a temperature-sensitive magnetic fluid. J. Magn. Magn. Mater. **6**(2–3), 363–365 (1987)
157. Maxwell, J.C.: A Treatise on Electricity and Magnetism. Clarendon Press, Oxford (1881)
158. McTaque, J.P.: Magnetoviscosity of magnetic colloids. J. Chem. Phys. **51**(1), 133–136 (1969)
159. Mezquia, D.A., Larranaga, M., Bou-Ali, M.M., Madariaga, J.A., Santamaria, C., Platten, J.K.: Contribution to thermodiffusion coefficient measurements in dcmix project. Int. J. Therm. Sci. **92**, 14–16 (2015)
160. Millan-Rodriguez, J., Bestehorn, M., Perez-Garcia, C., Friedrich, R., Neufeld, M.: Defect motion in rotating fluids. Phys. Rev. Lett. **74**(4), 530–533 (1995)
161. Montel, F.: Importance de la thermodiffusion en exploration et production petrolieres. Entropie **184–185**, 86–93 (1994)
162. Morozov, K.I.: On the theory of the Soret effect in colloids. In: Köhler, W., Wiegand, S. (eds.) Thermal Nonequilibrium Phenomena in Fluid Mixtures, pp. 38–60. Springer, Berlin (2002)
163. Morris, S.W., Bodenschatz, E., Cannell, D.S., Ahlers, G.: The spatio-temporal structure of spiral-defect chaos. Phys. D **97**(1–3), 164–179 (1996)
164. Moses, E., Fineberg, J., Steinberg, V.: Multistability and confined state, traveling-wave patterns in a convecting binary mixture. Phys. Rev.: Gen. Phys. **35**(6), 2757–2760 (1987)
165. Mukhopadhyay, A., Ganguly, R., Sen, S., Puri I., K.: A scaling analysis to characterize thermomagnetic convection. Int. J. Heat Mass Transfer **48**, 3485–3492 (2005)
166. Neel, L.: Influence of thermal fluctuations on the magnetization of ferromagnetic small particles. C. R. Acad. Sci. Paris **228**(6), 664–666 (1949)
167. Nkurikiyimfura, I., Wang, Y., Pan, Z.: Heat transfer enhancement by magnetic nanofluids—a review. Renew. Sust. Energ. Rev. **21**, 548–561 (2013)
168. Normand, C., Pomeau, Y., Velarde, M.G.: Convective instability: a physicists approach. Rev. Mod. Phys. **49**(3), 581–624 (1977)
169. Odenbach, S.: Drop tower experiments on thermomagnetic convection. Microgravity Sci. Tech. **6**(3), 161–163 (1993)
170. Odenbach, S.: Ferrofluids: Magnetically Controllable Fluids and Their Applications. Springer, New York (2002)
171. Odenbach, S.: Magnetoviscous Effects in Ferrofluids. Springer, New York (2002)

172. Odenbach, S.: Recent progress in magnetic fluid research. J. Phys.: Condens. Matter. **16**, R1135–R1150 (2004)
173. Odenbach, S.: Colloidal Magnetic Fluids: Basics, Development and Application of Ferrofluids. Springer Lecture Notes in Physics, vol. 763. Springer, New York (2009)
174. Odenbach, S., Müller, H.W.: On the microscopic interpretation of the coupling of the symmetric velocity gradient to the magnetization relaxation. J. Magn. Magn. Matter. **289**, 242–245 (2005)
175. Odenbach, S., Raj, K.: The influence of large particles and agglomerates on the magnetoviscous effect in ferrofluids. Magnetohydrodynamics **36**(4), 312–319 (2000)
176. Ogorodnikova, N.P., Putin, G.F.: Periodic and irregular convective self-oscillations in an ellipsoid (in Russian). Akademiia Nauk SSSR, Doklady **269**(1), 1065–1068 (1983)
177. Ohlsen, D.R., Hart, J.E., Weidman, P.D.: Waves in radial gravity using magnetic fluid. In: Proceedings of the Third Microgravity Fluid Physics Conference, Cleveland, Ohio, USA, pp. 717–721 (1996)
178. Orlov, D.V., Kurbatov, V.G., Silaev, V.A., Sizov, A.P., Trofimenko, M.I.: Ferromagnetic fluid for magnetofluidic seals. USSR Patent 516861 (1976)
179. Ostroumov, G.A.: Free convection as internal problem. Gostechizdat, Moscow, Leningrad, U.S.S.R. (1952)
180. Ovchinnikov, A.P., Shaidurov, G.F.: Convective stability of homogeneous fluid in a spherical cavity (in Russian). Hydrodynamics **1**, 3–21 (1968)
181. Padovani, S., Sada, C., Mazzoldi, P., Brunetti, B., Borgia, I., Sgamellotti, A., Giulivi, A., D'Acapito, F., Battaglin, G.: Copper in glazes of renaissance luster pottery: nanoparticles, ions, and local environment. J. Appl. Phys. **93**(12), 10058–10063 (2003)
182. Page, M.A.: Combined diffusion-driven and convective flow in a tilted square container. Phys. Fluids **23**, 056602 (2011)
183. Paliwal, R.C., Chen, C.F.: Double-diffusive instability in an inclined fluid layer. Part 1. Experimental investigation. J. Fluid Mech. **98**, 755–768 (1980)
184. Papell, S.S.: Low viscosity magnetic fluid obtained by the colloidal suspension of magnetic particles. US Patent 3215572 (1965)
185. Pareja-Rivera, C., Cuellar-Cruz, M., Esturau-Escofet, N., Demitri, N., Polentarutti, M., Stojanoff, V., Moreno, A.: Recent advances in the understanding of the influence of electric and magnetic fields on protein crystal growth. Cryst. Growth Des. **17**, 135–145 (2017)
186. Parekh, K., Lee, H.S.: Magnetic field induced enhancement in thermal conductivity of magnetite nanofluid. J. Appl. Phys. **107**(9), 09A310 (2010)
187. Peterson, E.A., Kruger, D.A.: Field induced agglomeration in magnetic colloids. J. Colloid Interface Sci. **62**(1), 24–34 (1977)

188. Philip, J., Shima, P.D., Raj, B.: Evidence for enhanced thermal conduction through percolating structures in nanofluids. Nanotechnology **19**, 305706 (2008)
189. Plapp, B.B., Egolf, D.A., Bodenschatz, E., Pesch, W.: Dynamics and selection of giant spirals in Rayleigh-B enard convection. Phys. Rev. Lett. **81**(24), 5334–5337 (1998)
190. Platten, J.K.: The Soret effect: a review of recent experimental results. J. Appl. Mech. **73**, 5–15 (2006)
191. Pocheau, A.A., Croquette, V.: Dislocation motion: a wavenumber selection mechanism in Rayleigh-Bénard convection. J. Phys. **45**, 35–48 (1984)
192. Polunin, V.M.: Acoustic Properties of Nanodisperse Magnetic Fluids (in Russian). Fizmatlit, Moscow (2012)
193. Pop, L.M., Odenbach, S.: Investigation of microscopic reason for the magnetoviscous effect in ferrofluid studied by small angle neutron scattering. J. Phys. Condens. Mater. **18**, S2785–S2802 (2006)
194. Pshenichnikov, A., Lebedev, A., Lakhtina, E., Kuznetsov, A.: Effect of centrifugation on dynamic susceptibility of magnetic fluids. J. Magn. Magn. Matter. **432**, 30–36 (2017)
195. Pshenichnikov, A.F.: Equilibrium magnetization of concentrated ferrocolloids. J. Magn. Magn. Matter. **145**(3), 319–326 (1995)
196. Pshenichnikov, A.F.: A mutual-inductance bridge for analysis of magnetic fluids. Instrum. Exp. Tech. **50**(4), 509–514 (2007)
197. Pshenichnikov, A.F., Elfimova, E.A., Ivanov, A.O.: Magnetophoresis, sedimentation, and diffusion of particles in concentrated magnetic fluids. J. Chem. Phys. **134**, 184508 (2011)
198. Pshenichnikov, A.F., Ivanov, A.S.: Magnetophoresis of particles and aggregates in concentrated magnetic fluids. Phys. Rev. E **86**, 051401 (2012)
199. Pshenichnikov, A.F., Shurobor, I.Y.: Segregation of magnetic fluids: conditions for formation and magnetic properties of droplet aggregates (in Russian). Izv. Akad. Nauk Phys. Ser. **51**(6), 1081–1087 (1987)
200. Putin, G.F.: Experimental investigation of the effect of a barometric distribution on ferromagnetic colloid flow. In: Proceedings of the 11th Riga Workshop on Magnetohydrodynamics (in Russian), vol. 3, pp. 15–18. Physics Institute of the Latvian Academy of Sciences, Riga (1984)
201. Rabinovich, M.I., Trubetzkov, D.I.: Introduction to the Theory of Oscillations and Waves (in Rusian). Nauka, Moscow (1984)
202. Rahman, H., Suslov, S.A.: Thermomagnetic convection in a layer of ferrofluid placed in a uniform oblique external magnetic field. J. Fluid Mech. **764**, 316–348 (2015)
203. Rahman, H., Suslov, S.A.: Magneto-gravitational convection in a vertical layer of ferrofluid in a uniform oblique magnetic field. J. Fluid Mech. **795**, 847–875 (2016)

204. Raja, M., Vijayan, R., Dineshkumar, P., Venkatesan, M.: Review on nanofluids characterization, heat transfer characteristics and applications. Renew. Sus. Energy Rev. **64**, 163–173 (2016)
205. Ramachandran, N.: Understanding G-jitter fluid mechanics by modeling and experiments. In: Proceedings of the 1st International Symposium on Microgravity Research and Applications in Physical Sciences and Biotechnology, Sorrento, Italy, vol. 2, pp. 925–930 (2000)
206. Ramachandran, N., Leslie, F.W.: Using magnetic fields to control convection during protein crystallization—analysis and validation studies. J. Cryst. Growth **274**, 297–306 (2005)
207. Rogers, J.L., Schatz, M.F., Bougie, J.L., Swift, J.B.: Rayleigh-Bénard onvection in a vertically oscillated fluid layer. Phys. Rev. Lett. **84**(1), 87–90 (2000)
208. Rosensweig, R.E.: Fluid dynamics and science of magnetic fluids. Adv. Electron. Electron Phys. **48**, 103–199 (1979)
209. Rosensweig, R.E.: Ferrohydrodynamics. Cambridge University Press, Cambridge (1985)
210. Rosensweig, R.E., Browaeyes, J., Bacri, J.C., Zebib, A., Perzynski, R.: Laboratory study of spherical convection in simulated central gravity. Phys. Rev. Lett. **83**(23), 4904–4907 (1999)
211. Rudakov, R.N.: Spectrum of perturbations and stability of convective motion between vertical plates. Appl. Math. Mech. **31**, 376–383 (1967)
212. Ryskin, A., Muller, H.W., Pleiner, H.: Thermal convection in binary fluid mixures with a weak concentration diffusivity, but strong solutal buoyancy forces. Phys. Rev. E **67**, 046302 (2003)
213. Sage, B.H., Member, A.I.M.E., Lacey, W.N.: Gravitational concentration gradietns in static columns of hydrocarbon fluids. Tans. AIME **132**(3), 120–131 (1939)
214. Sazaki, G., Yoshida, E., Komatsu, H., Nakada, T., Miyashita, S., Watanabe, K.J.: Effects of a magnetic field on the nucleation and growth of protein crystals. J. Cryst. Growth **173**, 231–234 (1997)
215. Schere, C., Figueiredo Neto, A.M.: Ferrofluids: properties and applications. Braz. J. Phys. **35**(3A), 718–727 (2005)
216. Schmidt, R.J., Milverton, S.W.: On the instability of a fluid when heated from below. Proc. Roy. Soc. Lond. A **152**, 586–594 (1935)
217. Schwab, L.: Field induced wavevector selection by magnetic Bénard convection. J. Magn. Magn. Mater. **65**, 315–316 (1987)
218. Schwab, L.: Konvektion in ferrofluiden. Ph.D. Thesis, University of Munich, Munich (1989)
219. Schwab, L., Hildebrandt, U., Stierstadt, K.: Magnetic Bénard convection. J. Magn. Magn. Mater. **39**, 113–114 (1983)
220. Shadid, J.N., Goldstein, R.J.: Visualization of longitudinal convection roll instabilities in an inclined enclosure heated from below. J. Fluid Mech. **215**, 61–84 (1990)

221. Shaidurov, G.F.: Stability of convective boundary layer in fluid filling the horizontal cylinder (in Russian). J. Eng. Phys. Thermophys. **2**(12), 68–71 (1959)
222. Shliomis, M.I.: Magnetic fluids. Sov. Phys. Uspekhi **17**, 153–169 (1974)
223. Shliomis, M.I., Smorodin, B.L., Kamiyama, S.: The onset of thermomagnetic convection in stratified ferrofluids. Philos. Mag. **83**, 2139–2153 (2003)
224. Si, S., Li, C., Wang, X., Yu, D., Peng, Q., Li, Y.: Magnetic monodisperse Fe_3O_4 nanoparticles. Cryst. Growth Des. **5**(2), 391–393 (2005)
225. Sidorov, A.S.: The influence of an oblique magnetic field on convection in a vertical layer of magnetic fluid. Magnetohydrodynamics **52**(1), 223–233 (2016)
226. Sidorov, N.I.: On the history of M. V. Lomonosov's mosaic recipes (in Russian). Proc. Acad. Sci. USSR, Ser. VII Phys. Math. **7**, 679–706 (1930)
227. Singh, J.: Energy relaxation for transient convection in ferrofluids. Phys. Rev. E **82**, 026311 (2010)
228. Skibin, Y.N.: Magneto-optical method of determining magnetic moment of magnetic fluid particles. In: Devices and Methods of Measurement of Physical Parameters of Ferrocolloids (in Russian), pp. 85–89. Ural Branch of the Academy of Sciences of the USSR, Sverdlovsk (1991)
229. Smorodin, B.L., Shliomis, M., Kaloni, P.: Influence of a horizontal magnetic field on free ferrofluid convection in a vertical slot. In: Book of Abstracts, 11 International Conference on Magnetic Fluids, Kosice, Slovakia, p. 6P22 (2007)
230. Soret, C.: Influence de la température sur la distribution des sels dans leurs solutions. C. R. Acad. Sci. Paris **91**, 289–291 (1880)
231. Sprenger, L., Lange, A., Odenbach, S.: Thermodiffusion in concentrated ferrofluids: experimental and numerical results on magnetic thermodiffusion. Phys. Fluids **26**, 022001 (2014)
232. Sprenger, L., Lange, A., Zubarev, A.Y., Odenbach, S.: Experimental, numerical and theoretical investigation on concentration-dependent Soret effect in magnetic fluids. Phys. Fluids **27**, 022001 (2015)
233. Stasiek, J.A., Kowalewski, T.A.: Thermochromic liquid crystals applied for heat transfer research. Opto-Electron. Rev. **10**(1), 1–10 (2002)
234. Suslov, S.A.: Thermo-magnetic convection in a vertical layer of ferromagnetic fluid. Phys. Fluids **20**(8), 084101 (2008)
235. Suslov, S.A.: Two equation model of mean flow resonances in subcritical flow systems. Discrete Contin. Dyn. Syst. Ser. **1**(1), 165–176 (2008)
236. Suslov, S.A., Bozhko, A.A., Putin, G.F.: Thermo-magneto-convective instabilities in a verticallayer of ferro-magnetic fluid. In: Proceedings of the XXXVI International Summer School—Conference "Advanced Problems in Mechanics", pp. 644–651. Repino, Russia, July 6–10 (2008)

237. Suslov, S.A., Bozhko, A.A., Putin, G.F., Sidorov, A.S.: Interaction of graviational and magnetic mechanisms of convection in a vertical layer of a magnetic fluid. Phys. Procedia **9**, 167–170 (2010)
238. Suslov, S.A., Bozhko, A.A., Sidorov, A.S., Putin, G.F.: Thermomagnetic convective flows in a vertical layer of ferrocolloid: perturbation energy analysis and experimental study. Phys. Rev. E **86**(1), 016301 (2012)
239. Suslov, S.A., Paolucci, S.: Stability of mixed-convection flow in a tall vertical channel under non-Boussinesq conditions. J. Fluid Mech. **302**, 91–115 (1995)
240. Suslov, S.A., Paolucci, S.: Stability of natural convection flow in a tall vertical enclosure under non-Boussinesq conditions. Int. J. Heat Mass Transfer **38**, 2143–2157 (1995)
241. Suslov, S.A., Paolucci, S.: Nonlinear analysis of convection flow in a tall vertical enclosure under non-Boussinesq conditions. J. Fluid Mech. **344**, 1–41 (1997)
242. Suslov, S.A., Paolucci, S.: Nonlinear stability of mixed convection flow under non-Boussinesq conditions. Part 1. Analysis and bifurcations. J. Fluid Mech. **398**, 61–85 (1999)
243. Suslov, S.A., Paolucci, S.: Stability of non-Boussinesq convection via the complex Ginzburg-Landau model. Fluid Dyn. Res. **35**, 159–203 (2004)
244. Taketomi, S., Tikadzumi, S.: Magnetic Fluids (in Russian: Trans. from Japanese). Mir, Moscow (1993)
245. Tareev, V.M.: Thermal conductivity of colloidal systems (in Russian). Colloid J. **6**, 545–550 (1940)
246. Terekhov, V.I., Kalinin, S.V., Lehmanov, V.V.: The mechanism of heat transfer in nanofluids: state of the art (review). Part 2. Convective heat exchange (in Russian). Thermophys. Aeromech. **2**, 173–188 (2010)
247. Tsebers, A.O.: Thermodynamic stability of magnetofluids. Magnetohydrodynamics **18**, 137–142 (1982)
248. Vargaftik, N.B.: Handbook of Physical Properties of Liquids and Gases. Hemisphere, New York (1975)
249. Vecsey, L.: Chaos in thermal convection and the wavelet analysis of geophysical fields. Ph.D. Thesis, Charles University, Prague (2002)
250. Veprik, I.Y., Fedonenko, A.I.: Some peculiarities of thermal convection in electrically conducting magnetic colloids. In: Abstract book of the Fifth USSR Colloquium on Physics of Magnetic Fluids (in Russian), Perm, Russia, pp. 38–39 (1990)
251. Vest, C.H., Arpachi, V.S.: Stability of natural convection in a vertical slot. J. Fluid Mech. **36**, 1–15 (1969)
252. Völker, T., Blums, E., Odenbach, S.: Determination of the Soret coefficient of agnetic particles in a ferrofluid from the steady and unsteady part of the separation curve. Int. J. Heat Mass Trans. **47**, 4315–4325 (2004)

253. Völker, T., Odenbach, S.: The influence of a uniform magnetic field on the Soret coefficient of magnetic nanoparticles. Phys. Fluids **15**, 2198–2207 (2003)
254. Vonsovsky, S.V.: Magnetism (in Russian). Nauka, Moscow (1971)
255. Wakayama, N.I.: Magnetic promotion of combustion in diffusion flames. Combust. Flame **93**(3), 207–214 (1996)
256. Wakitani, S.: Formation od cells in natural convection in a vertical slot at large Prandtl number. J. Fluid Mech. **314**, 299–314 (1996)
257. Wang, X.Q., Mujumdar, A.S.: Heat transfer characteristics of nanofluids: a review. Int. J. Therm. Sci. **46**, 1–19 (2007)
258. Weiss, P.: L'hypothèse du champ moléculaire et la propriété ferromagnétique. J. Phys. Theor. Appl. **6**, 661–690 (1907)
259. Wen, C.Y., Chen, C.Y.Y.S.F.: Flow visualization of natural convection of magnetic fluid in a rectangular Hele-Shaw cell. J. Magn. Magn. Mater. **252**, 206–208 (2002)
260. Yih, C.S.: Convective instability of a spherical fluid inclusion. Phys. Fluids **30**(1), 36–44 (1986)
261. Yin, D.C., Wakayama, N.L., Harata, K., Fujiwara, M., Kiyoshi, T., Wada, H., Niimura, N., Arai, S., Huang, W.D., Tanimoto, Y.: Formation of protein crystals (orthorhombic lysozyme) in quasi-microgravity environment obtained by superconducting magnet. J. Cryst. Growth **270**, 184–191 (2004)
262. Zablotsky, D., Mezulis, A., Blums, E.: Surface cooling based on thermomagnetic convection: Numerical simulation and experiment. Int. J. Heat Mass Transfer **52**, 5302–5308 (2009)
263. Zavarykin, M.P., Zorin, S.W., Putin, G.F.: Vibrational Effects in Hydrodynamics, chap. Experimental study of regimes of thermal convection in a vertically oscillating vertical layer, pp. 71–79. Collection of scientific works (In Russian). Perm State University (2000)
264. Zharkova, G.M., Kovrizhina, V.N., Khachaturyan, V.M.: A study of the flow structure in the near wall region of a complex shaped channel using liquid crystal. In: Greated, C., Cosgrov, J., Buick, J.M. (eds.) Optical methods and data proceeding in heat and fluid flow, Trowbridge, pp. 143–150 (2002)
265. Zhong, F., Ecke, R.: Pattern dynamics and heat transfer in rotating Rayleigh-Bénard convection. Chaos **2**(2), 163–171 (1992)
266. Zhong, L., He, R., Gu, H.C.: Oleic acid coating on the monodisperse magnetic nanoparticles. Appl. Surf. Sci. **253**, 2611–2617 (2006)
267. Zhukhovitsky, E.M.: On stability of non-uniformly heated fluid in a spherical cavity (in Russian). J. Appl. Math. Mech. **21**(5), 689–693 (1957)
268. Zsigmondy, R.A., Thiessen, P.A.: Das kolloide Gold. Lpz. (1925)

Index

A. A. Bozhko, S. A. Suslov, *Convection in Ferro-Nanofluids: Experiments and Theory*, Advances in Mechanics and Mathematics 40,
https://doi.org/10.1007/978-3-319-94427-2

Printed in the United States
By Bookmasters